Solar Photovoltaic
Energy

Other volumes in this series:

Solar Photovoltaic Energy

Anne Labouret and Michel Villoz

Preface by Jean-Louis Bal
French Environment and Energy
Management Agency (ADEME)

Translated from French by
Jeremy Hamand

The Institution of Engineering and Technology

Published by The Institution of Engineering and Technology, London, United Kingdom

The Institution of Engineering and Technology is registered as a Charity in England & Wales (no. 211014) and Scotland (no. SC038698).

Fourth edition © 2009 Dunod, Paris
English translation © 2010 The Institution of Engineering and Technology

First published 2003
Second edition 2005
Third edition 2006
Fourth edition 2009
English translation 2010

The Institution of Engineering and Technology
Michael Faraday House
Six Hills Way, Stevenage
Herts, SG1 2AY, United Kingdom

www.theiet.org

British Library Cataloguing in Publication Data
A catalogue record for this product is available from the British Library

ISBN 978-1-84919-154-8 (paperback)
ISBN 978-1-84919-155-5 (PDF)

Typeset in India by MPS Ltd, a Macmillan Company
Printed in the UK by CPI Antony Rowe, Chippenham

Contents

Preface

At a time when photovoltaic (PV) solar energy – in France and throughout the world – is expanding at an explosive rate unaffected by the financial crisis, I cannot recommend too highly to all technicians and engineers who wish to expand their understanding of PV technology a careful reading of this book by Anne Labouret and Michel Villoz. These two experts, recognised for some decades for their knowledge of industry and engineering, share their knowledge and experience of manufacturing techniques and the design of many complete systems, both grid connected and stand-alone.

This book reminds us that PV systems, although these two experts may make them appear simple, must comply with rigorous engineering rules if they are to attain the main objective, which is to provide a quality service at an economic price to the user. ADEME has always maintained that the quality of products and services is indispensable to the ongoing development of new sectors such as renewable energy. Solar energy, in its photovoltaic and thermal forms, will probably be the largest contributor to the energy supply of the planet in the second half of the twenty-first century and beyond – provided that its current massive growth does not generate disillusion among consumers.

I am therefore pleased to underline the fact that the two authors of this book make an important contribution to the consolidation of the solar sector by recalling the strict rules to be followed in PV engineering and technology.

They remind us of another important truth – the vital service that PV energy can bring to the rural areas of developing countries. In the early years of the PV market in the 1980s and 1990s, developments were concentrated on rural electrification and water pumps for use by communities not on the electric grid. The emergence of grid-connected solar generators in the industrialised countries has meant that the more modest applications for the rural areas of the countries of the South have been neglected. There are, however, today several hundred million people who now have light, water and health services, thanks to solar PV – and more than a billion and a half human beings could see such benefits in the future. Shouldn't this be the first priority in terms of sustainable development?

Jean-Louis Bal
Director of Renewable Energy, French Environment and
Energy Management Agency (ADEME)

Foreword

It is only stating the obvious to say that solar photovoltaic (PV) energy has suddenly taken off. Even the experts have been surprised by the incredible expansion of this economic sector, which has exceeded the most optimistic forecasts. There has been dynamic progress at all levels: investments, technological development, cost reduction, establishment of new factories, sale and installation of PV systems, exploitation of power stations. Figures published by analysts suggest production of 7.9 GW of PV cells in 2008, an increase of 85% over 2007 and by a factor of almost 5 over 5 years (production in 2004: 1.4 GW). The PV arrays installed more than doubled annually, from 2.4 GW in 2007 to 5.5 GW in 2008.

The European Union largely contributed to this impressive progress: Spain alone, which had only 0.5 GW at the end of 2007, installed 2.6 GW out of the 4.6 GW installed in Europe in 2008, thanks to its sunny climate and generous feed-in tariff (since revised downwards). Germany installed 1.5 GW, France only 45 MW. By the end of 2008, the European Union had 9.5 GW installed – 80% of the global PV park. On the production side, Germany again, the leader in European PV industry, was alone responsible for producing 1.5 GW of solar cells – nearly 20% of world production, and an increase of 67% over 2007.

There are many reasons for this success. Of course, financial incentives such as the repurchase of PV electricity at favourable rates by electricity companies, subsidised loans and tax credits were very effective, in particular in the sunnier countries – Spain is the perfect example. But such provisions, although subject to modification by governments, are not new. More recent developments are the successful efforts to lower costs by some manufacturers, particularly the world's leading manufacturer of thin-film cadmium telluride (CdTe), as well as some recent major technological advances such as interdigitated crystalline cells with more than 20% efficiency, heterojunction with intrinsic thin-layer (HIT) cells and tandem micromorphous silicon cells; and finally, the appearance on the market of complete generating stations of 30, 60 MW or more, which enable highly automated production units to be installed rapidly almost anywhere in the world.

In 2008, two important psychological barriers were overcome: the cost price of a CdTe PV panel came down to less than $1/Wc, and the PV electricity cost fell to parity with traditional power generation in a number of US states and southern Europe. When module costs come down, when PV output is high in a sunny climate and systems are optimised, and when low interest rates encourage investment, a PV power generator can today compete with a traditional power station on cost.

There remain, however, a number of uncertainties blighting this booming industry. In 2005–06, following a shortage of silicon in 2004, many new manufacturers entered the field, actually resulting in overproduction in 2008–09, amplified by the financial crisis of 2008, which led to a cutback of investment in solar power stations. This situation, although temporary, led and will still lead in the future to a number of small manufacturers to withdraw, especially the most recent entries to the field (in China, for example) who had not so far consolidated their position in the market. Brand familiarity is of prime importance in a market where PV panels have a life expectancy of 20 years or more, in order to guarantee the profitability of operations. Also, manufacturers have to constantly face up to questions about the environmental credentials of each technology: crystalline silicon consumes the most energy, CIS (copper indium selenide) and CdTe cells contain cadmium, and the carbon balance of production factories is also under criticism. A final potential threat is the possible shortage of indium, much in demand for the manufacture of flat screens, which could make the manufacture of CIGS (copper indium gallium selenide) cells more expensive (recycling solutions exist, but are expensive).

In stand-alone applications, solar PV energy can also prove an excellent technical solution. Outside electrified areas, it has a large number of economic domestic and light industrial applications and often provides an irreplaceable service. The market for stand-alone systems, while exhibiting constant growth, is slower to develop (annual growth estimated at 15%), and its share of the global market continues to fall because of the exceptional vitality of the grid-connected market. Obviously, the installed power for each unit is much smaller than in systems connected to the grid, but it remains very useful and well implanted in a large number of different activity sectors.

More broadly, the current increase of interest in renewable energy is certainly linked to the necessity to revise energy policies, both to combat excess CO_2 emissions and to prevent major world energy shortages. Planned revisions of the Kyoto Protocol, political change in the United States and further engagement by the European Union are all signs of hope for a real evolution of global energy policy.

In this new era, renewables have a larger role than ever to play. Fossil fuels are concentrated natural energies formed very slowly in the early history of the Earth. Should we not value this resource at its true worth, compared to the cost of renewable energy, which we would be capable of producing on a regular basis to cover all our consumption needs? Biofuels have shown their limits, and there is no obvious solution – although a drastic reduction of our consumption of all forms of energy seems inevitable.

But when we consider the true cost of all energy sources, renewables become entirely competitive, and the new methods of 'clean' storage should be developed (hydro or compressed air, for example). Fossil fuels should only be used in exceptional circumstances, to provide energy, for example, during periods of low sunlight (when PV does not produce) and low wind speed (to replace wind power). In this way, the maximum potential of renewable energies could be used in grids drawing on many different energy sources.

In addition, it is estimated that more than 2 billion people are not connected to an electrical grid – and will not be in the foreseeable future. The situation is caused by questions of profitability for the energy companies, related to remote location, low population density, poverty and low energy demand. For people in these situations, stand-alone PV systems can play a very important role by providing a truly economical solution covering basic electricity needs.

This basic energy supply, providing light, refrigeration, water treatment or telecommunications, can bring important health improvements to remote communities and the possibility to expand employment in crafts and agriculture.

Solar PV energy, originally developed to provide power for space satellites, emerged as an alternative source of energy after the oil shocks of the 1970s. Originally a niche market serving communities sensitive to the environment, PV has today become a modern and ambitious industry. The quality of solar PV panels, manufactured in highly automated factories, has greatly improved, and most manufacturers now offer 20 year guarantees. Associated technologies such as inverters and batteries have also made considerable progress and improved the reliability of PV systems.

The energy landscape is changing fast, many new companies are jumping on the bandwagon and the technology is evolving: information and training are more than ever necessary.

Introduction

The direct exploitation of solar energy captured by solar panels uses two different technologies: one produces calories – this is thermal solar energy – and the other produces electricity, and it is this technology, solar photovoltaic energy, which is the subject of this book.

A successful photovoltaic installation is based on a rigorous design and installation and uses reliable components responding to the needs of the project. In order to help professionals achieve this result, this book introduces basic tools for use by designers and foremen involved in photovoltaic installations. Much information is included on components – solar panels, inverters, batteries, regulators and others – so that their characteristics and the methods for selecting, associating, installing and maintaining them are clearly understood.

Chapter 1 provides a summary of the possibilities and potential uses of photovoltaic energy, which should enable the reader to acquire straightaway basic knowledge and sizing guidelines. It also contains some ideas on the contribution of PV energy to sustainable development.

The nature of light, the energy received on Earth from the Sun and the mechanisms of converting light to electricity are described in Chapter 2. These fundamental concepts are not indispensable to the rest of the book but will interest those intrigued by solar energy and who wish to understand the phenomena that govern it.

The dynamism of the PV sector has been boosted by all the many recent developments on the solar panels themselves, with two main tendencies: increase in performance and reduction in cost. To provide an understanding of the different PV materials and to explain the advantages of the different technologies, the whole of Chapter 3 is devoted to PV cells and modules.

Since grid-connected and stand-alone installations call for different criteria, they are dealt with in two separate chapters: Chapter 4 is entirely devoted to grid-connected systems and Chapter 5 to stand-alone systems. Both these chapters contain numerous examples and complete case studies.

This open structure makes it easier for the readers to concentrate on the approach most appropriate for their own needs.

Chapter 1

Some basic questions on photovoltaics

This first chapter attempts to provide answers to basic questions on the nature of photovoltaic (PV) energy; on what it produces, at what cost and for what applications; and on the type of equipment needed in particular cases. All these points are dealt with in greater depth in the chapters that follow. In more general terms, we will consider the role of PV energy in sustainable development, for today this is a major challenge that can affect many of our choices.

1.1   What is solar PV energy?

1.1.1   Electricity or heat?

Photovoltaic exploitation of solar energy, the subject of this book, consists of converting directly radiated light (solar or other) into *electricity*. This transformation of energy is carried out using PV modules or panels made up of solar PV cells (see Chapter 2). Thermal solar energy, on the other hand, functions quite differently, producing *heat* from the Sun's infrared radiation to heat water, air or other fluids. This technology is simple compared to PV technology, and therefore less costly. It works by capturing calories through heat absorbing surfaces such as black metal sheets (Figure 1.1). The heat energy collected can be used to heat individual or collective water systems. These are in widespread use in sunny countries to produce hot water for houses and apartments without central heating (Greece, Israel, African countries) and, in the last 10–20 years, partly thanks to financial incentives and subsidies, these systems have also been in use in temperate countries like France and Britain.

Thermodynamic solar energy, however, is the system used in large power stations equipped with solar radiation concentrators in the form of curved mirrors that heat a fluid to a very high temperature (several hundred degrees), which is then used to generate steam by thermal exchange and produce electricity from a steam-driven turbine. Although not widespread and requiring a direct solar flux (countries with little cloud cover), this technology is very spectacular. It is the principle used in the solar power stations of Font-Romeu in southern France and Andasol 1 in central Spain.

These aspects of solar energy are not dealt with in this book, which is strictly limited to PV energy. The term 'solar collector' will thus not be used to avoid ambiguity between the different technologies.

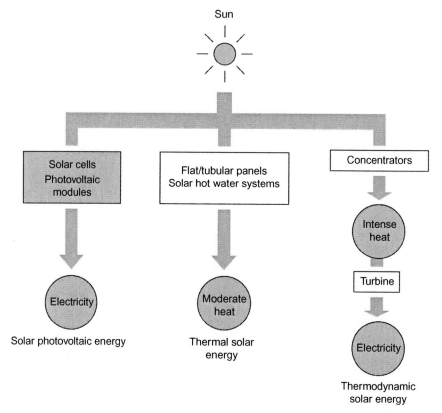

Figure 1.1   The different methods of exploiting solar energy

Note
Only thermal panels can be used for solar heating applications. It would be pointless to heat water with electricity generated by PV modules – this would have a very low efficiency and be much more expensive.

1.1.2   Is much sunshine necessarily needed?

The answer is no, of course. Otherwise PV would not be suitable for use in temperate countries. The term *solar energy* is somewhat ambiguous; in fact, any light source can be converted into electricity. For this reason, some people prefer the expression *light energy*. Having said this, since the Sun is the most intense source in our environment, production is always superior under sunlight. A much smaller luminous flux is available inside (in a building, under artificial light). Recoverable light energy is much weaker than outside and there are far fewer potential applications. The pupil of our eye can adjust to different levels of light and contrast. Briefly, if the value of 1000 is applied to maximum radiation in sunny weather, corresponding to a solar flux of 1000 W/m^2, a cloudy sky will radiate between 100 and 500 (100–500 W/m^2) and an interior only between 1 and 10 (100–1000 lx). Thus, up to 1000 times more PV energy can be produced outside than inside.

It is therefore out of the question to install a PV pump indoors. As soon as the power required exceeds 1 W, PV modules must be placed outside. In an internal environment, only electronic applications for timekeeping and others with very low electrical consumption can be used.

PV collectors also differ according to the applications they are used for, with the technology being adapted for both strong and weak radiation. This is dealt with in detail in Chapter 3.

1.1.3   Direct or alternating current?

Solar cells and PV modules produce electricity in *direct current* (DC), like batteries, and not the *alternating current* (AC) of the grid, which is usually provided at a voltage of 220–240 V and frequency of 50 Hz.

To supply equipment in AC or to connect to the grid and feed in electricity produced from PV energy, it is therefore necessary to have DC/AC converters that produce AC from DC (*inverters*).

Note

These inverters are similar to the uninterruptible power supply device used in a computer except that the computer device includes a battery providing backup energy. This inverter stores the energy and supplies it in case of loss of power. The PV inverter on the other hand is only a DC/AC converter.

The DC voltages generated by the PV panels available on the market vary according to their application: to charge lead-acid batteries, the panels are of 12 or 24 V, while for connections to the grid, voltage is often higher, 40 or 72 V for example, according to the size of the solar array being built and the input voltage of the inverter.

1.1.4   How much does a PV module generate?

For an explanation of units (W, Wh, Ah, etc.), see Appendix 1.

Outside (in direct sunlight), the electrical output of a solar panel depends on

- its dimensions,
- its technology,

- the radiation received,
- the duration of exposure.

With optimal solar radiation of 1000 W/m^2, a crystalline silicon PV module of 1 m^2 produces an instant power output of around 130 W (at 13% module efficiency).

Over a day, if the panel is well positioned, the following guidelines may be applied:

- In France/Switzerland/Belgium, 1 m^2 generates between 150 and 650 Wh/day between winter and summer and depending on the region.
- In Africa, 1 m^2 generates between 400 and 800 Wh/day depending on the country.

These values are further explained later in the book. The price of a 1 m^2 panel of this type is of the order of €250–€500 (or €2–€4/W according to the technology used and quantity required).

1.1.5   Should PV energy be stored?

Some applications can in fact be run without batteries, either when they use DC or by pumping water or when the installation is connected to the grid (see Section 1.2).

But stand-alone feeds derived only from PV must inevitably be able to provide current permanently, including when the current consumed is more than the current produced at the moment of use, and without the possibility of recourse to another source of energy. Most applications with storage fall into this category, notably isolated domestic appliances such as lighting, mainly needed at night, and refrigerators, which need to run round the clock. Storage must be adequate to ensure function during periods when the PV production is zero or below what is required.

The batteries most used in the PV domain are still lead-acid batteries, which, although large, offer the best quality to price ratio. Not all lead batteries are suitable for solar application, and some models have been specifically developed for this purpose (see Section 5.1.1).

A battery's capacity, expressed in ampere × hours (Ah), describes the amount of electricity that can be stored under a nominal voltage: for example, with 100 Ah, 10 A is available for 10 h or 4 A for 25 h. Batteries normally used for solar storage are either 2 V cells mounted in a series or in blocks of 6 or 12 V, and have a capacity of 1–4000 Ah.

There are other forms of storage: nickel and super-capacity batteries, for example, for low-power electronic applications (see Section 5.1.1). It is also possible to store solar energy in hydraulic rather than electrical form (by solar-powered pumping), and in the form of compressed air or hydrogen. In the latter case, electrolysis of water is carried out when the solar electricity is produced and resulting hydrogen is stored in tanks, which can be used, for example, to run fuel cells. This is not without risk, but much experience already exists in this field, and it is perhaps a more sustainable solution and more ecological than batteries.

1.2   The components of a PV generator

A PV module or group of modules is only very rarely used on its own: in the case of a grid-connected system, there is at least an inverter as we have seen, and in the case of a stand-alone feed, a battery in most cases. Other elements are often necessary, and the whole constitutes a PV system or PV generator, the functions of which are illustrated in Figure 1.2.

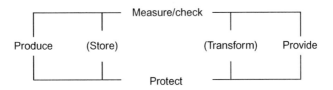

Figure 1.2   The functions of a PV generator

There are two major families of PV generators:

- *Grid-connected* installations, the electricity produced by which is fed into the collective grid.
- *Stand-alone* installations, intended to power certain functions on the spot, either without another energy source or with a complementary energy source, in which case they are described as *hybrid systems*.

Of the functions shown in Figure 1.2, those in brackets are not always present and are examined in detail later in this chapter. The functions of 'measure/check' and 'protect' are indispensable for verifying the functioning of the system and dealing with any malfunction, and preventing components from being damaged and ensuring that they last as long as possible. In a stand-alone system, the battery needs to be checked most often because it has the lowest life expectancy.

1.2.1   Stand-alone and hybrid systems

1.2.1.1   Stand-alone direct-coupled systems

These are the simplest systems since the PV energy is used directly from collecting panels without electrical storage. They are used in two versions.

Direct feed

In this case, functions are reduced to 'produce' and 'supply'.

The appliance receiving the current will only function when light is available at a sufficient level to provide the power required.

This is of interest for all applications that do not need to function during darkness and where the energy requirement coincides with the presence of daylight. If there is daylight, it works, if not, it stops.

But the panel or the solar cell has to be designed in such a way that there is sufficient power to work the application with the weakest illumination available, and this is often a serious constraint because high levels of illumination are not always available; there is no storage available, and so there is no recovery of any surplus solar energy.

Two concrete examples

Fans are widely used in hot countries. There is an obvious advantage of solar energy in this case: the overlap between the need for ventilation and the available supply of energy. The sunnier and hotter it is, the more the energy produced by the solar panel, and the faster the fan turns. (However, a protective device to prevent motor overheating is needed.) The correct functioning of the fan requires that it starts at 400 W/m^2. The PV element will be chosen to provide the starting energy of the motor at this radiation threshold. It may also be of use to have a short duration storage facility or an electronic booster (a sort of starter) to manage the calls for current at start-up.

The *pocket calculator* (one of the first, highly successful applications of the solar cell) also functions by direct feed. A condenser is placed as a 'buffer' between the photocell and the circuits to ensure the start-up current necessary for the circuits and to store the information in the memory in case of momentary loss of light. This is clearly a form of storage, but of very short duration. Regarding the choice of solar cell, it must be capable of powering the circuits at a level of illumination compatible with reading the screen (approximately 100 lx).

The direct-feed solar pump

Here, the requirement is to store water in a tank. The solar pump is directly connected to the solar panels via a regulator or a transformer. The tank is situated at a high point to receive the pumped water. The rate at which the water is delivered to the tank varies as a direct function of the solar radiation available. So there is storage in this case, but it is hydraulic storage (Figure 1.3).

The water supply is available at any time (see the solar pump presentation in Section 5.4.4 and the case study in Section 5.6.3).

1.2.1.2   Stand-alone systems with battery

This is the most usual configuration of stand-alone PV systems. The system normally functions with DC, which is preferable because it is simpler (although requiring considerable amount of cabling). But in the home, there are usually appliances that require AC because there are no DC alternatives (see Section 5.2).

The battery is at the heart of the system. It charges up during the day and acts as a permanent reservoir of energy, just like the water tank in the solar pumping system described in the earlier section. It can receive a charge current at any time and supply a discharge current of a different value. The appliances supplied are therefore connected to the battery through a charge controller. When the battery is

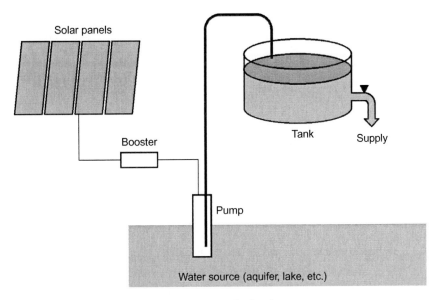

Figure 1.3   Direct-feed solar pump

full, the controller switches off the charge to prevent overcharging. (Charge controllers are described in detail in Chapter 5.) Consequently, the surplus energy produced (over and above that required for the application) is lost. This is what happens, for example, in temperate climates, if the consumption is constant throughout the year: the system is adjusted to balance consumption and production in winter and therefore there is a surplus production in summer, the surplus being absorbed by the charge controller. The huge advantage of grid-connected systems is to avoid this loss of energy (see Section 1.2.2). When an appliance requires AC (the form of electricity most widely used today), the battery power supply has to be converted from DC to AC. This is not without consequence since it increases the cost and size of the installation, reduces the energy efficiency (no inverter passes on 100% of the energy supplied) and increases the risk of total breakdown when the inverter fails if all the electric appliances are supplied through it.

Stand-alone systems and their detailed design are dealt with at greater depth in Chapter 5.

Cost example of a PV system with battery

We give later in this section an approximate budget for a *stand-alone system with battery* of the type represented in Figure 1.4, without DC/AC conversion. A more detailed budget is always necessary for such a project (see Chapter 5). It is often better to entrust such calculations to a solar energy professional, because the details always vary in individual cases and the energy balance must be correctly calculated so that the result is an efficient service at minimum cost.

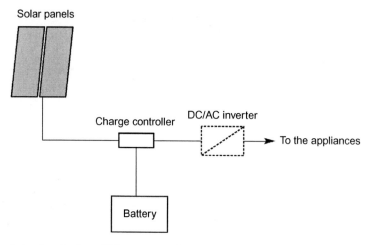

Figure 1.4   Stand-alone PV system with storage by a battery (with or without energy conversion)

Note

Contrary to what one might think, a 500 W panel should not be installed to supply an appliance that requires 500 W permanently. It's not so simple! Such a calculation does not include the 'time' parameter, which is essential. In fact, a panel rated at 500 W will only supply this power during the period of strong sunshine and for a variable period of time, at best just a few hours of the day. For the rest of the time it will produce less – and nothing at all during the night. Consequently, one could not rely on having a supply of 500 W from such a panel.

For a detailed budget, one needs to have solar radiation values provided by weather stations. Chapter 2 includes details on this point, but at this stage we are going to use data from the city of Nantes (western France), which are included in Appendix 2.

Stages of calculation
1. *Estimate the electrical consumption of the appliance over 24 h.* This is equal to the rated power multiplied by duration of operation over 24 h. For example, for an appliance rated at 20 W (at 24 V DC), the total daily consumption of the appliance will be 20 W × 24 h = 480 Wh.
2. *Estimate the output of the solar panels needed for the installation.* To do this, divide the consumption by the most unfavourable daily solar radiation over the period of use at the place of installation and the position of the panel (including orientation and angle of slope); see Further information panel.

For example, in Nantes in December, facing south and with an optimal angle of slope of 60° from the horizontal, the average total daily solar radiation value is 1.84 kWh/day (see Appendix 2).

Hence, the output of the solar panels required is 480 Wh/1.84 h = 261 W.

This result should be increased by a loss coefficient of 0.7 to give the preliminary estimate.

Total output required after allowing for losses is 261 Wc/0.7 = 373 Wc.

Bearing in mind that a solar panel of 100 Wc is approximately 0.8 m^2 in size, this application would require 3 m^2 of PV modules (four 100 Wc panels).

Further information

This method of calculation may seem strange. Why should the consumption be divided by solar radiation? The reasons is that, since sunshine is not constant during a day, the calculation is not based on what the panel produces at any given instant but over the course of a whole day. In order to do this, one assumes that the radiation during the day was constant at 1000 W/m^2 and that it lasts over a certain number of hours referred to as *number of equivalent hours*. In the above example, 1.84 kWh/m^2/day is based on the 1.84 h of solar radiation of 1000 W/m^2 (see detailed calculations in Chapter 5).

---

3. *Storage capacity.* Storage capacity is calculated for the number of days of autonomy necessary (generally taken as 7 days for France, to allow for successions of cloudy days). Values are then given in Ah.

   The capacity needed is therefore theoretically 480 Wh × 7 days/24 V = 140 Ah for 7 days.

   But the battery's capacity will be reduced by cold and other technical constraints, so this result needs to be divided by the loss coefficient, which is assumed to be 0.7 in this example.

   The true needed capacity is therefore 140 Ah/0.7 = 200 Ah.

4. *Charge control.* The charge regulator will be sized for 400 W of solar panels at 24 V, with a margin of around 50%, or a maximum current of 400 W/24 V × 1.5 = 25 A.

5. *Cost evaluation.* A PV system will therefore be composed of
   - four 100 Wc PV modules at 24 V,
   - one 200 Ah battery at 24 V,
   - one charge regulator at 24 V–30 A and
   - installation accessories: mechanical panel supports, cables junction boxes, etc.

A basic PV system of this type, without energy conversion, including PV modules, a charge regulator and solar lead batteries in open technology (the most recent), costs around €2400,[1] or €6/Wc, for the material, plus installation.

Now let us consider the main applications of stand-alone solar generators.

---

[1] Price excluding tax in 2008 for a professional consumer, but for small quantities.

Electricity in remote situations

Solar PV energy can be extremely valuable when there is no other source of energy and when panels can be installed outside. Often, the costs of connecting to the grid are much more than the cost of installing a PV system. Solar PV is most often used for

- rural electrification in developing countries: providing current for dwellings, health centres, agricultural uses, pumping, etc.;
- isolated dwellings, retreats, island communities, etc. and
- isolated professional installations (telecommunication relays for example).

Bearing in mind the millions of people in the world who are currently living off grid, there is no shortage of interest in this technology. However, there are sociological, technical and financial barriers to its widespread use. Installation budgets are considerable, and training and maintenance must also be financed. Many non-governmental organisations are active in this area, for example, in France by *Energies pour le monde* (Energy for the World – see Organisations and associations at the end of the book).

Portable electricity

Individual electronic devices, measuring equipment, leisure applications – electricity is everywhere. The use of solar energy in these portable applications depends on a number of factors, there is no general rule.

Although it depends largely on the situation, particularly the available light energy, it is by no means unusual that PV energy (often in the form of a very small solar cell) is competitive with disposable batteries, as in the case of the pocket calculator and some nautical applications.

The comparison of solar PV to rechargeable batteries depends on the consumption of the device in relation to its size and length of time without recharging. In the case of a small device, for instance a mobile telephone, there is often inadequate space for a solar cell large enough to provide a significant amount of the energy needed. This is especially the case when the device is normally used under artificial light, which is at a considerably lower energy level. It may therefore be worthwhile to couple two sources of energy, a solar cell, for example, providing or extending the standby consumption, and recharging through the mains providing the functional consumption.

1.2.1.3   Hybrid stand-alone systems

One of the limits of a purely PV stand-alone system of the sort described in the earlier section is that it supplies a limited amount of power, which varies according to the seasons. This means that one cannot use a higher consumption than is being produced, without risking damaging the battery by a deep discharge. And end users often have needs that change – and not necessarily in phase with the seasons!

With a hybrid system, another source of stand-alone electricity complements the PV energy source. This other source may be a generator or a wind turbine, for example.

A wind turbine will be favoured if the average wind speed is good, particularly if that is likely during seasons when sunshine is scarcer. But where a supply of diesel is available, a small generator is a better option (apart from the noise and exhaust!), because it can be switched on when needed. And if available it can also be used to recharge the battery when necessary (Figure 1.5).

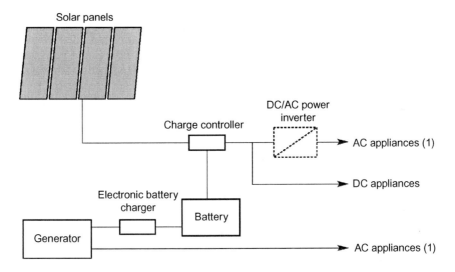

(1) The management of PC appliances running off the generator or the battery depends on the type of appliance and desired method of functioning.

Figure 1.5   Hybrid stand-alone system using PV and diesel generator

1.2.2   Grid-connected systems

Instead of directly powering appliances on the spot, grid-connected systems feed their electricity into a collective grid (Figure 1.6). They are installed on the ground or on buildings where there is space and good exposure to sunlight (Figure 1.7). The authorisation to feed into the grid and a contract from the electricity company to purchase the current are necessary pre-conditions for setting up such a system. They are dealt with in more detail in Chapter 4.

The huge advantage of this solution is that the grid fulfils the role of infinite storage and thus enables energy to be permanently available. Batteries and charge controllers are not needed. The electricity generated in DC must, however, still be converted to AC by means of an inverter according to the standards agreed by the electricity company receiving the current (sine wave quality and other parameters are essential). If the production site is also a user site (a dwelling for example), two

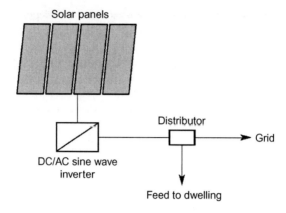

Figure 1.6   Grid-connected PV system

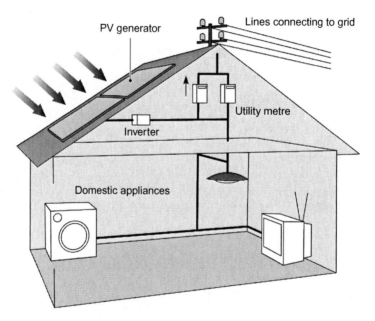

Figure 1.7   Solar PV roof scheme

solutions are possible: either all the current produced is sold and consumption is met by the electricity provided by the company or only the surplus current is sold. Often the first solution is more profitable for the owner of the grid-connected PV generator, because the feed-in tariff offered by the electricity company is considerably higher than the price they have to pay: in France, from an integral PV generator, electricity can be sold to the grid at €0.60/kWh, while the mains electricity bought directly from EDF (Electricité de France) costs only €0.08–€0.10/ kWh (2009 prices).

Note

These systems cannot provide emergency backup in the case of a grid power cut, because they have no energy storage capacity. For this reason, in some cases an emergency battery is built in to provide backup during short power cuts.

Compared to a stand-alone system, it has the following advantages:

- exploitation of all the PV energy generated by the panels (storage is infinite),
- an economy of around 40% on investment (batteries not needed), virtually maintenance free (in stand-alone systems the batteries require the most attention) and
- improved life expectancy.

1.2.2.1 The economics of the grid-connected system

Although only 4 or 5 years ago it was hard to imagine that PV could ever be competitive with price of electricity supply from the grid, today, with the price of modules in bulk falling to around €2/W or even less, PV has begun to be economic, with return on investment (time to amortise the investment) reducing all the time, from 5 to 10 years. In certain cases, with very low panel prices and very favourable sunshine, a PV generator can today produce electricity at the same cost as the grid. Parity with the price of the grid kWh has been reached in California and southern Spain, for example.

But the extension of PV energy production, from private integrated solar roofs of 3 kW to the largest PV power stations in the world, like the 40 MW Waldpolenz solar park near Leipzig in Germany, calls for heavy investment, whether public or private. For this investment to be amortised, the system must pay for itself, either in the form of savings on energy bills or through cash flow.

So the further deployment of PV systems depends today on the economics, and these in turn on three essential factors: first, the reduction of the costs of solar modules (and so the size of the installation); second, government policies encouraging this investment, and finally, favourable feed-in tariffs from the electricity companies. On this last point, several countries are passing legislation to push the energy companies to sign feed-in contracts.

Countries with an ambitious energy policy are introducing the following measures:

- an advantageous feed-in tariff of up to €0.60/kWh in France in 2009 (for building-integrated modules), which is considerably higher than the sale price of traditional electricity and is guaranteed for 20 years;
- low-cost loans to finance new installations;
- regional subsidies;
- tax credits.

In Europe so far, Germany in particular, and Spain, France and Italy are using these incentives.

To obtain a favourable return on investment, energy production needs to be maximised. Cost-effectiveness is thus higher in the sunnier countries, as is shown by the boom in the market in Spain: over 2.5 GW of PV generating capacity was installed in 2008 alone, up from only 0.5 GW at the end of 2007, and only 4.7 GW for the whole of Europe.

1.2.2.2   Example of the cost-effectiveness of a roof-integrated PV generator connected to the grid

This example is based on a 3 kWc domestic system in the south of France. In this region, the output of a solar panel 1 kWc is estimated at 1300 kWh/year. Thus, 3 kWc will generate 3900 kWh/year. The map in Figure 1.8 shows the annual estimated generation for different regions of France, per installed kWc, assuming optimum orientation of the panels.

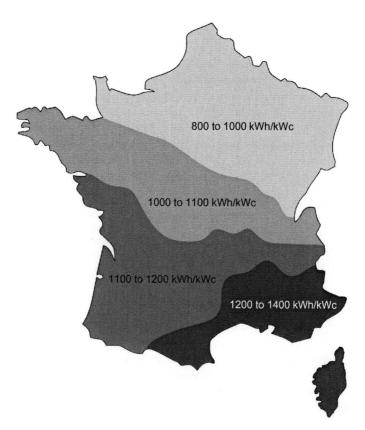

Figure 1.8   Estimated annual production of PV energy in France (in kWh/kWc with optimum orientation)

Bearing in mind that the price per kWh charged by EDF is around €0.10 (2009 price), the 3900 kWh generated would have cost €390/year if bought from the grid. The PV system costs in total around €5/W including installation, so a total of €15,000 in this case.

Self supply

If the user is producing energy only for their own use, without feeding into the grid, the time needed to amortise the user's investment would be €15,000/€390/year = 38 years, without aid, subsidies or resale of electricity.

The 'all PV' solution for stand-alone consumption is thus not justified, because it is not competitive with electricity supplied by the grid.

Resale of energy to the electricity company

Let us now look at all the costs and revenue, including the purchase of PV current by the electricity company (at €0.6/kWh in the case of roof-integrated modules), and the result is given in Table 1.1.

Table 1.1   PV current balance sheet

Expenditure	
Cost excluding tax of basic materials	€15,000
VAT at 5.5% (house in France more than 2 years old)	€825
Cost of connection to the grid	€1,000
Replacement of inverter after 10 years	€4,000
Total fixed cost	€20,825
Annual maintenance costs	€200/year
Income	
Annual sale of power produced	€2,340/year (€0.6 × 3,900 kWh/year)
Result	
Balance of annual income	€2,340–€200 = €2,140/year
Period of return on investment	€20,825/€2,140 = 9/10 years

This is only one example; with regional subsidies or a tax credit, the investment would be amortised more quickly. But with a lower PV output (for example, in northern France or with a less favourable orientation of the roof), the breakeven point would take longer to reach.

There are several other examples in Chapter 4, which is entirely devoted to grid-connected systems. In France, it is useful to consult the Perseus Guide intended for potential PV installers (see Bibliography). Also, ADEME (the French Environment and Energy Management Agency)[2] and the Hespul Association[3] give

[2] http://www.ademe.fr
[3] http://www.hespul.org

up-to-date information on the national and regional aid and subsidies available and set out the necessary steps in the installation process.

1.2.3   Tracker systems and concentrating systems

A tracker system is an array of panels that follows the trajectory of the Sun in order to maximise the power generated. The tracking may be done in one or two planes, and the systems need to be equipped with an articulated mechanism, more or less sophisticated, guided either by an astronomical clock or by a detection cell to ensure that the array is oriented towards the Sun at all times. The extra cost of the orientation equipment has to be covered by the extra production generated, and it is consequently rarely economic for a single panel: generally an array of several panels is mounted on the tracker, comprising total surface areas of at least 10 m^2. A number of tracker arrays can be mounted on the ground and linked to make a solar power station. An example is described in Section 4.4.3.

Additionally, on the model of the thermodynamic power stations described at the beginning of this chapter, solar radiation concentrators can be focused on very high efficiency PV cells (40% and more) to produce electricity with a total efficiency of up to 25%. Again, the Sun's trajectory must be tracked to ensure maximum efficiency of the concentrator, so these arrays must be able to track in two planes. The first PV concentrators were based on huge cylindrical or parabolic mirrors. Today they are increasingly made with smaller lenses, or with small juxtaposed parabolic mirrors, focused on solar cells.

For these systems to work efficiently, the Sun must be directly visible in the sky; as soon as it is cloudy, diffuse radiation dominates (see Section 2.2.2) and the concentrators and trackers are of no use. This is why they are mainly deployed in Spain and California, for example. Also, unlike fixed systems, they need mechanical maintenance and also cleaning when there are mirrors, both of which are constraints that have to be taken into account in budgeting. They are thus really only of interest in extremely sunny regions and where maintenance can easily be carried out.

1.3   The role of PV in sustainable development

Renewable energies in general, and PV in particular, are often looked on as sustainable alternative solutions to current energy and resource problems, at least as far as electricity is concerned. We give in the following sections the main arguments supporting this view.

1.3.1   Impacts on the planet

1. *The Sun's energy is the most renewable of all sources*
   The Sun is the primary source of energy present on Earth, except for geothermal whose energy derives from the Earth's core. The International Energy Agency[4]

---

[4] http://www.iea.org

has calculated that a surface area of 145,000 km², or 4% of the most arid deserts, would be sufficient to meet all the energy needs of the planet.

2. *Solar PV energy preserves natural resources*
   Silicon is one of the most abundant materials in the Earth's crust, so solar PV energy preserves natural resources. It is true that huge quantities of material have to be treated to obtain pure silicon, but despite that, the added value obtainable from each unit of the material is high, since silicon continues to function for more than 20 years. This argument is only valid for solar panels using silicon. Other technologies use heavy metals and sometimes rare or even toxic elements like cadmium, indium or gallium (see Sections 3.3 and 3.4), which raises questions about the long-term use of these materials.

3. *The use of PV technology reduces the amount of energy consumed to produce electricity*
   It is what is called 'grey energy', in comparison with other methods of production. It is estimated today that a solar panel generates the energy required for its manufacture in only a few years (4–6 depending on the technology used).

4. *The manufacture of solar panels uses mainly recyclable or recycled materials*
   Silicon often comes from recycled electronic waste, while the glass for mirrors and lenses and the aluminium used in arrays and mechanical supports are all materials that already benefit from a high rate of recycling.

5. *Electricity generated by a PV generator does not emit any greenhouse gases*
   No pollution comparable to traditional methods of generation is produced. PV energy is estimated to produce 600 g/kWh less CO_2 emissions than coal-fired power stations, and up to 900 g/kWh less in an isolated location when it replaces oil-fired generation. According to the European Photovoltaic Industry Association (EPIA),[5] by 2030, solar PV will enable global CO_2 emissions to be reduced by 1.6 billion tonnes per year, or the equivalent of 450 coal-fired power stations with an average generating power of 750 MW.

6. *It is a reliable and sustainable energy source*
   PV generators are modular and easy to install and maintain. They only have a minimal amount of wear and tear. Their life expectancy is 20–30 years.

1.3.2 Human impacts

1. *This industry minimises toxic waste*
   Pollution generated by the manufacture of solar cells is relatively small (except those cells that use certain risky materials like cadmium, see Section 3.3). And there is absolutely no toxic emission while electricity is being generated by the solar panels.

---

[5] http://www.epia.org

2. *The technology can help improve public health*
   Especially in countries with low population density, where degree of electrification is likely to be low, refrigeration from solar PV enables food, drugs and vaccines to be preserved, thus contributing to hygiene and health in the developing countries. In the same way, solar pumping and water purification systems improve access to clean drinking water.

3. *The technology fosters human development*
   By bringing electricity to remote rural areas, PV makes a notable improvement to local standards of living: education is better when a school has lighting and is equipped with fans and a TV set; farming is improved by the introduction of irrigation pumps and mechanical appliances such as grain mills; craftsmen's work is made easier when electricity is available for sewing machines, tools and lighting in the evening.

4. *The technology reduces rural depopulation and excessive urbanisation*
   Consequently, the availability of solar PV reduces rural depopulation and the drift to the cities, which is a major headache for poor countries, unable to provide jobs and decent accommodation for all these rural migrants.

5. *PV contributes indirectly to the prevention of global overpopulation*
   Insofar as there is generally a direct link between an increase in living standards and a reduction in the birth rate, PV technology contributes indirectly to the prevention of global overpopulation.

6. *PV generates economic activity and jobs*
   In countries where solar panels are produced, but also more or less everywhere where they are sold, installed and maintained, PV technology generates economic activity and additional jobs.

Chapter 2
Light energy and photovoltaic conversion

2.1 Light in all its forms

When discussing the phenomenon of light in physics, the first constant that comes to mind is the speed of light, which can never be equalled and even less exceeded. Nothing can travel faster than light, and it is on this axiom that Einstein based his famous theory of relativity. He discovered that matter (m) is energy (E) and vice versa, and that these constants are linked by the square of the speed of light. Hence, the famous formula $E = mc^2$. In a vacuum, the speed of light c is 299,792,458 m/s, which means light can travel from the Earth to the Moon in just over a second.

Light appears to our eyes like a ray travelling in a straight line, and obeying certain laws of geometric optics: reflection from a surface, diffraction (the deviation of a beam entering a transparent body), focussing by a lens, diffusion on a rough surface, etc. All these phenomena come into play in the capturing of light in a solar cell (see Section 2.3.1).

But they do not explain everything – far from it. Why do we need light for our eye to see our surroundings? How can light travel through glass? How is a rainbow formed? There are many such questions, and to try to answer them, scientists since the Middle Ages have sought to describe the profound nature of light and developed many theories, some of them contradictory.

2.1.1 Wave–particle duality

If a beam of light passes through two slits close together onto a screen behind, it appears not as two spots of light but as a group of alternately dark and light bands, an *interference pattern*. This phenomenon can only be explained if light is a wave moving in space. When two waves arrive at the same point, they can either reinforce each other or cancel each other out, creating the alternating dark and white bands seen on the screen. Many experiments have confirmed this wave theory of light, in particular, the observations of the astronomer Huygens and the work of Young, Fresnel, Arago and Maxwell, who devised formulae to describe the movement of light waves.

Other physicists like Newton argued for a different view of light, that it consists of a beam of particles, to explain reflection: for example, grains of light 'bounce' off the mirror.

It was only in the twentieth century that these two series, the wave and corpuscular theories of light, were finally reconciled following the discovery of the *photon* by Planck and Einstein. For, in practice, light does actually have a double nature.

- It is an electromagnetic wave or a periodic oscillation characterised by its wavelength λ (spatial periodicity, Figure 2.1) or its frequency v: the higher the frequency, the shorter the wavelength and vice versa; $v = c/\lambda$, where c is the speed of light. In the visible part of the solar spectrum (see Section 2.2.2), the wavelength is shown by the 'colour' of the light.
- It is also a bundle of photons that are like 'particles of light' carrying energy, each according to their wavelength, according to the formula of Louis de Broglie (1924):

$$E = hv = \frac{hc}{\lambda} \tag{2.1}$$

where h is Planck's constant.

It is this energy carried by the photons that enables PV conversion by liberating electrical charges in the material (see Section 2.3.2).

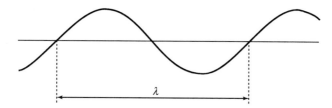

Figure 2.1   Definition of wavelength (λ)

2.1.1.1   Spectral distribution of electromagnetic waves

The dividing of light waves according to their wavelength is called *spectral distribution* or the *spectrum*. For visible light, this is demonstrated by the fact that white light is in fact made up of many colours as can be seen when it is refracted through a prism (Figure 2.2) or in a rainbow. But electromagnetic waves are not all contained in the visible spectrum, which in reality only represents a tiny proportion of all known electromagnetic radiation. Obviously, the term 'light' refers to this visible part seen by man, but by extension, it is often applied to the entire solar spectrum, which ranges from near ultraviolet (250 nm) to near infrared (10 µm) (see Section 2.2.2 for details).

Table 2.1 gives a brief description of all electromagnetic waves, their wavelength domain and some of their applications.

2.1.2   Sources of light

Now let us consider these sources of light in our environment. The main source of natural light and by far the strongest is of course the Sun, to which we will return in

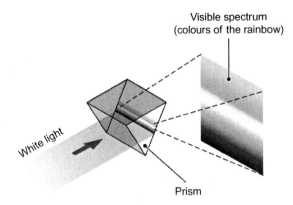

Figure 2.2   Dispersal of white light through a prism

Table 2.1   Wavelength distribution of electromagnetic waves

	Wavelength	**Frequency**	**Examples of use**
Gamma rays	<0.01 nm		
X-rays	0.01–10 nm		Radiography
Ultraviolet	10–400 nm		Suntanning, water purification
Visible	400–800 nm		Daytime vision, photosynthesis
Near infrared	800 nm–10 μm		Nocturnal vision
Thermal infrared	10 μm–1 mm		Heating, cooking
Microwaves	1 mm–10 cm		Microwave ovens
Radar waves	10 cm–1 m	3 GHz–300 MHz	Mobile telephone, speed detectors
Radio waves	>1 m	<300 MHz	Radio, TV, telecommunications

detail in Section 2.2. PV and solar thermal appliances were by definition developed to convert this energy of solar origin.

But this is not the only source of light, and since discovering fire, mankind has invented and manufactured many sources of artificial light.

Fire, torches, candles, oil and paraffin lamps that produce light from combustion are concrete illustrations of Einstein's energy/matter equivalence. In these applications, the decomposition of the material used as fuel produces the emission of light. These are the oldest traditional sources used by mankind to provide light at night. And when early physicists attempted to measure light, they naturally took the amount emitted by a candle as a unit. Candlepower was used as a measuring unit and later the candela, defined several times and finally fixed in 1979.[1]

---

[1] The candela is the luminous intensity, in a given direction, of a source that emits monochromatic radiation of frequency 540×10^{12} Hz and that has a radiant intensity in that direction of 1/683 W/sr.

The discovery of electricity led to many different sources of electric light: incandescent bulbs, then halogens, fluorescent tubes, discharge and semiconductor lights (electroluminescent diodes, LED).

2.1.2.1   Types of sources of light

All these sources of light may be classified into four categories, according to the type of spectrum they emit, or in other words according to the distribution of luminous energy emitted in different wavelengths.

Continuous spectrum
In this type of spectrum, luminous energy is emitted continuously over each wavelength. They are mainly thermal sources that use heat as a source of energy. Examples are incandescent and halogen light bulbs, the Sun or a candle (Figure 2.3).

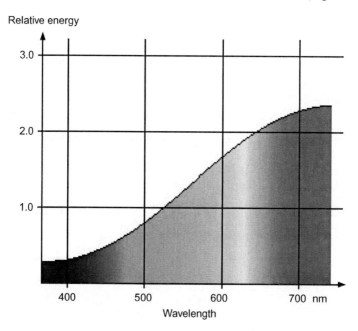

Figure 2.3   Continuous spectrum emitted by the halogen bulb

Discontinuous spectrum
In this type of spectrum, the band is broken by numerous gaps, from which no luminous energy is emitted. Light sources using an electrical discharge in an ionised gas generally emit a discontinuous spectrum (Figure 2.4).

Combined spectrum
This is a combination of continuous and discontinuous spectra. This particular type is emitted by light sources with modified electrical discharge, such as fluorescent tubes (Figure 2.5).

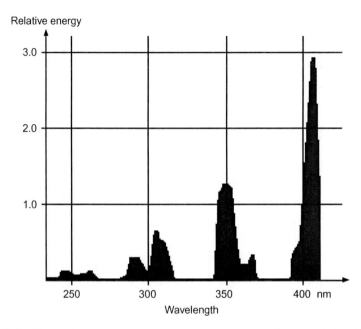

Figure 2.4 Discontinuous spectrum of a mercury vapour lamp, emitting in the ultraviolet wavelength

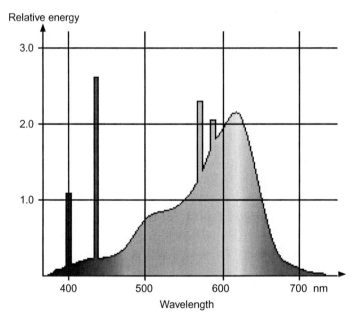

Figure 2.5 Combined spectrum of a 'warm white' type fluorescent tube

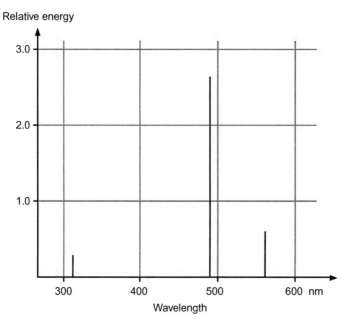

Figure 2.6   The three main emission wavelengths of the argon-ion laser

These tubes are the most common source of artificial lighting in public places today. They are sometimes wrongly referred to as neon tubes because of an earlier version, which was filled with neon gas. The form of these spectra is important for PV. For example, they show that combined sources like fluorescent tubes include an important element of blue light, which is well absorbed by amorphous silicon, and this enables solar modules using this material to produce current when exposed to this type of lighting.

Laser spectra
Some light sources like lasers or laser diodes emit only in a few wavelengths (Figure 2.6). When associated with the narrow bandwidth filters, these sources become practically monochromatic.

2.1.2.2   Colour temperature

By comparing the continuous spectrum emitted by a thermal source to that of a 'blackbody', an ideal object whose emission depends only on temperature, each thermal source can be assigned a colour temperature value, expressed in Kelvin, which defines the spectral distribution of this source. This temperature value describes the apparent colour of the light source, which varies from the orange red of a candle flame (1800 K) to the bluish white of an electronic flash (between 5000 and 6500 K, depending on the manufacturer). Paradoxically, a blue coloured light,

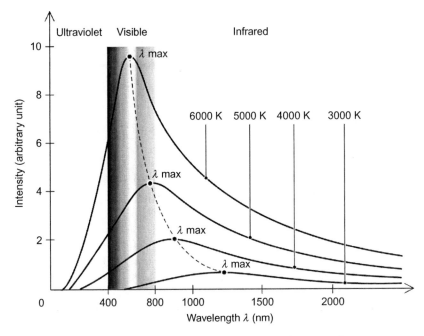

Figure 2.7 Emission spectrum of a blackbody according to colour temperature

which appears colder to the eye, in fact corresponds to a higher colour temperature, and vice versa, with a red colour having a lower colour temperature (Figure 2.7).

- candle: 1800 K
- Sun on the horizon: 2000 K
- sodium vapour lamp: 2200 K
- incandescent bulb: 2400–2700 K
- warm white fluorescent tube: 2700–3000 K
- metallic halogen lamp: 3000–4200 K
- halogen lamp: 3000–3200 K
- neutral white fluorescent tube: 3900–4200 K
- midday sunshine (cloudless sky): 5500–5800 K
- solar spectrum AM 0 (see Section 2.2.2): 5900 K
- daylight fluorescent tube: 5400–6100 K
- electronic flash: 5000–6500 K
- cloudy sky: 7000–9000 K

It can be seen that the Sun's spectral distribution varies according to the time of day, ranging from a reddish sauce at dawn to sunset to a much bluer source in cloudy conditions. This is important in understanding PV phenomena and we will return to the subject when we compare the different technologies in Chapter 3.

2.2   Terrestrial solar radiation

2.2.1   Geometry of the Earth/Sun

The Sun is a pseudo-spherical star with a maximum diameter of 1,391,000 km. Its average distance from the Earth is 149,598,000 km. Composed of gaseous matter, mainly hydrogen and helium, it undergoes permanent nuclear fusion reactions, and its core temperature reaches 10^7 K.

2.2.1.1   Movements of the Earth

The Earth revolves around the Sun on a slightly elliptical trajectory with the Sun as its focus (Figure 2.8). In fact, the distance separating them varies by ±1.69% during the year because of the slight eccentricity of the terrestrial orbit ($e = 0.017$).

The Earth's rotational axis is inclined by 23°27′ from the *ecliptic plane* (terrestrial orbit plane). The angle formed by the axis of Earth/Sun with the equatorial plane at a given moment of the year is called *declination* δ. Thus, the declination value is +23°27′ at the summer solstice, −23°27′ at the winter solstice and zero at the equinoxes. This declination is responsible for the seasons, since in the northern hemisphere the Sun's rays reach us from a higher angle in summer and a lower angle to the horizon in winter (the reverse is true in the southern hemisphere). It also explains why the seasonal differences are more marked in higher latitudes. We also know that solar activity is not constant and that it varies according to solar eruptions, but the variation of radiation intensity does not exceed 4%.

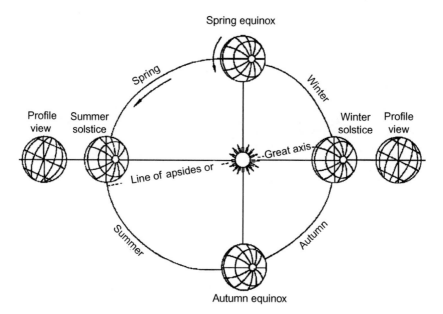

Figure 2.8   Plane of the ecliptic: the terrestrial orbit and the seasons

2.2.1.2   Apparent trajectory of the Sun

For an observer on the surface of the Earth, the Sun describes an apparent trajectory that depends on the latitude and longitude of the place of observation, latitude being the angular distance from any point on the globe compared to the equator (from 0° to 90° in the northern hemisphere), and longitude the angle measured from east or west of the Greenwich Meridian (the arc of the circle passing through the two poles and the town of Greenwich in England). The position of the Sun is defined by two angles: its angular height h – the angle between the direction of the Sun and the horizontal plane of the location – and its azimuth a – the angle between the meridian of the location and the vertical plane passing through the Sun, measured negatively towards the east (Figure 2.9).

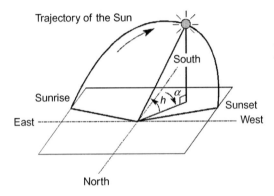

Figure 2.9   Definitions of the Sun's position (height and azimuth)

Figure 2.10 shows the trajectories of the Sun at a given place, and the Sun's height and azimuth for a particular moment in the year.

2.2.2   Solar radiation characteristics

2.2.2.1   Renewable energy

The energy that reaches us from the Sun represents virtually all of the energy available on Earth. Apart from direct contributions in the form of light and heat, it creates biomass (photosynthesis) and drives the cycles of water, winds and ocean currents, and millions of years ago, it laid down the world's reserves of gas, oil and coal.

The only non-solar energy resources are the heat of the Earth (geothermal), the tides and nuclear energy.

The Sun's energy is produced by thermonuclear fusion reactions: hydrogen nuclei (protons) fuse with helium nuclei (two protons + two neutrons). This energy is launched into space from the Sun's surface, mainly in the form of electromagnetic waves in the visible, ultraviolet and infrared spectrum.

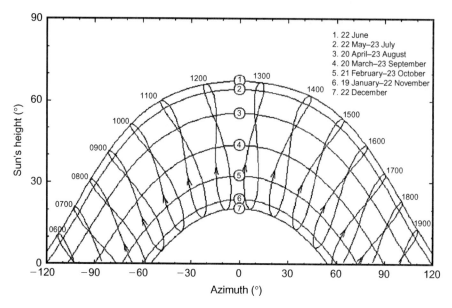

Figure 2.10   Sun's trajectories at Geneva (latitude 46°12′ N, longitude 6°09′ E)

2.2.2.2   Role of the atmosphere

The Sun's light energy before it reaches the Earth's atmosphere has been precisely measured by NASA at 1357 W/m². This is the instantaneous solar radiation (irradiance) received at a given moment above the Earth's atmosphere at normal incidence (at a plane perpendicular to the Sun's direction). This value is called the *solar constant*, although it does change slightly because of the small variations in the distance between the Earth and the Sun and in solar activity.

But the full force of this energy does not reach the surface of our planet because it undergoes transformations due to absorption and diffusion while passing through the atmosphere.

The atmosphere contains a majority of nitrogen and oxygen (78% and 21%, respectively) and also argon, CO_2, water vapour and the famous stratospheric ozone layer, which has an important role in filtering the most dangerous ultraviolet rays. Dust and clouds (formed from tiny droplets of water, and not to be confused with water vapour, which is a gas) are also important in diffusing solar radiation.

2.2.2.3   Airmass

The closer the Sun is to the horizon, the greater the thickness of the atmosphere that sunlight must pass through, and the greater the diffusion effect.

The ratio between the thickness of the atmosphere crossed by the direct radiation to reach the Earth and the thickness vertically above the location is called *airmass* (Figure 2.11).

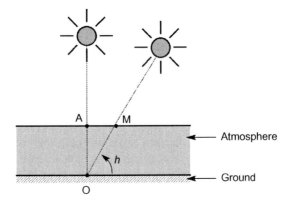

Figure 2.11   Definition of airmass

This principally depends on the angular height h of the Sun as previously defined (Figure 2.9). Using the points O, A and M and the angle h shown in Figure 2.11, the length of the Sun's trajectory through the atmosphere can be measured:

$$OM = \frac{OA}{\sin h}$$

Therefore, airmass

$$\frac{OM}{OA} = \frac{1}{\sin h} \tag{2.2}$$

In the expression AM x, x designates the ratio OM/OA.

Examples

AM 0: conventionally used to designate solar radiation outside the atmosphere
AM 1: Sun at the zenith (sea level)
AM 1.5: Sun at 41.8° – chosen as reference for PV
AM 2: Sun at 30°

2.2.2.4   Direct, diffuse and total radiation

As the Sun's radiation passes through the atmosphere, it is partially absorbed and scattered. At ground level, several different components can be distinguished.

Direct radiation is received from the Sun in a straight line, without diffusion by the atmosphere. Its rays are parallel to each other; direct radiation therefore casts shadows and can be concentrated by mirrors.

Diffuse radiation consists of light scattered by the atmosphere (air, clouds, aerosols). Diffusion is a phenomenon that scatters parallel beams into a multitude of beams travelling in all directions. In the sky, the Sun's rays are scattered by air molecules, water droplets (clouds) and dust. The degree of diffusion is thus mainly dependent on weather conditions. In cloudy weather, diffuse radiation is described as *isotropic*, that is, identical radiation is received from all directions of the sky. When the sky is clear or misty, as well as when it is relatively isotropic blue (diffusion by the air), there is a more brilliant ring around the Sun (component known as *circumsolar brightening*) and often a brighter band on the horizon, called *horizon brightening*.

Albedo is the part reflected by the ground, which depends on the environment of the location. Snow, for example, reflects a massive amount of light radiation whereas asphalt reflects practically none. This has to be taken into account in evaluating radiation on sloped surfaces.

Total radiation is simply the sum of these various contributions as shown in Figure 2.12.

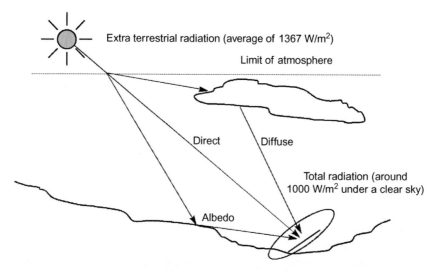

Figure 2.12   Components of solar radiation on the ground

Normal direct radiation is defined as the direct radiation measured perpendicularly to the Sun's rays. Measured at an angle, the same radiation covers a larger surface and is reduced in proportion to the cosine of the angle: this is called the *cosine effect* (Figure 2.13).

This effect explains why direct radiation onto a horizontal plane is always inferior to radiation received on a plane perpendicular to the Sun. However, the diffuse radiation received may be greater because the horizontal plane 'sees' a larger part of the sky.

This phenomenon leads designers of solar generators to use a horizontal installation of solar panels normally receiving diffuse radiation, in geographical

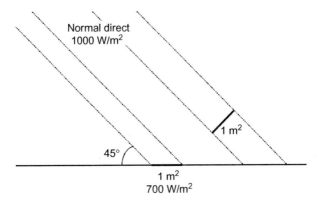

Figure 2.13 The cosine effect

sites where the sky is often cloudy. When the Sun is rarely visible, direct radiation is less intense, and it is better to expose the panel to the whole sky to recover a maximum of diffuse radiation (see case study in Section 5.6.1).

2.2.2.5 Solar spectrum

The solar spectrum is defined as the distribution of sunlight in wavelengths or colours as described in Section 2.1.1.1. Sunlight is made up of all kinds of rays of different colours, characterised by their wavelength.

Photons, the particles that make up this electromagnetic radiation, carry energy that is related to their wavelength according to the formula:

$$E = h\nu = \frac{hc}{\lambda}$$

where h is Planck's constant, ν the frequency, c the speed of light and λ the wavelength (see Section 2.1.1).

The standard curve of the spectral distribution of extraterrestrial solar radiation AM 0, compiled from data received by satellites, is as follows:

Ultraviolet UV	$0.20 < \lambda < 0.38$ µm	6.4%
Visible	$0.38 < \lambda < 0.78$ µm	48.0%
Infrared IR	$0.78 < \lambda < 10$ µm	45.6%

For units of measurement, see Appendix 1.

Figure 2.14 shows the attenuation of solar radiation observed after passing through an atmospheric density corresponding to AM 1.5, the equivalent of a solar elevation of 41.8° at sea level. Spectral irradiance is the solar flux for a given wavelength (and thus for a given colour as far as visible light is concerned).

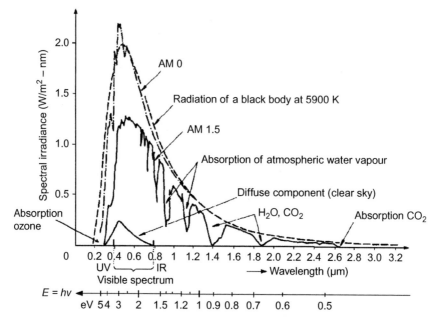

Figure 2.14   Spectral distribution of solar radiation: (a) outside the atmosphere (AM 0); (b) at an incidence of 41.8° (AM 1.5)

This clear sky spectrum, described as AM 1.5, is used as a reference to measure PV cells. The simulators used to measure the modules try to reproduce it as accurately as possible. Note also the spectrum of fine weather diffused radiation, which shows a bias towards blue because of the effect of Rayleigh scattering.

On the AM 1.5 spectrum, the absorption bands corresponding to atmospheric gases, notably CO_2 and water vapour, can be clearly seen. Also represented is the spectrum of a blackbody with a colour temperature of 5900 K, very close to the AM 0 solar spectrum. As we have seen earlier in this chapter, the Sun is taken as equivalent to this blackbody, which enables physicists to develop models to explain its behaviour and its radiation (see Section 2.1.2).

In practice, it should not be forgotten that this spectrum defines a reference solar radiation, not one that is at all permanent. The reality is much more diverse and complex. For example, as we have seen in Section 2.1.2.2, the colour temperature and thus the spectrum of sunlight can vary from 2000 to 10,000 K.

2.2.3   Solar radiation on Earth

The design of PV systems requires an accurate knowledge of the solar radiation available for the solar panels on the installation site. This is one of the essential parameters for the preliminary study: for a given requirement of electricity, the higher the amount of solar energy available means fewer solar panels to install, and vice versa.

One might assume that because extraterrestrial solar energy is a known value, as is the Sun's trajectory at all points of the globe, it would be easy to work out the solar energy received at any point on the ground. But this would be to overlook the influence of the atmosphere, which is responsible for the scattering and absorption of part of the incident radiation. Diffusion represents more than 50% of the annual available radiation in temperate regions, as against 30–45% in sunnier countries and in the mountains, and 15–20% on the finest days, even in the tropics.

The modification by the atmosphere of solar radiation is subject to complex and largely random phenomena. The state of the sky and thus the light flux received at ground level at any given time depends on a large number of parameters:

- gases present in the atmosphere,
- clouds,
- albedo (ground reflectivity),
- the ambient temperature,
- wind,
- relative humidity, etc.

  In turn, these parameters depend on

- the geographical location,
- the season,
- the time of day,
- the weather conditions at the time.

Of course, scientists, and in particular climatologists, have developed models to describe and predict atmospheric phenomena, but the surest means of obtaining reliable data is still to consult statistics accumulated in earlier years by measuring. In time, this may change, and account must be taken of that, but these changes are relatively slow compared to the safety margins, which will, in any case, be built in.

Let us now review the instruments used to measure solar radiation received on the ground.

2.2.3.1 Measuring instruments

The *heliograph* is the oldest instrument used for this purpose. It records the duration of sunshine, or more precisely, the period of the day during which solar radiation exceeded a certain threshold. The solar radiation is concentrated by a lens to burn a record of the day's sunshine on slowly moving paper. This information is mainly of use to horticulturists and farmers, since the growth of some plants depends on this duration, and chickens lay more eggs if daylight is artificially prolonged.

For PV, this device is not very useful because it provides no information on radiation intensity.

Note

It is important to avoid a common error that consists of thinking of an 8 h day as 8 h of standard solar radiation at 1000 W/m² AM 1.5, for example. In fact, solar radiation is never constant (see Appendix 1 for useful units of measurement).

The *pyranometer* is the most useful measuring device because it can estimate, with the help of a thermopile, the total solar irradiance (direct and diffuse) on a given surface, over a very broad spectrum, from 0.3 to 3 μm of wavelength (Figure 2.15). Its glass dome provides very wide angular acceptance, almost a hemisphere (it can collect radiation coming from all directions, even horizontal rays).

Figure 2.15   A second-class pyranometer according to ISO classification 9060 – DeltaOhm (Italy)

It can also, uniquely, measure the diffuse part of solar radiation. This is done by masking the direct radiation with a curved cover that follows the Sun's trajectory to eliminate direct radiation.

The *pyrheliometer* only measures direct irradiance. It tracks the Sun's movements and measures radiation by means of a detector placed at the bottom of a tube with a narrow opening.

These instruments are installed by professionals in meteorological recording stations and research centres. They are expensive and somewhat difficult to use. For day-to-day but accurate measurement, solar energy professionals use reference solar cells calibrated by professional laboratories (LCIE, Ispra, Fraunhofer Institut, etc.). This enables manufacturers to calibrate the electrical measuring devices of their PV modules and to verify their behaviour on the ground against solar radiation at any given moment.

An even more economical solution for the installer is a simple solarimeter with a small crystalline silicon cell, which is 95% accurate. Their spectral response is narrower than that of a thermopile – only from 400 to 1100 nm – but by definition similar to that of silicon modules. Some solar panel manufacturers wrongly refer to these silicon cell solar radiation sensors as 'silicon pyranometers'.

2.2.3.2 Meteorological databases

By using the instruments described in the previous section in different positions and orientations, meteorological observation stations can compile solar radiation statistics from billions of data collected. Databases are compiled with this information along with other useful data such as minimal and maximal temperatures, humidity rates, etc. Unfortunately, access to this information is not easy to obtain and almost never free.

These irradiance data are indispensable in the sizing of a PV installation. For Europe, the information can be found in the *European Solar Radiation Atlas*.[2] It contains numerous maps (including the map in Figure A2.1 in Appendix 2) and average values over 10 years of total and diffuse solar irradiance, for fine or average weather on all days, for different orientations and inclinations, recorded by several dozen European weather stations.

In Appendix 2, some of the data tables for the main towns of Europe are reproduced, as well as some for other places in the world, supplied by international databases. It is important to note the exact conditions in which these data are valid (see caption to tables).

Other data can be obtained from irradiance databases accessed through sizing software such as Meteonorm 2000 (version 4.0)[3] and PVSYST[4], both produced in Switzerland. The meteorological database of NASA[5] is also very comprehensive, and part of this can be downloaded from the Canadian free software site RETscreen[6] (more information on sizing software is given in Chapters 4 and 5).

2.2.3.3 Using data

Useful measures

There are two types of irradiance data: instant values and cumulative values.

Cumulative radiation

For most of the time (as is shown in Chapter 5), the sizing of a PV system is based on the monthly averages of daily solar energy of the region. For this, it is enough to have *12 cumulative daily values* (for the 12 months of the year) for the plane of the panels. More precisely, the physical value used is the *total solar irradiance averaged over 1 day*, more simply referred to as *total daily irradiance*: this is the total

---

[2] W. Palz, J. Greif (eds.), *European Solar Radiation Atlas: Solar Radiation on Horizontal and Inclined Surfaces*, Springer–Verlag, Berlin; 2008.

[3] http://www.meteotest.ch

[4] http://www.pvsyst.com

[5] http://eosweb.larc.nasa.gov

[6] http://www.retscreen.net

(direct and diffuse) radiation received during the day, the cumulative radiation in Wh/m^2/day (see units in Appendix 1). These values are accumulated from year to year, and from these, averages can be compiled for each month of the year.

As it is impossible to arrange pyranometers facing all directions, the meteorological databases mentioned in Section 2.2.3.2 will provide these values finally, with certain orientations and inclinations. Sometimes only the horizontal radiation is measured. It is possible to calculate, using mathematical models and some statistics, the radiation on inclined surface from the horizontal radiation, according to the latitude and longitude of the location and the albedo coefficient, etc.; these calculations can nowadays be made using good software such as PVSYST.

Some typical values:

• In France, for horizontal exposure, the daily cumulative irradiance ranges typically from 0.5 (in winter at Lille) to 7 kWh/m^2/day (in summer at Nice).
• In Côte d'Ivoire, for horizontal exposure, it ranges from 4 to 6 kWh/m^2/day.

Instantaneous radiation
There are, however, cases where detailed profiles of instantaneous radiation during the day are needed, especially when obstacles near the panels are likely to cause shadows for several hours at certain times of the year. Then hourly data are needed (graphs of the intensity of radiation according to the hour of the day) to quantify the losses caused by this. More detail is given in Chapters 4 and 5.

Variations in total daily irradiance
Without going into the detail given in Appendix 2, we will now look at the main tendencies to see how solar radiation varies according to location in the world and months of the year.

Influence of latitude
In Europe, irradiance falls fairly quickly in latitudes above 45° N. Comparing Scotland and Spain, for example, daily irradiance in Spain is twice as high in average over the year and three or four times higher in December (for horizontal exposure). Yet these two countries are not far apart when considering the size of the whole planet. The differences are due to the lower angle of the Sun's rays, which means that the PV modules must be inclined at a lower angle at higher latitudes: an incline equal to the latitude +10° is generally the best choice in order to recover the maximum amount of solar energy in winter for a stand-alone system. Further details are given in Section 5.5.3.

The best sites for solar energy are situated in the subtropical regions (latitudes 25–30°). Equatorial zones are in general more likely to be cloudy (monsoon and storm phenomena).

But the latitude has the most influence on seasonal distribution of irradiance, especially on horizontal radiation. In the equatorial and tropical zones, there are only small variations throughout the year, which is extremely advantageous for the use of PV energy. The higher the latitude, the more marked the differences. Figure 2.16 shows the annual changes in solar radiation on a fine day according to latitude.

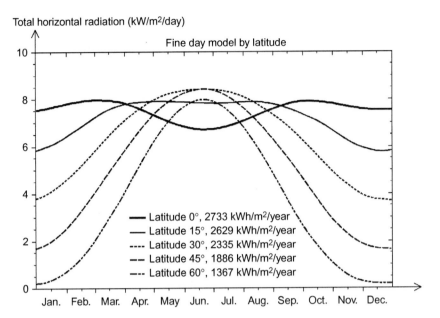

Figure 2.16 *'Ideal' daily total irradiance (fine day without cloud model) by season and latitude*

This has direct consequences on the design of stand-alone PV systems. In low-latitude countries (between 15° S and 15° N), daily irradiance is relatively constant, and the electric output of a PV generator varies little throughout the year. On the other hand, when the summer/winter contrast is increased at higher latitudes, sizing has to be carried out with regard to the lowest irradiance of the year, and excess energy produced during the sunny season must be carefully managed. This means that charge regulators have to be carefully designed so as not to damage the batteries, and priorities may have to be established by control systems for appliances.

Influence of situation
In temperate and cold countries, the ideal curve shown in Figure 2.16 is further affected by the fact that there is usually more cloud in winter. Low values of total winter irradiance are certainly a disadvantage for the expansion of PV in temperate climates. Fortunately, the situation can be improved a bit by keeping the modules oriented due south and, when possible, inclining them at an angle suitable for the latitude.

Figure 2.17 illustrates this attenuation of the summer/winter imbalance for the city of Paris (48.8° N): winter solar radiation values are improved by inclining the panels at 60° from the horizontal, facing due south.

It can be seen that the two curves cross: in summer, horizontal exposure is better, in winter 60° S exposure gives the best values. So for temperate countries in the northern latitudes, while orientation to the south is always desirable, the best

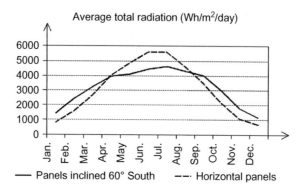

Figure 2.17   Daily total radiation at Paris, measured with two different exposures

angle of inclination must be decided according to the location and use. If the use is mainly in the summer months, a low inclination is to be preferred (10–30° from the horizontal), whereas for annual use, it is best to follow the rule of latitude of location +10°.

Altitude, albedo and other factors
Obviously, the latitude, season and exposure are the main factors determining the solar radiation available on the ground at a given location. But other factors may also play a role, particularly in the mountains, on the coast or in cities where atmospheric pollution is high.

At higher altitudes, solar irradiance is almost always higher than in the plains at any given time. Often this is simply because you are above the clouds. In general, cloudiness reduces with altitude. Also snow cover can play a major role in winter, increasing ground reflectivity considerably (there is a fourfold difference between ordinary cultivated ground and ground covered with fresh snow). Radiation reflected from the ground does not strike horizontal PV modules, but affects all inclined panels and, above all, vertical ones. This is explained further in the case study of a Swiss chalet in Section 5.6.2.

These two elements combined (albedo and reduced cloudiness) modify the radiation received in the mountains, especially in winter. It will be seen that in Figure A2.1 in Appendix 2, which gives total irradiance for the month of December for an exposure of 60° S, there are important variations in the curves for locations in the Alps. The statistical values given for the location of Davos in Switzerland (altitude 1590 m), compared to cities not far away (Zurich and Milan, for example), are a clear example of this (see table for Europe in Appendix 2).

As for localised pollution and other disturbances giving rise to microclimates, only local meteorological data can take account of these. The designer of a solar installation should obtain data from local meteorological services to detect any possible microclimates, which could lead to divergences from general recognise statistics. Sometimes, images obtained by satellite are also a useful source of infor-mation on the state of the atmosphere. The cloudiness of a sector can be deduced from the values of brilliance observed, for example, by the METEOSAT satellite. An international programme has stored and made use of these data since 1984.[7]

2.3 Photovoltaic conversion

We will now deal with the core of the PV phenomenon: the conversion of light into electricity.

The word 'photovoltaic' comes from the Greek word *photos* meaning *light* and 'Volta', the name of the Italian physicist who discovered the electric battery in 1800. However, in 1839, it was the French scientist Antoine Becquerel who was the first to demonstrate the conversion of energy through the variation in conductivity of a material under the effect of light.

The first *photo-electric resistors* were used in cameras to measure light levels. These are resistors whose values depend on the luminous flux received. It was only later that active cells that generate current, *PV cells*, were used. The first cells used on cameras were made of selenium. All PV devices that are *energy converters* transform light into electric current. Using a hydraulic analogy, the electric battery could be compared to a tank of water and the PV cell to a natural spring, the flow of which is proportional to the amount of sunlight at a given moment.

The mechanics of this energy conversion make use of three physical phenomena closely linked and simultaneous:

- the absorption of light into the material,
- the transfer of the energy of the photons into electrical charges and
- collection of the current.

It is thus clear that PV material must have specific optical and electrical properties to enable PV conversion.

2.3.1 The absorption of light

As we have seen in an earlier section, light is made up of photons, 'particles of light', each carrying energy dependent on its wavelength (or colour). These photons can penetrate certain materials and even pass through them: objects that are transparent for our eyes allow visible light to pass through them.

More generally, a light ray striking a solid can undergo three optical events (Figure 2.18):

- *reflection*: the light bounces off the surface of the object,
- *transmission*: the light passes through the object,
- *absorption*: the light penetrates the object and remains within it, its energy converted to another form.

The optical properties of the material struck by the light ray govern the distribution of these various contributions, in intensity and wavelength.

---

[7] ISCCP (International Satellite Cloud Climatology Project) – http://isccp.giss.nasa.gov

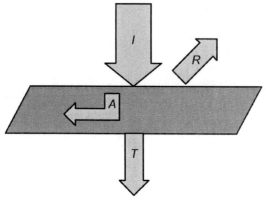

I (incident flux) = R (reflected) + A (absorbed) + T (transmitted)

Figure 2.18   Reflection, transmission and absorption

Take the example of a piece of red coloured glass. It transmits red light, seen by our eye. The part reflected will be as much as 8% of the luminous flux (including all colours) because of the refractive index of the glass. The remaining blue and yellow light is absorbed by the material.

This absorption will be perceptible to touch under intense illumination because the glass will warm up. What happens in most materials is that the absorbed portion of light is converted into heat, that is, infrared radiation (wavelength between 1 µm, the red limit of the visible spectrum, and 1 mm, the starting point of radio waves).

In a PV material, a part of the absorbed luminous flux will be returned in the form of electrical energy (Table 2.2). Therefore, at the start, the material should have the capacity to absorb visible light, since this is what we are trying to convert, whether sunlight or artificial light sources. Care should also be taken to minimise the purely optical losses by reflection or transmission.

Further information

When a material absorbs light, energy is subject to a law of exponential reduction, because the part remaining to be absorbed reduces as it penetrates into the material. If E_{inc} is incident energy, the energy remaining at depth d is described thus:

$$E = E_{\text{inc}}e^{-\alpha d}$$

Thus, energy absorbed in thickness d is equal to

$$E_{\text{abs}} = E_{\text{inc}} - E_{\text{inc}}e^{-\alpha d} = E_{\text{inc}}\left(1 - e^{-\alpha d}\right)$$

The *coefficient of absorption* α depends on the material and the wavelength of the incident energy. It is expressed as cm^{-1}, with the thickness d in cm.

Table 2.2 Optical absorption of some photovoltaic
materials (thickness 0.59 μm)

Material	a (cm^{-1})
Crystalline silicon	4.5×10^3
Amorphous silicon	2.4×10^4
Gallium arsenide	5.4×10^4

The various PV materials and their properties are given in detail in Chapter 3, but it should be pointed out straightaway that there cannot be transmission of light in crystalline silicon cells on account of the thickness of silicon (0.2 μm). In a thin-film cell of amorphous silicon, with active thicknesses of less than 1 μm, the part passing through the active material is significant, especially at the red end of the spectrum where absorption is lower. These cells are therefore optimised to improve the quantity of absorbed light. A backing electrode with reflective qualities such as aluminium ensures that the light is passed back through the active layers. Diffusion is another way of improving absorption: when the layers are rough, part of the diffused light is 'trapped' in the cell and is forced to pass several times (Figure 2.19). This structure gives a more brownish appearance to PV cells made of amorphous silicon (instead of the red colour without optical trapping).

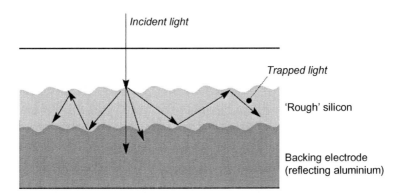

Figure 2.19 Principle of trapping by diffusion in PV cell

As for reflection, this will mainly depend on the refractive indices of the materials used. The greater the difference in the index between one side of a surface and the other, the more light it reflects.

The reflection ratio is expressed thus: $R = [(n_2 - n_1)/(n_2 + n_1)]^2$, if the materials in contact are rated n_1 and n_2.

Thus, raw silicon with the index equal to 3.75 at $\lambda = 0.6$ μm, in contact with air ($n = 1$), will reflect 33% of the light it receives. For this reason, it is essential not to lose a further third of the luminous flux.

In practice, the silicon is not directly exposed to air, as can be seen in Figure 2.20. The crystalline silicon is coated in EVA (ethylene-vinyl acetate) resin, covered in turn with a sheet of glass to protect it. The EVA and the glass each have an index of 1.5, so there remains a considerable contrast with the silicon. A layer with an intermediate index is therefore placed on the silicon, an oxide with an index close to 2. Its thickness is optimised to play the role of anti-reflector to a fairly central wavelength (0.6 μm for crystalline silicon).

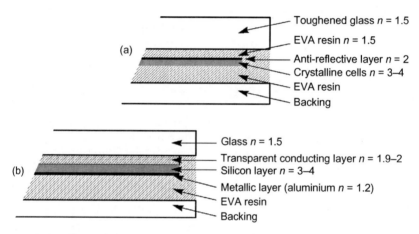

Figure 2.20   Optical stacking of a crystalline silicon cell (a) and an amorphous silicon cell (b)

In the case of amorphous silicon, the stacking of refraction indices is more favourable, and the transparent electrode situated between the glass and the silicon already plays an anti-reflective role since its index is 1.9–2.1 (between the glass with index 1.5 and the silicon with index 3–4). However, its thickness should be optimised to favour the entry of visible light in the amorphous silicon (whose response is basically centred on 0.5 μm).

2.3.2   The transfer of energy from protons into electric current

We shall now look at the light absorbed into the PV material and explain how its energy is converted into electricity.

The charges that produce the electric current under illumination are electrons, elementary negative charges, contained in the semiconductor matter. All solids are made up of atoms, which each comprise the nucleus (formed by protons and neutrons) and a group of electrons circulating around it.

The absorbed photons simply transfer their energy to the peripheral electrons (the furthest from the nucleus) of the atom, enabling them to liberate themselves from the attraction of their nucleus. These liberated electrons produce an electric current if they are attracted to the exterior (see Section 2.3.3).

In the ongoing process, the liberated electron leaves a 'hole' that translates into a positive charge. If this electron is attracted outside the atom, the electron from a neighbouring atom will move to fill this hole, leaving in turn another hole, which is filled by a neighbouring electron and so on. In this way, a circulation of elementary charges is generated, of electrons in one direction and holes in the other direction, which results in an electric current.

The simplest analogy is the 'parking spaces' one. The driver takes his car from one parking space to another, which suits him better. Another car does the same thing and comes to take the place the first driver has liberated, leaving another free space that will be taken by a third car, etc. Imagining the scene, one can easily see a 'car current' in one direction (electrons) and a 'space current' in the other direction (the holes).

This physical phenomenon, called *photoconductivity*, is specific to semi-conductors because they comprise 'unbound' electrons, unlike an insulator, where all the electrons are strongly bound, and an electric conductor, in which there is a high density of totally free electrons.

Depending on the material, there is a threshold of the minimum energy necessary for the liberation of the electrons by the photons. This threshold depends on the material because the electronic structure is different for each type of pattern (number of orbits and number of electrons pattern) and so the energies released are also different.

This threshold is called the *optical gap* of the material or the *forbidden bandwidth*. If the photon has a lower energy, it will not be able to create the electron–hole pair and will not be absorbed. Optical and electronic properties are thus closely linked.

If the photon has an energy superior or equal to the optical gap, it means that it has a wavelength below a certain value, since these two measures are inversely proportional, as shown here:

$$E = hv = \frac{hc}{\lambda}, \text{ which translates by } E(\text{eV}) = \frac{1.24}{\lambda} \text{ (nm)}$$

(See details on measurement units in Appendix 1.)

The optical gap of crystalline silicon is $E_g = 1.1$ eV. The photon possessing this energy has a wavelength of 1.13 μm (in the near infrared). For amorphous silicon, $E_g = 1.77$ eV. The photon possessing this energy is in the red part of the spectrum, with a wavelength of 700 nm (= 0.7 μm).

All photons with energy above these thresholds, and so with lower wavelengths, are suitable for PV conversion.

Figure 2.21 shows the portion of the solar spectrum that can be converted into electrical energy in the case of crystalline silicon. The part of the spectrum shown in dark grey is impossible to convert because it is not absorbed into the material.

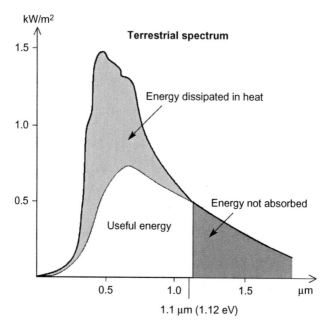

Figure 2.21   AM 0 solar spectrum and parts usable by crystalline silicon

Let us now look at what happens to the portion shown in light grey. To make this clearer, we give in Figure 2.22 another representation of the energy transfer from photons to charged particles. This diagram represents the different energy states in the semiconductor material.

In the energy domain situated below the optical gap, we find valence electrons of the material – those which are linked to atoms. In the conduction band are those which have been released and are free to circulate in the material. This domain is therefore empty when the semiconductor is not illuminated. When a photon has sufficient energy, it is absorbed and causes an electron to pass from the valence band to the conduction band. What happens if it has an energy superior to E_g? Photon 2 in Figure 2.22b generates an electron–hole pair at a superior level, but the excess is lost by a process of spontaneous de-energisation, which produces heat and reduces its energy to E_g. Thus, whatever its energy, provided it is superior to E_g, *each photon absorbed creates only a single electron–hole pair of energy E_g.*

Since the available energy at each wavelength of a given solar spectrum (AM 0 or AM 1.5, for example) is known, the quantity of photons can be determined (total solar energy at this wavelength divided by the energy of the photon), and by adding all the photons, the current and the total power that they can generate can be calculated, according to the optical gap of the material (Table 2.3). These are purely

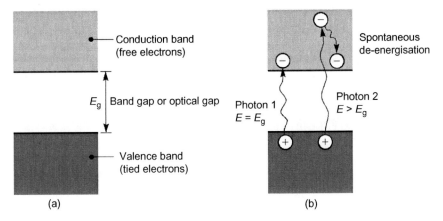

Figure 2.22 Energy diagram of a semiconductor: (a) in darkness; (b) under illumination

theoretical electrical performances that do not take into account losses by reflection, and assume that all the electron–hole pairs generated are collected, which is not the case as we shall see in a later section.

Table 2.3 Maximum theoretical PV performances of semiconductors for irradiation AM 0 with the power of 1350 W/m²

	Crystalline silicon $E_g = 1.1$ eV	Gallium arsenide $E_g = 1.4$ eV
Current (mA/cm²)	53.5	39
Power (mW/cm²)	58.8	55
Efficiency	0.44	0.41

Further information

The maximum theoretical electrical power P_{th} is calculated from the theoretical current I_{th} and the optical gap of the material as follows: $P_{th} = (1/q)I_{th}E_g$, q being the charge of the electron. The electrical efficiency is the ratio between the electrical power generated and the power of the solar radiation (here 1350 W/cm²).

For example, the theoretical efficiency of crystalline silicon for AM 0 is

$$R = 58.8/135 = 0.44$$

These data are interesting because they give the maximum theoretical efficiency, which will never be able to be improved upon with the PV materials available today, and with the light energy available on Earth, coming from the Sun. It can thus be seen that as things are, it is impossible to convert more than 44% of the extraterrestrial solar spectrum.

This takes into account two types of inevitable losses:

- the impossibility of converting photons with energy below the optical gap and
- the loss of energy of photons, which exceeds those of the optical gap.

To convert a higher rate of light energy it would be necessary for all the photons of the light source to have the same energy (a red sun, for example!), and a material be available whose optical gap corresponds exactly to this energy level.

2.3.3 Charge collecting

For the charges liberated by illumination to generate energy, they must move. They therefore have to be 'attracted' out of the semiconductor material into an electrical circuit. Otherwise they will recombine: the negatively charged electrons neutralising the positively charged 'hole'. In other words, the liberated electrons would revert to their initial state at the periphery of their atom: this would generate some heat energy but no electrical energy.

This charge extraction is achieved by a *junction* created in the semiconductor. The aim is to generate an electrical field within the material, which will align the negative charges on one side and the positive charges on the other.

This is possible through the *doping* of the semiconductor. The junction of a silicon photo cell is made up of at least one part doped with phosphorus, called type 'n', joined to a part doped with boron, called type 'p'. An electric field is created at the junction of these two parts, which separates the positive and negative charges. Let us look at this in more detail.

2.3.3.1 Doping of semiconductors

The doping of a pure semiconductor enables it to receive higher charges, which will improve the conductivity of the material.

Figure 2.23 shows a two-dimensional schematic view of silicon atoms (with four electrons in the external layer) that are each linked to four other silicon atoms.

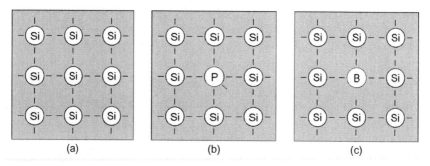

Figure 2.23   (a) Pure (intrinsic) silicon; (b) n-type silicon; (c) p-type silicon

When the silicon is doped with phosphorus atoms, which have five electrons in their external layer, one electron in each phosphorus atom cannot bind with its silicon counterpart, there will then be an excess of negative charge in the crystal (Figure 2.23b). The material then becomes a potential 'donor' of electrons, available for conducting electricity, and the silicon doped in this way is called *type n silicon*.

On the other hand, silicon can also be doped with boron, which has only three electrons per atom in its valence band. The result is the appearance of excess holes, in other words, positive charges, since there is one electron missing in each boron atom to match the four silicon electrons (Figure 2.23c). The material is then the opposite of the example above, an 'acceptor' of electrons. Material doped in this way is called *type p silicon*.

2.3.3.2   p–n and p–i–n junctions

When two regions doped in opposite ways in a semiconductor are placed into close contact, the result is a diode. At the interface where the concentrations of extraneous atoms create a junction between p-type and n-type silicon, a so-called *depletion region* appears, which arises because of the tendency of the excess electrons from region n to try to pass to region p to which they are attracted by the excess holes and the tendency of the holes to do try to pass to region n by reciprocity (Figure 2.24). This exchange of charge carriers in the spatial charge region creates an electric field that will counterbalance the charge exchange and re-establish equilibrium.

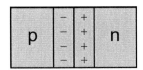

Figure 2.24   Schematic representation of a pn junction

A solar cell is therefore usually a wafer of silicon doped 'p' on one side and 'n' on the other, to which electrical contacts are added to collect the current generated. This junction thus has the electrical characteristics of a classic silicon diode and, when exposed to light, causes the appearance of a photocurrent independent of the voltage and proportional to the luminous flux and to the surface of the cell.

But the simple p–n structure, adapted to crystalline silicon, is not sufficient in all cases. For example, amorphous silicon with p type doping is not a very good photoconductor, and it is preferable that the PV conversion is produced in a non-doped material, called *intrinsic* and described as 'i'. The classic amorphous silicon cell is therefore made up of three layers: p–i–n. The 'i' layer at the centre of the device is the thickest and carries out the charge conversion. The p and n layers enable the creation of an internal electrical field that extends throughout the i layer, which encourages the separation of charges.

This p–i–n junction can also be doubled or tripled to form multi-junctions (see Section 3.2.4).

2.4   The function of the PV junction

2.4.1   *Current–voltage characteristic*

As we have seen, the junction at the core of a PV cell is a diode. When the light falls on it, this diode produces a photocurrent that depends on the quantity of incident light. Hence the term *photodiode*, also used to designate a PV cell.

Figure 2.25 shows the two current–voltage characteristics of this photodiode, the broken line showing it under dark conditions and the solid line under illumination.

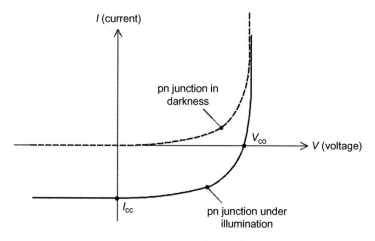

Figure 2.25   Current–voltage characteristics of a silicone diode in darkness and under illumination

2.4.1.1   Short-circuit current and open circuit voltage

It will be noticed that the curve produced under illumination is simply shifted from the other curve by a value I_{cc}, which represents the constant generation of current by light. This value is called a *short-circuit current*, because it is the current generated by the photo cell under the light at zero voltage (in short circuit). The value V_{co}, on the contrary, is the *open circuit voltage*, voltage of the photo cell under illumination at zero current.

Further information

To determine the characteristic curve of this PV cell, we start with the main characteristic of a silicone diode, which is defined as (pn junction in darkness, Figure 2.25, broken curve)

$$I = I_s(e^{V/V_t} - 1) \text{ or, more simply, } I = I_s e^{V/V_t} \text{ when } V \gg V_t \quad (2.5)$$

where
  V = voltage imposed on the diode,
  V_t = kT/q = 26 mV at 300 K,
  $k = 1.38 \times 10^{-23}$ Boltzmann constant,
  $q = 1.602 \times 10^{-19}$ electron charge,
  T = absolute temperature in K,
  I_s = saturation current of the diode.

Under illumination, with a change of sign for the current,[8] this equation becomes

$$I = I_p - I_s(e^{V/V_t} - 1) \tag{2.6}$$

where I_p is the photocurrent.

With the help of this equation, the following parameters can be quantified:

The *short-circuit current*, I_{cc}, value of the current when the voltage $V = 0$, is

$$I_{cc} = I_p \tag{2.7}$$

And the *open circuit voltage*, V_{co}, when the current is zero, is

$$V_{co} = \frac{kT}{q} \text{Ln} \left(1 + \frac{I_p}{I_s} \right) \tag{2.8}$$

or, when I_{cc} is greater than I_s

$$V_{co} = \frac{kT}{q} \text{Ln} \frac{I_p}{I_s} \tag{2.9}$$

It is important to note that this voltage increases with the log of I_p, in other words, with the log of illumination. On the other hand, it reduces with temperature despite the function kT/q. In reality, the saturation current, I_s, depends on the surface of the diode (and so the surface of the cell) and the characteristics of the junction: it varies exponentially with temperature, and this temperature dependence largely compensates for kT/q. Thus, the open circuit voltage, V_{co}, falls with temperature, and this is an important consideration in the sizing of systems.

The diagram of a solar cell (Figure 2.26) can be completed by adding two resistances to take account of internal losses. R_s represents the series resistance, which takes into account only losses of the material, the bonding and the metal/semiconductor contact resistance. R_p represents a parallel

---

[8] This is conventional in the PV field. The sign of the current is changed so that the current–voltage curves can be drawn above and not below the voltage axis, which is more convenient (Figure 2.27).

resistance (or leakage resistance) arising from parasitic currents between the top and bottom of the cell, from the frame, in particular, and within the material from inhomogeneities or impurities.

The current–voltage characteristic equation then becomes

$$I = I_\mathrm{p} - I_\mathrm{s}\left(e^{\frac{q[V+(I \cdot R_\mathrm{s})m]}{kT}} - 1\right) - \frac{V + I \cdot R_\mathrm{s}}{R_\mathrm{p}} \tag{2.10}$$

and it will be noticed that the short-circuit current I_cc, when $V = 0$, is no longer strictly equal to I_p.

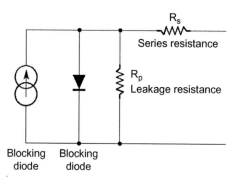

Figure 2.26   Diagram of a solar cell

Summary

- The current of the solar cell is proportional to the illumination and the surface of the cell. It increases with temperature.
- The open circuit voltage varies with the log of illumination and falls with temperature.

2.4.1.2   Power and efficiency

The most useful part of the current–voltage characteristic for the user is the part that generates energy, rather than the point of open circuit voltage or the point of short circuit, which do not generate any energy, since power is the product of current and voltage. Figure 2.27 shows the characteristic of a photo cell under illumination and the theoretical curves of constant power (broken lines).

At point P_m, situated at the 'elbow' of the curve, the power of the photo cell is at its maximum for the illumination under consideration. This point of maximum power is associated with a *maximum voltage* V_m and a *maximum current* I_m. It should be noted, however, the photo cell can also be used at a lower power, for example less than V_m (point P_2 in Figure 2.27).

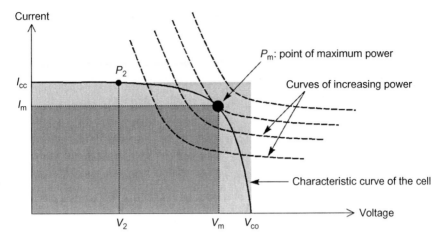

Figure 2.27   Maximum power on a current–voltage characteristic

It can be clearly seen that the 'squarer' the curve, the higher is the maximum power. This property is measured by the *fill factor*, which is defined as follows:

$$FF = \frac{P_m}{V_{co} \times I_{cc}} \tag{2.11}$$

The *energy efficiency* is defined as the ratio between the maximum power produced and the power of solar radiation striking the PV module.

If S is the surface of the module and E the illumination in W/m², the energy efficiency is described as

$$\eta = \frac{P_m}{E \times S} \tag{2.12}$$

This efficiency is often measured under reference conditions: with a solar irradiation of 1000 W/m², at a temperature of 25°C and under a spectrum of AM 1.5. These standard conditions are described as STC (*Standard Test Conditions*). The maximum power output (P_m) of a panel under STC conditions is peak power expressed in *watt-peak* (Wp).

2.4.2   Spectral response

The *spectral response* is the response curve of the solar cell according to the colour of the incident radiation. As we have seen, white light is made up of various colours ranging from ultraviolet to infrared and passing through all the colours of the rainbow (see Section 2.1). The spectrum is different for a clear sky or a sky with a lot of diffuse radiation, etc.

In anticipation of topics covered in Chapter 3, which describes all PV materials, we show in Figure 2.28 the responses of crystalline silicon and amorphous silicon (there is no notable difference between the response of monocrystalline and multicrystalline silicon) (see Section 2.2.2 for details on the solar spectrum).

Amorphous silicon has a better response in blue and green (short wavelengths, 350–550 nm), but crystalline silicon has a better response in the red and near infrared (700–1100 nm). This property explains the better performance of crystalline silicon in sunlight, and the preference for amorphous silicon for artificial and diffuse lighting, which is richer in blue light (higher colour temperature).

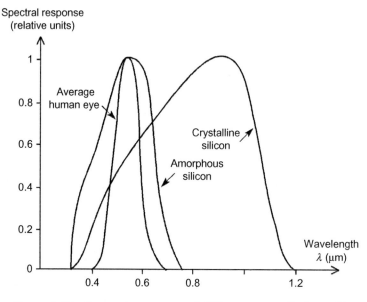

Figure 2.28   Spectral responses of different types of solar cells

Manufacturers seek to improve the spectral response by finding ways of increasing the absorption of different colours in the cell, on the upper surface for blue light, absorbed in the first layer of the material, and for red light for the core or back of the cell (in the case of thin-film cells).

Some examples of possible improvements in spectral response are as follows:

- reducing the reflection on the front surface by an anti-reflection layer;
- using a backing mirror as metallic electrode (of silver, which reflects better than aluminium);
- stacking of cells with different optical gaps (see Section 3.2.4).

Chapter 3
Solar panel technologies

In Chapter 2, certain PV materials such as crystalline and amorphous silicon have been described. Other materials apart from silicon can, however, be used. We consider here their nature, properties and potential uses.

There are two major families of PV materials:

- Solid crystalline materials
- Solid thin films (on a substrate)

We also cover other technologies, in particular *Grâtzel-type* dye-sensitised solar cells, polymer solar cells, cells containing small organic molecules and photoelectric chemical sensors. These technologies, while promising in terms of cost and adaptability, have yet to show their competitiveness against the traditional technologies. Most have stability problems and have low efficiency ($<5\%$). This is why, despite much promise, they have only had a few recent commercial applications.[1]

The material most widely used in solar PV is silicon, a type IV semiconductor. It is tetravalent, meaning one atom of silicon can bind with four other atoms of the same type. Solar silicon is either crystalline or amorphous.

In the amorphous state, it is used as a thin film with thicknesses of around 1 μm and above, deposited on a backing, while crystalline cells are solid and around 0.1–0.2 mm thick.

Other semiconductors used are types III–V like gallium arsenide (rare and expensive) and thin films like CdTe (cadmium telluride) and CIS (copper indium selenide).

3.1 Crystalline silicon cells and modules

3.1.1 Preparation of the silicon and the cells

Crystalline silicon cells are still the most widely used. They are used in the form of square, trimmed square or sometimes round wafers. The material is described as crystalline because the constituent silicon is a crystal with an orderly arrangement of atoms in the tetrahedral type of atomic structure.

---

[1] http://www.konarka.com

If the cell consists of a single crystal, it is described as *monocrystalline silicon*, and it has a uniform bluish grey or black appearance (Figure 3.1). If it is made of *multi-crystalline* (also called *polycrystalline*) *silicon*, it is made up of several assembled crystals and has the appearance of a compact 'mosaic' of bluish metallic fragments of a few millimetres to a few centimetres in size, called 'grains' (Figure 3.2). However, new crystallisation procedures mean that the grain is sometimes too fine to

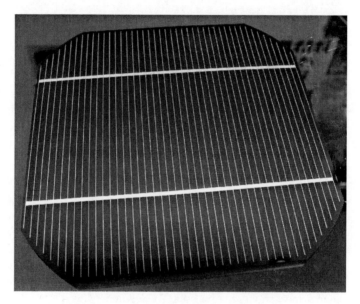

Figure 3.1   Monocrystalline silicon cell

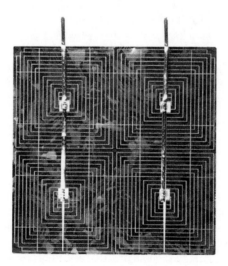

Figure 3.2   Polycrystalline silicon cell [Photowatt International]

be seen. They can be recognised by their true square format, the corners not being cut-off as with the monocrystalline cells obtained from circular ingots.

3.1.1.1 Preparation of metallurgical silicon

Silicon exists in large quantities in nature in its oxidised form, since it is the basic constituent of sand, in the form of silica (SiO_2). To make pure silicon, sand in the form of crystallised quartz is refined by a reduction process by carbon in an arc electric furnace. The reaction follows the equation:

$$SiO_2 + 2C \rightarrow Si + 2CO \tag{3.1}$$

This method is used to produce millions of tonnes of so-called metallurgical-grade silicon every year. Its purity is around 98–99%, the main impurities being aluminium and iron. The main use of silicon is as an additive for aluminium and steel manufacture. A proportion of the total production is purified for the electronic and solar industries. To obtain material sufficiently pure to manufacture electronic or solar components, the Siemens process is used, which transforms silicon into trichlorosilane, by using hydrochloric acid:

$$Si + 3HCl \leftrightarrow SiHCl_3 + H_2 \tag{3.2}$$

As this reaction is reversible, it can also be used to recover silicon after purification, which is achieved by fractional distillation of trichlorosilane. The silicon obtained by reduction with hydrogen is finally deposited on a wafer of heated silicon in the form of small polycrystalline grains. This polycrystalline silicon with a purity of 99.999% is expensive, because this last stage has a low material efficiency (approximately 37%) and requires large amounts of energy.

Note

This polycrystalline silicon is not the same material as that used in modules called polycrystalline because the grain is too fine (see the following section).

3.1.1.2 Manufacture of silicon wafers

Silicon wafers are produced from this purified silicon for use in solar cells.

Monocrystalline silicon

To obtain these wafers, pure granulated silicon has to be converted to a solid material. The classic and still widespread method of obtaining monocrystalline silicon is the *Czochralski process* (Figure 3.3). This consists of drawing cylindrical bars from molten silicon.

The granulated polycrystalline silicon is melted in a crucible with a doping material, for example, boron to obtain a basic p-type material. (For the principle of doping, see Section 2.3.3.) During this process, a seed of monocrystalline silicon is placed in the precise orientation required and the crystal is made to grow, the

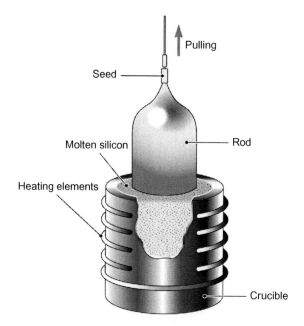

Figure 3.3   Czochralski process

temperature being controlled very accurately (Figure 3.4). In this way, crystals 1–2 m long and up to 30 cm in diameter are grown with the same orientation as the seed inserted.

Another process (the 'zone fusion process'), similar to the Czochralski process, consists of pulling more rapidly and then melting a zone of the rod by an electro-magnetic bobbin to enable it to crystallise regularly from the seed.

Finally, to obtain silicon wafers of around 150–300 μm, rods are sliced with a wire saw.

Figure 3.4   Monocrystalline rods and wafers

A steel wire of around 0.2 mm diameter coated with an abrasive compound, for example, silicon carbide, slices through the silicon at high speed. This procedure enables about a hundred wafers to be sliced at once, the wire being wound round the rod many times.

The whole process has a fairly low efficiency (15–25%) and uses a fair amount of energy.

Polycrystalline silicon

By the mid-1970s, it was understood which elements were detrimental to the efficiency of solar cells and how a cheap silicon crystal of 'solar' quality could be manufactured. The result was the development of solar quality polycrystalline silicon that appears as the juxtaposition of small monocrystalline crystals of different orientations and sizes ranging from millimetre to centimetre. To manufacture this material, waste from the pulling of monocrystals, or purified metallurgical silicon, is re-melted in a square crucible, at a temperature close to 1500 °C, in a controlled atmosphere. Several thermal and chemical processes are used at this stage to 'push' the main impurities to the edge of the crucible, forming a crust, which is eliminated after solidification. The correct method of cooling is essential and determines the size of the crystals and the distribution of remaining impurities, which are mainly concentrated at the edges of the crystals, called *grain boundaries*. It is even possible to orient these grains parallel to the surface to improve the diffusion of electrical charges in the cells (see, for example, the Polix process developed by the Photowatt company). The ingot obtained in this way is then cut into square rods (12.5 × 12.5 or 15.6 × 15.6 cm), which are then sliced into wafers with a wire saw, like the rods of crystalline silicon. This process is economical: the wafers are directly produced in squares, the material efficiency is good and the 'filling' of the PV module is denser.

Silicon ribbon

To completely eliminate the sawing stage, which uses a lot of energy and results in considerable material losses, many methods have been tried since the end of the 1980s to produce PV panels directly from molten silicon, processes described under the generic term of silicon ribbon. The molten silicon is drawn directly in the form of a flat or tubular ribbon. The main problems in the process arise in the difficulty of finding a suitable support for the ribbon, how to remove the heat that arises and the treatment of the edges.

The EFG (*Edge-defined Film-fed Growth*) ribbon technology consists of drawing an octagonal tube from a bath of molten silicon up to 6 m long, the ends of which are subsequently sliced by laser to form wafers of today's standard size of 156 × 156 mm.[2] The mechanical behaviour of wafers obtained by this method is a critical parameter because the laser slicing makes the edges of the cells fragile. The degree of crystallisation depends on the speed of the pulling, and a slow speed can lead to silicon ribbon that is virtually monocrystalline with an efficiency of 15–16%. With another method, the SR (*String Ribbon*) technique, developed in the United

[2] *Photovoltaic International*, 2nd edn, 2008. Available from http://www.pv-tech.org

States, a single ribbon is pulled from a bath of silicon, supported on either side by high temperature wires (or *strings*). This simpler method results in better productivity, especially because the speed of pulling can be higher. The promoters of this impressive technique claim that material efficiency is doubled in comparison with traditional slicing,[3] but its detractors maintain that it is too limited in terms of wafer sizes.

Other methods consist of producing the ribbon on a backing, which is subsequently removed, according to the technique called RGS (*Ribbon Growth on Substrate*), in which the material is grown on a moving substrate.

Similarly, the technique called CDS (*Crystallisation on Dipped Substrate*) consists of dipping substrates in a bath of molten silicon. The promoters of this recent technique claim that it is the only method compatible with future mass production because it combines larger wafer size and high productivity.

However, it should be said that with the market as it is in 2009, with the price of silicon having fallen considerably, these technologies have lost part of their attraction, their main aim being to reduce the costs of the raw material when they were developed during the period of silicon shortage in 2003–04.

3.1.1.3   From the wafer to the classic cell

Once the silicon wafers – usually p type – have been formed, they must undergo the following stages in solar cell manufacture:

- Cleaning of the surface with caustic soda to repair damage caused during the sawing process, and etching to create a rough texture to increase their light-gathering capacity.
- Treatment with phosphorus to create the PV junction, by forming an n+ layer on the surface and n at the junction.
- Doping of the base with aluminium (silkscreening and firing), which creates a diffusion p+ layer and a surface that improves the collection of charges.
- Deposition of an anti-reflective layer on the face.
- Deposition of a metallisation grill on the face (− electrode).
- Deposition of a solderable metal on the back (+ electrode).
- The testing and grading of all the cells manufactured.

Figure 3.5 shows the cross section of a monocrystalline silicon cell (not to scale).

To economise on energy and reduce manufacturing costs, manufacturers aim to set up production lines using dry processes as far as possible and avoiding any manipulation of wafers (which is a source of wastage). The different heat treatments and sometimes even the diffusion are carried out in tunnel furnaces, and the process can be continued without intermediate stock, which facilitates its automation. Another tendency is to slice increasingly finely, 100 μm being adequate to capture the whole of the spectrum received on Earth, and some manufacturers are trying to treat wafers of 130–180 μm, but this thickness is currently the minimum

[3] http://www.evergreensolar.com/technology

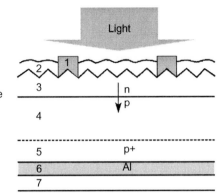

1   Front metallised grid
2   Anti-reflective layer
3   n-doped and textured front surface
3-4 Junction and electric field
4   p substrate
4-5 BSF (black surface field)
5   p+ doping
6   Aluminium metallisation
7   Metal contact

Figure 3.5   Cross section of a monocrystalline silicon cell

that can be used to limit breakage during manipulation and possible thermal shocks. Also, cells are being made increasingly large to reduce subsequent assembling stages. From cells of 125 × 125 mm, manufacturers have now gone to 156 × 156 mm, and even larger sizes are being developed currently in the most modern factories.

To improve performance still further, manufacturers are increasingly using sophisticated processes, such as very close diffusion of the front surface to improve the collection of short-wavelength photons (very energetic blue photons that do not penetrate far into the silicon), an extremely fine front metallised grid deposited on a laser scribed pattern, two anti-reflective layers, etc.

These 'classic' crystalline cells currently available on the market (early 2009) have an energy conversion rate of 15–18%.[4]

We will now examine two recent crystalline silicon technologies, which give even better results.

3.1.1.4   Interdigitated back-contact solar cells

This highly original process first developed in 1994[5] for use in concentrated solar generation is complicated and expensive. It has now found its place among the classic types of solar panel, produced industrially with monocrystalline silicon, with more dual conversion rates as high as 20%, among the best on the market.[6] More recently, sales of this type have also been produced using polycrystalline silicon with a prototype efficiency of 18.5%.[7]

---

[4] Examples are the German cells manufactured by Q-cells and Ersol, the Japanese Kyocera cells, or the French Photowatt and Tenesol. For a directory of all manufacturers, see http://www.solarbuzz.com.

[5] P.J. Verlinden, R.M. Swanson, R.A. Crane, 'High efficiency silicon point-contact solar cells for concentrator and high value one-sun applications', *Proceedings of the 12th EC Photovoltaic Solar Energy Conference*, Amsterdam, April 1994, pp. 1477–80.

[6] Sunpower Corporation.

[7] Kyocera Corporation. European PVSEC conference, Milan September 2007.

In these cells, there is not just one p–n junction, placed on the upper surface perpendicularly to the light source (Figure 3.5), but several n–p junctions interconnected in the form of combs, located on the back (Figure 3.6). These modules are remarkable for the fact that they have no contacts on the upper face: all the contacts are placed at the back of the cell and are thus invisible on the module.

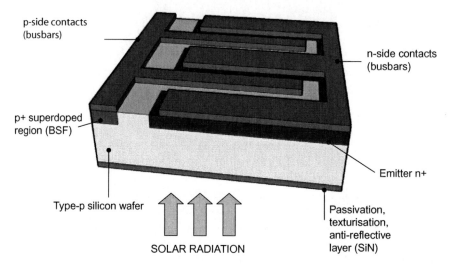

Figure 3.6   Cross section of an interdigitated back-contact cell

This result is obtained by very accurate etching and diffusion methods, with many stages based on microelectronic techniques, which explains the high cost; also, the silicon wafer used must be of very high quality.

But there are many advantages, particularly:

- The 'shadow' current of the cell is reduced because the junction surface is small.
- There is no shading by the contacts because there are none on the upper surface.
- As the contacts are on the back surface, they can be large, which reduces the series resistance of the cell.
- The connections between the cells in a module can be made more easily by the location of all the contacts on the back surface.

With this technology, an efficiency of up to 24% can be obtained in laboratory conditions and 21–22% in factory conditions, which is obviously much higher than with classic technologies.

3.1.1.5   Heterojunction cells called HIT or HIP

A heterojunction is a PV junction composed of different materials. HIT (*heterojunction with intrinsic thin layer*) cells are composed of a wafer of crystalline silicon and thin slices of amorphous silicon. The idea is to take advantage of the better efficiency of crystalline silicon in bright illumination and its spectral response to near-infrared (see Section 3.1.2), and of the higher sensitivity of amorphous silicon to low light and blue light and its lower loss of power with temperature. According to the Sanyo process, a thin wafer of type-n crystalline silicon is bonded with layers of amorphous silicon of type p and i on the upper surface and i and n on the rear surface (Figure 3.7). The manufacturer claims cells with an efficiency of 20%, and offers modules from 16% to 17% and a temperature coefficient reduced to 0.4–0.3%/°C[8] (see Section 3.1.2 for details of this temperature influence).

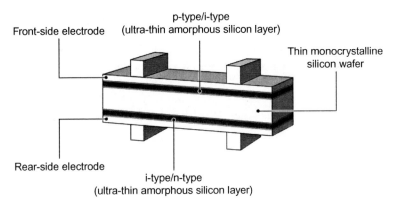

Figure 3.7

HIT cells also have their detractors: the temperature advantage is not always borne out on the ground, and more results are needed as the technology is still recent.

3.1.2 Properties of crystalline cells

3.1.2.1 Spectral response

As described in Chapter 2, crystalline silicon material has a spectral response (sensitivity to different colours of the light source) that ranges from blue (400 nm) to near-infrared (1100 nm); in other words, it is well suited to the solar spectrum, with a weak point in blue (Figure 2.28). This lack of current in the blue part of the solar spectrum is essentially due to reflection (the raw cells appear blue), and so a good anti-reflective layer improves this response.

---

[8] http://www.solarelectricsupply.com/Solar_Panels/Sanyo/HIT-190BA19.html

3.1.2.2   Current–voltage performances

Imperfection of the definition of 'efficiency' according to STC standards

As seen earlier, when we defined energy efficiency and the conditions of measuring this efficiency (see Section 2.4.1), solar cell technologies are often compared between each other only in sunlight, with an irradiance of 1000 W/m² (Standard Test Conditions, or STC) and at a temperature of 25 °C, which is far from the usual case.

How could a panel exposed to full sunlight be maintained at 25 °C? But when these standards were defined, all the panels were optimised for bright sunlight, no doubt because it was not expected at the time that solar applications would be used in temperate climates or under artificial light.

Today, this standard is widely criticised, because it does not allow developers to plan the installation of an array of panels on the ground, which also depends on the instant radiation, often much weaker, the state of the weather (diffuse or direct radiation), the orientation of the panels and their operational temperature, etc. Instead, real production values in kWh produced per Wc installed are used.

Figure 3.8 shows some current–voltage characteristics of a monocrystalline cell with an efficiency of 15% and a surface of 10×10 cm (100 cm²). The top curve represents an incident radiation of 1000 W/m² in STC, and those below represent weaker solar radiation intensities, all at 25 °C.

In STC, a cell of this type typically has an open circuit voltage of 0.58 V, and a short-circuit current of 33 mA/cm². In operation, that is, at the elbow of the curve, it produces 30 mA/cm² at 0.5 V (which results in 3 A $\times$ 0.5 V = 1.5 W at an efficiency of 15% over 100 cm²). In practice, these values vary according to the crystalline technology used (see the sample of modules in Table 3.2).

Influence of illumination

Obviously, the PV production of a solar panel depends directly on the luminous flux received, since this is a source of energy. But all the parameters of the characteristic are not affected in the same way. In the left part of the curves in Figure 3.8, it will be seen that the current is directly proportional to radiation at these levels of illumination (>200 W/m²). Voltage, on the other hand, is less effective by lower light levels, since, as we have seen in Section 2.4.1, the voltage of the cell varies with the logarithm of illumination. In a monocrystalline cell, the parasitic shunt resistance remains fairly high and the cell can supply a correct voltage even at low light levels. For a polycrystalline cell, which has a lower shunt resistance, the voltage can sometimes drop significantly once illumination has fallen below 30–50 W/m² (3–5% of maximum insolation). This property is disadvantageous for the use of crystalline cells in temperate countries.

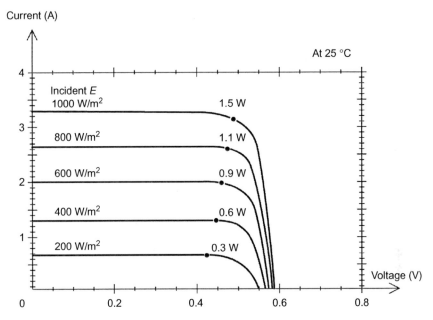

Figure 3.8   Curves I/V of a monocrystalline cell under various radiation intensities

A further disadvantage is that crystalline silicon cannot be used under artificial light, where illumination is typically between 100 and 1000 lx, the equivalent of 0.1–1% of standardised solar radiation (the percentage depending on the spectrum of the lamp).

Amorphous silicon is used in PV applications for internal use, operating under reduced illumination: watches, calculators, measuring devices, etc.

Influence of temperature
Temperature has an important impact on the performance of crystalline cells, and thus on the design and production of panels and systems.

The voltage of a crystalline cell drops fairly sharply with temperature, as we have seen in Section 2.4.1. This effect is shown in Figure 3.9 in curves of a crystalline cell between 10 and 75 °C within the radiance of 1000 W/m^2.

The voltage lost is typically 2 mV/°C per cell (which is –0.4%/°C for 500 mV), which results in a fall of around 80 mV between 25 and 65 °C, for example. For a 12 V module with 36 cells, this reduces the operational voltage U_m by 16%: from 16 to 13.1 V (36 cells × 0.08 V = –2.9 V). Obviously, this can have consequences on the charging of batteries if the panel is submitted to a temperature of 65 °C. It may be asked whether a cell can reasonably be expected to reach such a high temperature on the ground. This is unfortunately possible, and it can even exceed 70 °C: the actual operational temperature of a cell is always higher than the ambient temperature, the difference depending on the construction of the module in which it is implanted (see Section 3.1.4).

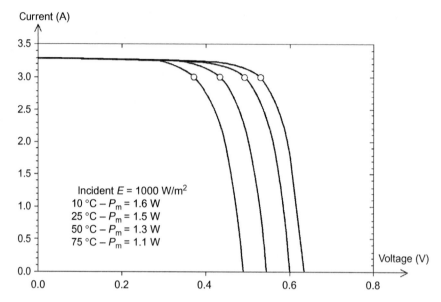

Figure 3.9   Curves I/V of a monocrystalline cell at various temperatures

Under the effect of a rise in temperature, the current does increase slightly. This is explained by a better absorption of light, the gap reducing as the temperature increases. But the increase in current is negligible at the point of maximum power, and the total behaviour of the crystalline cell at higher temperatures is a loss of 0.4–0.5% per degree, sometimes only 0.3% for HIT cells, according to the manufacturers. This translates in practice to losses of around –15% for cells at 60 °C.

3.1.3   From cell to PV module

The PV module is by definition a collection of cells assembled to generate a usable electric current when it is exposed to light. A single cell does not generate enough voltage, around 0.6 V for crystalline technology. Almost always, several cells have to be mounted in series to generate a useful voltage.

Additionally, this series of cells has to be protected to enable the module to be used outside. The cells are fragile objects and sensitive to corrosion, which need to be mechanically protected and sheltered from climatic excesses (humidity, temperature variations, etc.).

So modules of different powers are produced according to the surface area to be used (typically from 1 to 300 Wc/module), capable of generating direct current when they are exposed to the light. These modules constitute the energy producing part of a PV generator.

Increasingly powerful modules are available on the market, particularly those designed for connection to the grid, limited only by weight, handling problems and maintenance constraints. So to make up a high-power generator, several PV modules are linked with cabling before being connected to the rest of the system. Series assembly requires panels of the same current, and parallel assembly requires panels of the same voltage (see Section 3.1.6).

Figure 3.10 shows the structure of a crystalline silicon module (mono- or polycrystalline). It is made up of a number of cells arranged in a row, connected to each other in series, and assembled in a weatherproof frame.

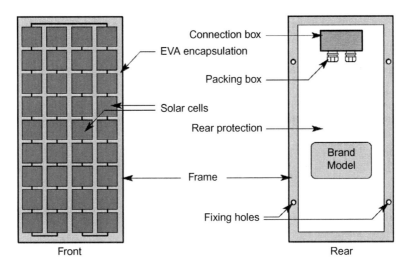

Figure 3.10   Structure of a crystalline silicon PV module

3.1.3.1   Series connection

It is essential for the cells to be mounted in series to produce a usable voltage. Electrical and mechanical rules govern how the cells can be assembled.

Number of cells per module

As we have seen in earlier section, the crystalline silicon cell produces an open circuit voltage of around 0.6 V and a maximum power voltage of around 0.45–0.5 V.

Let us now look at the assembly of cells to make up, for example, a PV module designed to charge a 12 V battery.

Remembering that we need to have around 14.5 V for an effective charge and that 2–3 V will be lost in cabling and through rises in temperature (see Section 3.3.2), we will need a panel supplying at least 17.5 V at its maximum power. If we then divide 17.5 by 0.48, we arrive at the round figure of 36 cells, the usual number for 12 V panels on the market. This allows the cells of such a module to be arranged in four rows of nine or six rows of six. In practice, 12 V panels are made up of 32–44 cells: this depends on the exact value of the voltage of each cell and the temperature of utilisation.

For panels intended to be connected to the grid, it is better to plan for a higher voltage, 40 V or even 60 or 72 V, partly because they have to feed into inverters with an increasingly high entry voltage (100 V at least), but also because the higher the voltages, the lower the currents, which simplifies cabling (thinner cables, lower amperage protective devices). These higher powered panels comprise a large number of cells, the quantity mainly determined by the size of standard cells (today 156 × 156 mm) and the constraints of panel dimensions. Thus, panels for grid connection may be made up of 72 or 96 cells in series (see modules in Table 3.2.)

Series connection of cells
Except for back-contact cells where all the contacts are at the back, the (−) contact on the front of the first cell must be connected to the (+) contact on the back of the following cell, the contact of the latter to the contact of the following, etc. These connections require a solderable contact on each side of the cells, usually tin or silver. The connections are made with ribbons of tinned copper, which is both supple, extra flat and solderable. Figure 3.11 shows these internal connections. In manufacturing, this stage is often increasingly automated: the machine holds the cells by means of suction cups and solders them two by two in strips called '*strings*', which are subsequently soldered together at their ends (Figure 3.12).

This stage is quite delicate because the cells are fragile. Manipulation and soldering requires special equipment built to precise formats. The series connection of thin-film cells is much simpler and more adaptable in terms of format. It is described in Section 3.2.

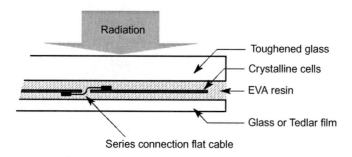

Figure 3.11   Cross section of crystalline silicon module

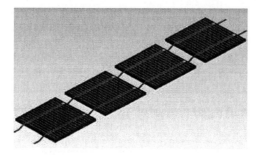

Figure 3.12   A 'string' of cells connected in series [credit ECN]

What happens from an electrical point of view when the cells are placed in a series? It is the same as with batteries or other generators: the voltages of all the cells are combined and the current is the same as that of a single cell. This is why cells must always be of the same current when they are mounted in series; in manufacturing, this is called pairing: the cells are sorted according to their current in order to connect them together. If one of them had a lower current, it would impose its current on the whole series, which would penalise the whole module.

If the cells are connected in parallel instead of series, the current increases and the voltage remains constant. In this case, the voltages of the cells would have to be paired and not the currents. This is what happens when PV modules are arranged in parallel to set up a more powerful generator (see Section 3.1.5).

It would be easiest to have completely identical cells, but manufacture is too widely dispersed (there is a gap between the best and less good cells). The sorting of cells into grades is a constraint for manufacturers, who often choose to manufacture identical modules in terms of size, but with different outputs according to the characteristics of the cells used (Figure 3.13). See, for example, the monocrystalline modules of 165–185 Wc made by Suntech (second row of Table 3.2).

Figure 3.13   Polycrystalline silicon (left) and monocrystalline silicon (right) modules [photo http://www.photovoltaique.info]

3.1.3.2   Encapsulation and framing

Once the connections have been made, the cells are encapsulated in polymer resin and sealed in a sandwich between two surfaces ensuring that the cells are at least 1 cm from the sides to avoid corrosion.

Front

The panel is faced with a transparent and resistant material, normally a sheet of 'high-transmission' toughened glass.[9] Such glass has a lower concentration of iron oxide than ordinary sheet glass, and is therefore more transparent. It is normally 3 or 4 mm thick depending on the size of the module, and many different qualities can be found on the market today.

Rear

Sometimes glass is also used as a backing, following the technique of double-sheet safety glass used for car windscreens. The solution is the best mechanically, and results in partially transparent modules that are attractive for architectural applications. But it is more economical and sometimes just as effective to use polyvinyl fluoride (PVF) plastic film, marketed under the brand Tedlar by Dupont de Nemours, or other films providing a barrier against humidity, such as plastic/metal laminates. In recent years, the supply of these materials has become a major challenge, bearing in mind the spectacular growth of world production of solar panels. This has led to shortages (particularly of Tedlar), which laminate manufacturers are now getting round by using alternative materials such as polyethylene, polyester and other innovative materials, usually combined with each other. The validation of this type of material is not easy and requires numerous calibrated tests in the climate tank and validation on the ground, because it directly affects the life expectancy of the panels.

The polymer resin used for encapsulation is normally *EVA (ethylene-vinyl acetate)*. It is supplied in the form of whitish sheets, which are placed between the outer covering and the cells. 'Solar' EVA, specially produced for this purpose, includes additives to promote hardening and to improve adherence to the glass. This sandwich of glass/EVA/cabled cells/EVA/glass (or plastic film) is heated to 100–120 °C to liquefy the resin, pressure is then applied to remove any air and ensure adhesion, then the temperature is maintained at around 140 °C for about 15 min to solidify the EVA. It then becomes transparent, with a refractive index close to that of glass, which avoids optical loss. The technology employed is similar to that used for safety windscreen glass, with a difference that since the glass is flat and not curved, only 1 bar of pressure is necessary to effect the bonding (as against 3–5 bars used in the autoclaves of the automobile industry). This operation is carried out in a laminator composed of a heating sheet, a vacuum pump and an air pocket to apply atmospheric pressure on the stacks. The PV module is then operational and fit to face all climatic challenges (see Section 3.1.4).

Often a fixing frame is added. The electric outlet must be well designed because the opening for the electrical cables must not allow water or water vapour to penetrate the interior of the panel. A *junction box* is usually fitted to the frame to connect the panel with devices compatible with its output amperage.

---

[9] The toughening of glass by chemical or thermal processes makes it much more resistant, and certainly virtually impervious to hail. As a consequence, it cannot be easily cut, and the toughening process is normally applied to ready cut sheets.

For connection to the grid, since direct current voltages are very high, typically 200–600 V, special cables with double installation and integrated connectors are used.

Waterproofing or not?

There are differences of view on the waterproofing of modules themselves or their junction boxes, irrespective of the technologies used. Should the aim be complete waterproofing to prevent any humidity entering the panel, whether in gaseous form (water vapour) or liquid? Today most manufacturers agree that, since total waterproofing is very difficult to achieve, it is better to allow the modules to 'breathe' so that any humidity can evaporate rather than stagnate. There is also agreement that junction boxes should have a hole to allow water to escape, rather than perfect waterproofing, which would trap any water vapour and allow it to condense to liquid and cause damage. There must, however, be mesh round the ventilation holes to prevent insects entering. For grid-connected panels, junction boxes are no longer used, only cable connectors are used for the series mounting, which are already in place.

3.1.4 Electrical and climatic characteristics of modules

The manufacturing stages from cell to module, which have been reviewed in Section 3.1.3, do have a bearing on PV performance. A module has characteristics slightly inferior to those of the cells that make it up, because of

- glass and EVA on the upper surface, which cause optical loss (approximately 4%);
- dispersal between the cells;
- losses due to geometrical arrangement: spaces between the cells, border and frame;
- small electrical losses in series: soldered joints, flat cables, etc.

Typically, these losses represent in total around 10%, and cause the efficiency to drop by 15% on cells and 13.5% on modules, for example.

The technical specifications of a PV module naturally include its physical characteristics: dimensions, weight, method of fixing, output connections and, especially, the electrical characteristics, which we will now deal with one by one.

Before we do this, it should be recalled that PV modules are measured and guaranteed according to reference conditions known as STC, which are

- solar radiation of 1000 W/m^2,
- solar spectrum AM 1.5,
- ambient temperature 25 °C.

These conditions are normally mentioned on the module technical specification sheets. When the panel is also checked and/or guaranteed for performance under weaker illumination, these data are sometimes provided, for example, at an irradiation of 200 W/m². This is an undeniable plus, because the STC are not representative of all situations encountered, far from it, for a solar radiation of 1000 W/m² is very high (remember that the extraterrestrial intensity AM 0 has only 1360 W/m²). In France, this level is only seen at midday on a fine spring day, with a cloudless sky (see the note, Imperfection of the definition of 'efficiency' according to STC standards, in Section 3.1.2.2).

3.1.4.1   Electrical parameters under illumination

A PV module exposed to the Sun or another form of illumination produces power continuously according to the characteristics previously described (see Figure 2.27 for a definition of the parameters of current, voltage, etc.).

Peak power, operational voltage and current
Peak power or the maximum output power of the PV module under solar radiation is the essential parameter. It is the ideal point of the current–voltage characteristic in STC.

The values of voltage (V_m) and current (I_m) under load are also important, especially for charging batteries: what is the use of producing a lot of watts if they do not charge the battery? If the STC voltage V_m is too weak (13 or 14 V, for example, for a 12 V solar panel), battery charging will be possible under strong illumination without any online losses, but impossible if these conditions are not fulfilled. A good 12 V panel should have a V_m under STC of 17 V minimum. This is especially true in regions where solar radiation is not always maximum. The panel can of course be invoked below this value, since the battery will impose the operational voltage, at 13.5 months, for example. But the voltage reserve will be useful in cases where it is less, for example,

- solar radiation below STC,
- temperatures below 25 °C,
- voltage losses along cables.

See Section 3.2.2.2 on the effects of irradiation and temperature on the electrical performance of cells.

The load current I_m is also important: this is the peak current that the panel can produce in operation. It affects the specification of all the components at the back of the panel, output cable, diodes, charge controller and inverters.

Open circuit voltage
The open circuit voltage (V_{co}) is easy to measure since it is the panel's no-load voltage without any current circulation, simply read by a voltmeter connected to its terminals. A value of 22–24 V is normal for a 12 V panel of good quality. Although it is not directly useful, this voltage can provide some information. For example, recording the variation in voltage according to solar radiation can provide a good idea of the panel's capacity to charge a battery under moderate irradiation. Knowing that the ratio between V_m and V_{co} is approximately 0.8, in a radiation situation, which would give an open circuit voltage of at least 16 V, the panel

would have no chance of correctly charging a 12 V battery, because its operational voltage would be below $16 \times 0.8 = 12.8$ V.

Measuring this voltage is also the simplest way during maintenance operations of verifying that the panel still has its internal electrical continuity. Finally, the measure V_{co} enables the temperature of the cells to be checked rapidly once its value in STC is known.

Short-circuit current and form factor
The no-load current, described as I_{cc} (see Section 2.4.1), as measured by an amperometer connected directly to the panel terminals, is not a very useful parameter taken on its own. At best, it can give an indication of the solar radiation at the time, since it is proportional to it. However, when the open circuit voltage is also known, it can be used to calculate the form factor (FF).

$$FF = \frac{P_m}{V_{co} \times I_{cc}} \tag{3.3}$$

As we have seen in Chapter 2, this parameter, with a value between 0 and 1, or expressed as a percentage, describes the more or less square shape of the current–voltage characteristic. If the shape was square, the form factor would be equal to 1, and the power P_m would be equal to $V_{co} \times I_{cc}$. But this would be without allowing for inevitable losses: in series, on account of the not-zero resistance of the cell constituents, and in parallel, on account of slight current leakages. Generally, this form factor is between 0.6 and 0.85.

Note
Putting a panel into short circuit cannot damage it, since this current is very close to the operating current I_m. On the other hand, if the panel has a power that is more than negligible, the object that caused the short circuit may be damaged and this can prove dangerous (fire risk) (see Section 5.5.7 for safety advice concerning the use of direct current).

NOCT
One might imagine that the actual site temperature of the PV cell within its module would depend on its immediate environment: front cover, rear cover, ambient climatic conditions and ventilation. The operating temperature of the cell is higher than that of the ambient air. To describe it, scientists have defined the nominal operating cell temperature (NOCT). This dictates the way the module is built and influences the operating temperature of the cells it contains. It is defined as the temperature that the cell reaches within its module in open circuit, under an irradiance of 800 W/m^2, with an ambient temperature of 25 °C and a wind speed of 1 m/s.

The values normally encountered today are between 40 and 50 °C. As we have seen, this high temperature affects the operation of the module. In order for the NOCT not to be too high, the colours on the back should be light to shed heat (a white rear panel heats up less than a black panel), and maximum ventilation should be provided. Panels with Tedlar encapsulation are therefore a priori better from this point of view than panels using two sheets of glass.

But this definition of the NOCT is questionable: the real temperature of the module is very frequently well above this, as soon as the ambient temperature exceeds 25 °C, for example, which it often does. It may be applicable to a free-standing module, ventilated back and front. In all other cases, the real temperature of the module depends on how it is integrated.

3.1.4.2   Life expectancy and certification

Good quality modules today have a life expectancy of over 20 years in any climatic conditions. There may be a slight loss in characteristics over time, because it is now known that the materials undergo some ageing in the long term. Most crystalline modules are today guaranteed for between 20 and 25 years to 80% of their minimum nominal power, and sometimes for a minimum of 90% after 10 years.

Apart from the quality of the cells themselves, their life expectancy depends on the protection techniques (encapsulation, termination system), which are the subject of much work to optimise the quality/price ratio of this essential manufacturing stage. So as not to have to wait 20 years for results, laboratories have worked out accelerated tests to simulate real climatic conditions. The principle is generally to assume that by submitting modules to higher temperatures, any damage produced over time will be accelerated (which is correct if the deterioration is caused by heat and the tests do not exceed the temperatures at which irreversible deformation or destruction of the materials takes place).

Current international standards concerning PV modules are issued by the International Electrotechnical Commission (IEC) based in Geneva[10]: IEC 61215, IEC 61646 and IEC 61730, which in turn became European standards.

These standards are based on earlier work, in particular by the research centre of the European Commission at Ispra in Italy, which carried out numerous investigations to define the best test procedures for PV modules. The specification number 503, which they developed, 'Terrestrial PV modules with crystalline solar cells', was adopted in 1993 as IEC standard IEC 1215, which has today become IEC 61215 and was ratified in 1995 as European standard EN 61215.

In Europe, standards on the construction of solar panels must be respected according to their technology as follows:

- EN 61215: 'Terrestrial photovoltaic (PV) modules with crystalline solar cells – Design qualification and type approval'
- EN 61646: 'Thin-film terrestrial photovoltaic (PV) modules – Design qualification and type approval'

---

[10] International Electrotechnical Commission (IEC): http://www.iec.ch

There is also a safety standard, EN 61730 'Photovoltaic (PV) module safety qualification', and another dealing with panel measurements, EN 60904 'Photovoltaic devices. Measurement of photovoltaic current–voltage characteristics'.

Standards 61215 and 61646 comprise a sequence of module tests that review all the events liable to affect ageing of PV modules in their natural exposure on the ground. The standards describe in detail the test conditions and acceptance criteria. They may be classified as follows:

- exposure to solar radiation, including UV[11];
- climatic testing: the effects of temperature, including rapid changes, and humidity;
- mechanical testing: hail, wind and snow;
- electrical testing: insulation and leakage measurement;
- in the case of thin-film modules, tests of degradation under exposure to light (see Section 3.2.3).

Table 3.1 lists the tests to be carried out and the conditions to be respected. The panels are approved if, after these tests, no major visual faults appear and the power of the solar panel is not significantly degraded. Approved inspection agencies such as TUV will then deliver a certificate of approval. Further detail of these test sequences can be found on the TUV website.[12]

Table 3.1   *Summary of test requirements for PV modules according to IEC standards 61215 and 61646*

Clause	Measurement/testing	Testing conditions
10.1	Visual inspection	According to defined inspection list 10.1.2
10.2	Determination of maximum power	See IEC 60904-1
10.3	Insulation test	1000 V DC + twice the open circuit voltage of the system at STC for 1 min For modules with a surface smaller than $0.1\ m^2$, the minimum resistance is 400 MΩ. Larger modules are required to have the measured resistance times the area of the module greater than 40 MΩ/m^2. Testing is performed using a test voltage of 500 V or maximum voltage of the system, whichever is higher
10.4	Measurement of temperature coefficients	According to details provided in 10.4 For further information, see IEC 60904-10

(Continues)

[11] There is an independent standard on exposure to UV, IEC 61345, which stipulates the minimum doses of UVA and UVB the panels must be submitted to.

[12] http://www.tuv.com/uk/en/pv_module_certification.html

Table 3.1   (Continues)

Clause	Measurement/testing	Testing conditions
10.5	Measurement of NOCT	Total solar irradiance $= 800$ W/m^2 Ambient temperature 20 °C Wind speed $= 1$ m/s
10.6	Performance at NOCT	Cell temperature $=$ NOCT/25 °C Irradiance $= 1000$ and 800 W/m^2 and solar spectrum distribution according to IEC 60904-3
10.7	Performance at low irradiance	Cell temperature $= 25$ °C and NOCT Irradiance $= 200$ W/m^2 and solar spectrum distribution according to IEC 60904-3
10.8	Outdoor exposure	60 kWh/m^2 total solar irradiation
10.9	Hotspot test	5 h exposure to 1000 W/m^2 irradiance in worst-case hotspot condition
10.10	UV pre-conditioning test	15 kWh/m^2 UV radiation (280–385 nm) with 5 kWh/m^2 UV radiation (280–320 nm)
10.11	Thermal cycling	50 and 200 cycles -40 °C to $+85$ °C with peak current at STC for 200 cycles
10.12	Humidity freeze	10 cycles at -40 °C to $+85$ °C, 85% relative humidity
10.13	Damp heat	1000 hours at $+85$ °C, 85% relative humidity
10.14	Robustness of terminations	As in IEC 60068-2-21
10.15	Wet leakage test	See details in 10.15 For modules with a surface smaller than 0.1 m^2, the minimum resistance is 400 MΩ. Larger modules are required to have the measured resistance times the area of the module greater than 40 MΩ/m^2. Testing is performed using a test voltage of 500 V or maximum voltage of the system, whichever is higher
10.16	Mechanical load test	Three cycles of 2400 Pa uniform load, applied for 1 h to front and back surfaces in turn Optional snow load of 5400 Pa during the last cycle
10.17	Hail test	25 mm diameter ice ball at 23 m/s, directed at 11 impact locations
10.18	Bypass diode thermal test	1 h at I_{sc} and 75°C 1 h at 1.25 times I_{sc} and 75°C
10.19*	Light soaking	Light exposure of cycles of at least 43 kWh/m^2 and module temperature of 50 ± 10 °C, until P_m is stable within 2%

*Only for thin-film PV modules (IEC 61646)

Nowadays, such certification is virtually essential to obtain any kind of aid, whether from governments or international programmes. In France, it is indispensable to obtain feed-in contracts for current sold to EDF (Electricité de France).

It is, however, important not to overestimate the value of these PV module qualification standards. They give an indication of good quality, but are not infallible. The reality is always more complex than laboratory tests, and nothing can match experience on the ground. There have been cases of highly certificated modules that showed erosion problems after some years of use, and conversely uncertificated panels that were still perfect after 15 years of service. Relying on a well-known brand is often the best guarantee.

3.1.4.3   Hotspots and bypass diodes

It can happen that a crystalline silicon module is not evenly exposed to light; there may be patches of shadow, or in the worst case, a dead leaf completely covering a cell. What happens in this case? As the cells are connected in series, the total current is reduced to the lowest cell output (the weakest cell imposes its current on the others), so when a cell has no output because it is not exposed to radiation, the current of the whole chain tends to fall to zero.

Worse than that, the shaded cell becomes the receiver of all the others in the series, receiving in inverse voltage the sum of all their voltages. It therefore starts heating up, hence the well-known term 'hotspot' to describe this phenomenon. It is essential to protect the panels from this happening, since the damage can even cause a fire. On panels with a voltage of 24 V and more and unprotected, the negative voltage applied on the shaded cell can easily exceed its breakdown voltage (30–35 V). This happens when the charge regulator is of the shunt type (see Section 5.1.2) and short-circuits the panels when the battery is full (Figure 3.14).

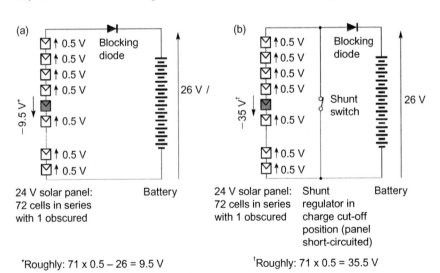

*Roughly: 71 x 0.5 – 26 = 9.5 V †Roughly: 71 x 0.5 = 35.5 V

Figure 3.14   Hotspot phenomenon on unprotected 24 V panel: a shaded cell receiving negative voltage: (a) during normal charging; (b) during panel short-circuit by a shunt regulator

And on the lowest voltage panels, the heating can cause irreversible damage (deterioration of terminals and anti-reflective layer). This can be seen directly with the naked eye by the colour of the cells on the damaged panels: cells that have been submitted to a hotspot will have turned brown (which can also happen to badly ventilated cells, placed in front of the junction boxes).

There is fortunately a fairly simple means of avoiding this phenomenon, which is systematically applied by most manufacturers. A diode has to be connected in parallel by a group of cells. By placing a bypass diode on each series of 18 cells, or two on a 36-cell panel, the reverse voltage applied to the shaded cell is limited to under 10 V, which causes limited heating in the case of a hotspot, generally below 60 °C, which is easily tolerated by modern modules.

These diodes are generally placed in the junction box at the panel outlet (Figure 3.15). For these diodes to be connected, it is essential that the panel has an electrical output accessible from the exterior at its mid-point (between the two series of 18 cells).

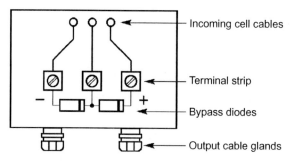

Figure 3.15   Mounting of bypass diodes at module output

3.1.5   Crystalline modules and manufacturers today

In the crystalline silicon industry, which still dominates the market with an 87% market share in 2008, it is not always the same companies that produce both the cells and the modules, although some do. So care must be taken not to calculate cells and modules from published data, since the modules are manufactured from cells.

Also the figures of MW produced in a country by its factories should not be confused with the MW installed, which is completely different. For example, in 2008, Spain installed 2.7 GW of PV power, but is only a small producer, with 170 MW produced in 2008 in the Isophoton and BP Solar factories. Obviously, a major proportion of the modules installed came from Germany and elsewhere.

In 2008, the world leader in crystalline cell manufacture was Q-cells with a total production of 574 MW (mainly crystalline silicon but also 11 MW of thin film), closely followed by the Chinese company Suntech (497 MW, again mainly crystalline) and the Japanese firm Sharp (with 473 MW in total), which has long been the market leader.

Most of the manufacturers following the use are in Asia and reflect the dramatic entry of China into this market, capable of developing an industry in only a few years. The main companies are Kyocera (Japan), Yingli (China), JA Solar (China), Motech (Taiwan), Sanyo (Japan), Trina (China), but there are also American companies such as Sunpower, with its 20% interdigitated cell technology. European companies that follow are mainly German: Solar World (190 MW) and Schott Solar (149 MW). France is way behind with Photowatt (58 MW).

Overall, the European PV industry is holding its place with another year of strong growth (+68%) in 2008: it comprises 28% of world production, led by Germany, which alone represents 65% of the European PV industry.

More detailed figures are given, for example, in the magazine *Photon International*,[13] the Solarbuzz consultancy website,[14] the EurObserv'er Barometer[15] or in French by the Cythélia newsletter.[16] This information generally has to be paid for (except for EurObserv'er).

3.1.5.1   A selection of crystalline modules

To show in concrete terms what today's crystalline silicon PV panels can achieve (in 2009), we list in Table 3.2 a selection of modules with data from the manufacturers' technical specifications.

In terms of efficiency, clearly Sunpower stands out, with its interdigitated cells giving a module with more than 19% efficiency, while the classic polycrystalline and monocrystalline cells all return between 11% and 15% efficiency.

It must be remembered that a module always has an efficiency slightly lower than that of the cells that constitute it, on account of the glass on the front, inactive surfaces and small electrical losses (mismatch of cells in series particularly).

Note also the range of power output (for example, 165–185 Wc for the same model), mainly caused by the variations of current between the cells (see the I_{cc} column). This leads the manufacturers to grade their cells and to produce modules that are physically identical but with an increasing power output, also sold at different prices, the market being based today on the price of the Wc. When installing a system, it is an advantage to have a batch of panels guaranteed within 3%; this will avoid losses on the ground and improve the production of the PV array (see Chapter 4).

3.1.6   Panel assembly

To ensure an installed power of several hundred watts, kilowatts or even megawatts, PV modules must be assembled in a *PV array* of varying area. On paper, series and parallel assembly respond to known laws of electricity: when the modules are mounted in series, voltage increases and the current remains constant, and when they are mounted in parallel, the reverse is true, the current increases and the

---

[13] http://www.photon-magazine.com
[14] http://solarbuzz.com/Marketbuzz2009.htm
[15] http://www.eurobserv-er.org/
[16] http://www.cythelia.fr/photovoltaique.php

Table 3.2 Examples of crystalline silicon PV modules (performance as given on the manufacturers' technical specifications)

	Type	No. of cells	Cell dimensions (mm)	Module dimensions (mm)	U_{oc}	I_{cc}	P_{max}	Module efficiency STC	Temperature coefficient
					Power > 150 Wc				
Sharp (Japan)	Mono	48	156 × 156	1318 × 994	29.4–30.2 V 0.610.63 V/cell	8.37–8.54 A 34.2–34.9 mA/cm²	170–185 Wc	13.0–14.1%	−0.485%/°C at P_{max}
Suntech (China)	Mono	72	125 × 125	1580 × 808	44–44.8 V 0.61–0.62 V/cell	5.05–5.29 A 32.3–33.8 mA/cm²	165–185 Wc	12.9–14.5%	−0.48%/K at P_{max}
SolarWorld (Germany)	Mono	72	125 × 125	1610 × 810	43.8–44.8 V 0.61–0.62 V/cell	5.00–5.5 A 32–35.2 mA/cm²	160–185 Wc	12.3–14.2%	−0.33%/K at V_{oc}
Sunpower (USA)	Mono back-contact	96	125 × 125	1559 × 1046	64.6 V 0.676 V/cell	6.14 A 39.3 mA/cm²	315 Wc	19.3%	−0.38%/K at P_{max}
Photowatt (France)	Poly	72	125 × 125	1237 × 1082	43–43.4 V 0.60 V/cell	4.8–5.3 A 30.5–33.6 mA/cm²	155–175 Wc	11.6–1.31%	−0.43%/K at P_{max}
Kyocera (Japan)	Poly	54	156 × 156	1500 × 990	33.2 V 0.61 V/cell	8.58 A 35.2 mA/cm²	210 Wc	14.1%	−0.46%/K at P_{max}
Heckert Solar (Germany)	Poly	54	156 × 156	1480 × 990	31.8–34.1 V 0.58–0.63 V/cell	8.2–8.45 A 33.7–34.7 mA/cm²	190–215 Wc	13–14.7%	−0.31%/K at V_{oc}

voltage remains constant. The current of the different panels therefore must be identical in a series array, and the same with voltage in a parallel array. The first rule to be remembered is that

- mount in series only those panels having the same operating current (and they need not have the same voltage)
- mount in parallel only those panels having the same operating voltage (but they need not have the same current).

In reality, as panels are not all absolutely identical, they can be paired in voltage or in current as required. This consists of connecting panels whose values are the closest.

Even when paired, panels may not always output the same power on the ground, simply because they do not all receive the same solar radiation. A shadow falling on one part of the array can cause the output of the whole array to drop significantly for a time, according to the hotspot principle described in Section 3.1.4.3. The simplest way of avoiding any problem of this kind is to place anti-return diodes of adequate power at the output of each series of panels (Figure 3.16).

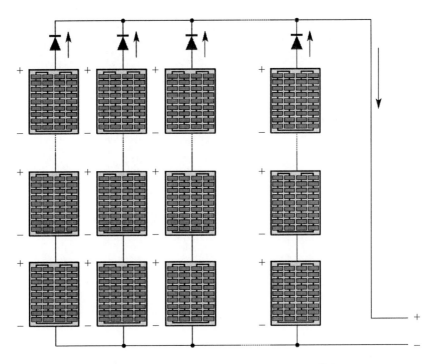

Figure 3.16   Array of panels mounted in parallel series

These diodes are often placed in the connection boxes, which also serve to connect the cables coming from the panels, and to output the total power through a cable of thicker diameter to the charge regulator (Figure 3.17).

If the PV array is reduced to a single panel or one series of panels, the simplest solution is to place this diode at the input of the charge regulator. It will also block any nocturnal current that may flow from the battery towards the panels.

The reduction in voltage should be as low as possible because it will directly affect the working voltage of the panel (a Schottky diode only causes a voltage reduction of 0.5 V as against around 1 V with a silicon diode[17]). In the majority of modern charge regulators, this diode is no longer fitted externally: the series regulator incorporates it or incorporates two MOS transistors mounted alternate ways up, controlled by a processor that cuts off the transistors' command at night. The series voltage of the regulator is often reduced to less than 0.2 V in this way.

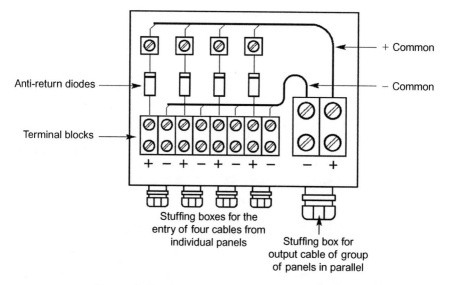

Figure 3.17   Junction box for panels in parallel

When the panels are mounted in parallel, their current is combined as we have seen. Thus, the total current of an array can rise considerably, especially at low voltage.

Examples

300 Wc of panels at 12 V:25 A; 1500 Wc of panels at 24 V:62 A; 20 kWc at 60 V:333 A!
This is why the larger the PV array, the more important it is to increase the voltage (more detail on the sizing of stand-alone systems is given in Chapter 5).

---

[17] When choosing a Schottky diode, attention should be paid to its reverse voltage, which could be too low: select double the nominal voltage of the panels.

3.2 Thin-film silicon cells and modules

3.2.1 The special properties of thin films

We shall now deal with thin-film technologies, which concern several PV materials – amorphous and polycrystalline silicon, CIGS and CdTe. All these materials use much less material than solid silicon, and they are produced in layers of around 1 μm thickness on rigid or flexible substrates. In the production of panels, these technologies also use similar methods: vacuum deposition and laser patterning, in particular.

Currently, these thin-film technologies represent less than 15% of the global market: 12.7% in 2008 with 1 GW produced, more than double of the 2007 production, which was only 433 MW. In a generally booming market this is not a bad result, but there is likely to be further expansion within a few years. The United States is already ahead on this front with more than 40% of all thin-film panels installed there.

These technologies are certainly cheaper (especially CdTe, which we will return to in Section 3.3) and offer a number of advantages on the ground: for example, it has now been demonstrated that amorphous silicon produces more kWh per kWc installed than crystalline silicon on account of its better response to diffuse irradiation and its lower temperature coefficient. But let us first consider the design of these cells and modules.

3.2.2 Simple junctions with amorphous silicon

It will be recalled that a simple junction, in the case of amorphous silicon, is a cell with three thin layers stacked, of p, i and n types: one layer doped with boron, one intrinsic layer (non-doped) and one layer doped with phosphorus (see Section 2.3.3). This structure is designed to produce an electric current to collect the charges produced under the effect of light. Several p–i–n structures can be stacked in this way to form multi-junctions, as we shall see later.

3.2.2.1 The manufacture of hydrogenated amorphous silicon cells

The silicon produced in thin films is basically *amorphous* in nature because it has a disordered, glass-like structure. The atomic organisation is not regular as in a crystal, but is deformed, and the crystalline structure is only maintained at very short distance (2–3 atomic bonds). The material therefore contains distortions and small cavities, and when the atoms are bonded to only three other atoms instead of four, this creates unsatisfied bonds or *dangling bonds*. Pure amorphous silicon is therefore not really a semiconductor – it contains too many defaults and cannot be doped.

But when it is produced from the gas silane (SiH_4), which it normally is, it contains a significant proportion of hydrogen (5–10%), which will bond with these dangling bonds, considerably reducing the density of faults and permitting the collection of charges and the doping of the material. Figure 3.18 shows the principle of the arrangement of amorphous silicon atoms with hydrogen.

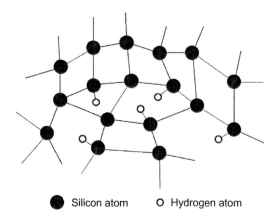

● Silicon atom   O Hydrogen atom

*Figure 3.18   Diagrammatic representation of a network of hydrogenated
amorphous silicon*

This material is thus a sort of amorphous alloy of silicon and hydrogen, which is described by scientists as a-Si:H (hydrogenated amorphous silicon) (Figure 3.19). This alloy has a higher optical gap (1.77 eV) than crystalline silicon, and absorbs light much more strongly (a layer 1 μm thick is sufficient to capture radiation at ground level). In practice, the thickness of the junctions can be as little as 0.2–0.3 μm.

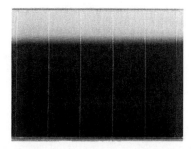

Figure 3.19   Amorphous silicon cell [SOLEMS ref 05/048/032]

The usual technique for manufacturing amorphous silicon cells is plasma-enhanced deposition. The layers are deposited directly from silane gas in a partial vacuum. The glass substrates are introduced to the machine, then heated to 150–200 °C. The silane introduced into the environment is decomposed by a radio frequency discharge. In the plasma thus created, the liberated silicon and hydrogen reform a solid but disordered material on the substrates.

The main advantage of this technique is that all kinds of different layers may be superimposed, just by modifying the gaseous mixture during the deposition, and even without halting the plasma. This can be used to form simple or multiple junctions. Doping is carried out by adding to the gaseous mixture elements in the form of gaseous hydrides: diborane (B_2H_6) for boron (p doping) and phosphine (PH_3) for phosphorus (n doping).

To complete the cell, two electrodes have to be fixed on either side of the silicon. When the cell is deposited on glass, which is the usual case, the (+) electrode is a transparent conducting layer deposited on the glass before the silicon. This is a metallic oxide such as SnO_2, tin oxide doped with fluoride, or ZnO, zinc oxide doped with aluminium. The quality of this front electrode is important, and if it is rough, it contributes to creating diffusion in the cell for a better absorption of light (see Section 2.3.1).

At the back, the (–) electrode is most often made of aluminium or silver, also in thin film, and sometimes nickel, to allow the soldering of the output conductors. Aluminium is a good reflector of light, so photons crossing the junction have a second chance of absorption. Another technique to increase diffusion is to make the rear conductor in zinc oxide transparent and not metal, and to place a diffusing material on the back of this layer to promote the trapping of light.

Figure 3.20 summarises the complete structure of a classic amorphous silicon cell (single junction).

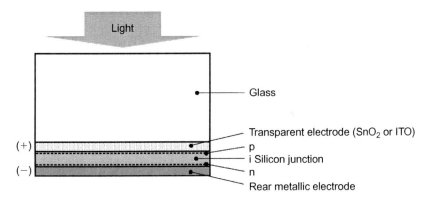

Figure 3.20   Structure of a hydrogenated amorphous silicon cell (not to scale)

3.2.2.2   Performance of simple junction amorphous silicon cells

Optical absorption and spectral response

What differentiates amorphous silicon from crystalline silicon from the optical point of view is

- its higher optical gap of 1.77 eV (see definition in Section 2.3.2) and
- its higher absorption of visible light: factor 4 at a wavelength of 590 nm (see Table 2.2).

Consequently, the cells are much finer (0.2–0.5 μm thick); they therefore use less material. The spectral response of a simple junction, illustrated in Figure 2.28, shows high values in the blue-green and yellow parts of the visible spectrum up to 600 nm, but cuts off earlier than crystalline silicon in the red, around 700 nm, the wavelength corresponding to the cut-off frequency of the optical gap of 1.77 eV. Thus, a fraction of the red light of the spectrum is not correctly absorbed in amorphous silicon, but is reflected by the rear electrode. This is why amorphous silicon cells often have a dark reddish appearance. This response can be improved in various ways: by increasing the optical reflection of the rear contact (to generate a second passage of light through the silicon); by trapping by diffusion as we have seen (Figure 2.19); or, particularly, by the use of multi-junction cells described in 3.2.4.

Performance under strong illumination

Figure 3.21 compares the typical performances of a crystalline silicon cell and a simple junction of amorphous silicon cell in STC (1000 W/m², 25 °C, AM 1.5 solar spectrum).

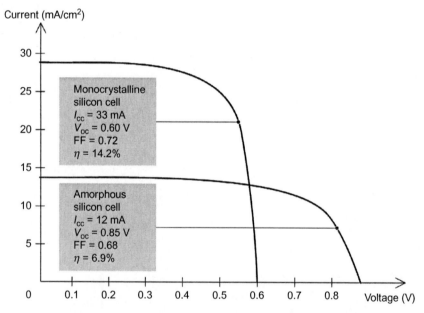

Figure 3.21   Current–voltage characteristics of an amorphous silicon cell compared with a crystalline silicon cell (1000 W/m² STC)

The amorphous cell, because its optical gap is higher (1.77 eV) than the crystalline silicon cell (1.1 eV), has a higher open circuit voltage (0.85 V against 0.6 V for crystalline silicon) and its operating voltage is also higher (0.7 V instead of 0.5 V). But its current is significantly lower on account of its inferior charge collection: 13 mA/cm² at maximum, as against 30–35 mA/cm² for the crystalline cell. The result is that, in industrial production, simple junction amorphous panels have an STC efficiency of 6–7%, which is considerably less than crystalline silicon. This is certainly a handicap: at identical peak power (measured in STC), an amorphous silicon panel is typically twice as large as a crystalline silicon panel.

But amorphous silicon has other advantages, in non-standard conditions, notably:

- Its voltage falls significantly lower than crystalline in low illumination.
- Its voltage falls generally lower than crystalline in high temperatures.
- It is more sensitive to blue light.
- It is more sensitive to diffuse irradiation.

Performance in low illumination
The fall in voltage with low illumination described for crystalline silicon in Section 3.1.2 is much less pronounced with amorphous silicon, which can function even at very low illumination. Its open circuit voltage only falls by 100 mV per log-scale decade of illumination.

This means that if its output is 850 mV/cell at 1000 W/m^2, this only falls to 750 mV/cell at 100 W/m^2, and 650 mV/cell at 10 W/m^2, equivalent to what a crystalline cell supplies at 1000 W/m^2. It is also able to function in very dull weather, corresponding to an illumination of 10 W/m^2, around 1000 lx, which is very low for outside irradiation.

In an indoor environment and artificial lighting, typical illumination ranges from 100 to 1000 lx, and the amorphous cell is still capable of providing a voltage of 0.5–0.55 V at 100 lx. Figure 3.22 shows the performance of an amorphous cell at these very low levels of illumination, inside and outside, down to a level of 0.1–1% of normal solar radiation.

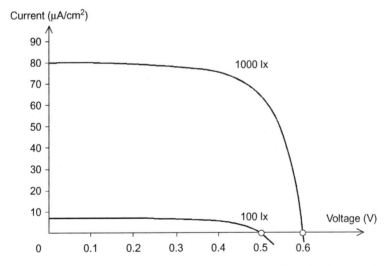

Figure 3.22 I/V curves of an amorphous silicon photocell under low-level fluorescent illumination

This property has led amorphous silicon to be used for the manufacture of cells for use indoors, as they will work under artificial light even at very low levels of illumination. These small cells are made in various voltages suitable for contemporary electronic circuits (Figure 3.19). Laser mounting in series (see Section 3.2.5) also facilitates the manufacture of amorphous silicon cells for this market, because it allows great freedom of current/voltage design for any format.

They generate micro-currents, sufficient for watches and portable devices such as calculators, measuring instruments, organisers, etc.

Sensitivity to blue light and diffuse irradiation

This indoor performance is also due to the good spectral response of amorphous silicon to short wavelengths. The fluorescent lighting widely used today provides a light with high colour temperature and a spectrum that is stronger in blue compared to classic incandescent light bulbs, which emit more red and infrared light (see Section 2.1.2). This factor reinforces the suitability of amorphous silicon cells on products for inside use.

Another interesting element working in the same direction is the disordered atomic structure of amorphous silicon, which is more sensitive to diffuse solar radiation coming from all directions (see Section 2.1). This diffuse radiation also contains a larger share of blue light on account of the spectral distribution of diffusion phenomena. There is consequently less loss of amorphous silicon when there is a little or no direct solar radiation, and when the panel is not accurately oriented towards the south. This is certainly an advantage for temperate climates (the panels work better in the winter) and for less than optimal orientations.

It is quite possible to place an amorphous silicon panel horizontally on a box: it will capture radiation from all directions of the sky.

See the example of telemetering in Section 5.6.1.

Influence of temperature

We have seen in Section 3.1.2.2 that crystalline silicon loses around 0.4% power/°C, or −16% for a temperature gap of 40 °C, between 25 and 65 °C, for example. In the case of amorphous silicon, this effect is smaller: because of the higher optical gap of 1.77 eV, the temperature effect is only −0.2%/°C on peak power.

This effect has a very important impact on energy production: it explains why, even in very sunny climates, amorphous silicon produces more kWh per installed Wc than crystalline silicon; the latter is affected by the bigger reduction of its power with temperature (−0.4%/°C). Numerous recent studies have demonstrated this,[18] and Figure 3.23 compares the outputs of amorphous silicon and crystalline silicon, showing that even in a very sunny desert climate, because of the high temperature, amorphous silicon has a higher annual output, for an identical installed power.

[18] K.W. Jansen, S.B. Kadam and J.F. Groelinger, 'The advantages of amorphous silicon photovoltaic modules in grid-tied systems', *Photovoltaic Energy Conversion, Conference Record of the 2006 IEEE 4th World Conference*, May 2006, vol. 2, pp. 2363–66.

S. Adhikari, S. Kumar and P. Siripuekpong, 'Comparison of amorphous and single crystal silicon based residential grid connected PV systems: case of Thailand', Technical Digest of the International PVSEC-14, Thailand, Bangkok, 2004.

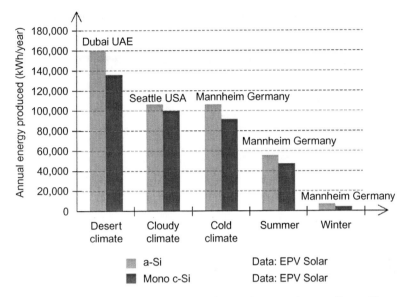

*Figure 3.23   Energy production of amorphous silicon and crystalline silicon
compared in different climatic situations [EPV Solar]*

3.2.3   Stabilisation under light

This stabilisation is a phenomenon specific to amorphous materials, the performance of which falls when they are first exposed to light. Known to scientists as the
Staebler–Wronski effect, it is often wrongly referred to as 'ageing'. It is not in fact a
permanent degradation, but more a 'running in' phenomenon: the material as we
have seen has a number of defects at the atomic scale and suffers a loss of efficiency when it is first exposed to the Sun (a simple junction of 0.3 µm showing a
loss of 20–25%) but its performance subsequently stabilises. This deterioration
arises because of certain defects in stability, which appear under illumination, weak
atomic bonding, notably. But the scale is limited, and the deterioration stops fairly
quickly (after some months in daylight). Manufacturers deal with this phenomenon
by improving the quality of the material, but it is not possible to totally eliminate it.
However, as the degree of stabilisation depends on the thickness of the junctions,
the use of multi-junctions is a good workaround, and enables the deterioration to be
limited to 10–15% (see the next section). The potential user has the right to know
the stabilised performances of amorphous silicon PV components. It sometimes
happens that documentation is not very clear on this point.

3.2.4   Thin-film silicon multi-junction cells

Another advantage of thin-film silicon technology is that it enables structures with
several junctions of different gaps to be made, each junction being specialised in
the conversion of a particular band in the light spectrum. As we have seen, it is easy

to stack layers simply by modifying the gaseous mixture in the plasma environment during the deposition of the cell.

The optical gap of amorphous silicon is 1.77 eV, and it does not absorb the red part of the visible spectrum ($\lambda > 0.7$ μm). Germanium, which is also tetravalent (with four bonds), has a much smaller gap (1.1 eV for amorphous germanium). Unfortunately, amorphous germanium on its own is a poor semiconductor, but a good silicon-germanium alloy absorbs part of this red light. It is therefore advantageous to place behind the silicon junction a silicon-germanium junction. Double-junction cells (tandem cells), or even triple-junction cells, therefore have a higher efficiency than a simple amorphous silicon junction: with solar radiation of 1000 W/m², module efficiencies of 7–9% stabilised, as against 6% for a simple amorphous silicon junction (see products of Schott Solar and Unisolar, in particular) (Figure 3.24).

These devices also have the advantage of increasing voltage of the cell: since there are two junctions in series, double the voltage is obtained. They can also reduce the stabilisation effects described in Section 3.2.3. Tandem cells have a stabilisation loss of 10–15% as against 20–25% for a simple junction.

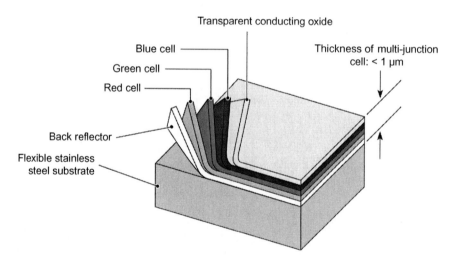

Figure 3.24   Example of triple junction cell: stabilised efficiency of 13% on flexible substrate [ECD – Ovonics]

3.2.4.1   Microcrystalline and polymorphous silicon, micromorph cells

By introducing a high proportion of gaseous hydrogen in the silane plasma for producing amorphous silicon, it is possible to create a certain proportion of crystallised micro-grains in the material being grown. This material, described as microcrystalline silicon and abbreviated as μc-Si:H, possesses some of the characteristics of crystalline silicon: it is more photoconductive than amorphous silicon and has a smaller optical gap, which makes it more suitable for the conversion of the solar spectrum than amorphous silicon (particularly in the red). Unfortunately, the speed of deposition is generally low. At first it was used in a very thin film as an interface in multi-junction cells, and now increasingly in micromorph cells.

Using the VHF-GD (*very high frequency glow discharge*, discharge of the plasma-enhanced chemical vapour deposition (PECVD) type, but at a higher frequency, 70 MHz) technique, developed at the University of Neuchâtel,[19] deposition is quicker and can produce under laboratory conditions microcrystalline cells only a few microns thick (Figure 3.25). By associating such a cell with one or even two thin amorphous silicon cells, this same laboratory has perfected tandem or triple cells known as *micromorph*. This process is now being industrially exploited by the Kaneka and Mitsubishi companies in Japan and Inventux in Germany, with module efficiencies between 8% and 11%.

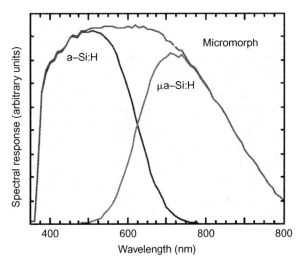

Figure 3.25   Spectral response of a micromorph cell [IMT, Neuchâtel]

Polymorphous silicon was developed in France at the Ecole polytechnique.[20] It is a material that could be described as 'nanocrystalline' because it contains crystals smaller than microcrystals. Its advantage lies in the fact that it could combine speed of deposition with photoconductivity properties close to microcrystalline, including at moderate temperatures compatible with deposition on plastic.

3.2.5   Thin-film silicon modules

3.2.5.1   Manufacture of modules

The manufacture of a PV module using amorphous silicon and other thin-film technologies is rather different from that of a crystalline module, especially on

[19] Photovoltaic and electronic thin-film laboratory, Institute of Microtechnology, University of Neuchâtel, Switzerland.

[20] Interface and thin-film physics laboratory, Ecole polytechnique, Palaiseau, France.

Figure 3.26   Amorphous silicon module (40 W–24 V)

account of the series arrangement of the cells, which is completely different (at least on glass substrates) (Figure 3.26).

Let us now look at the structure of an amorphous silicon module on a glass substrate (Figure 3.27): the cells are not physically separated like crystalline cells. The module appears as a uniform surface with only fine lines separating the cells. These lines in fact mark the links in the series mounting of the cells: each rectangular 'stripe' is a cell, and these stripes are connected in a series by three patterns.

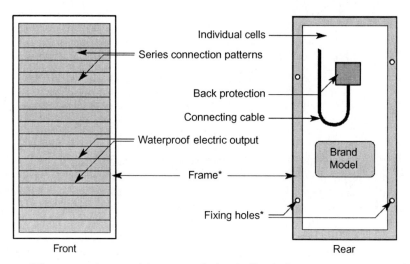

Front   Rear

* Some amorphous models are manufactured without a frame.

Figure 3.27   Structure of an amorphous silicon PV module

Let us examine these integrated series connections more closely.

Charges are created under illumination in the silicon layer, a phenomenon already described in detail in Chapter 2. These charges are subsequently collected

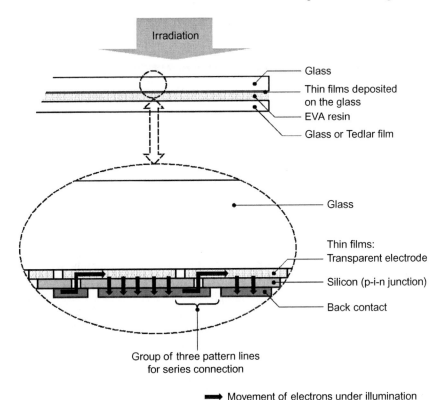

Figure 3.28   Series connection of the cells of an amorphous silicon module

(see the arrows on Figure 3.28) by the means of two electrodes on either side: transparent electrode (the (+) pole of the cell) and the back contact (the (−) pole). The patterns scribed in three layers, offset from one another, enable the adjacent cells to be connected in series. The (−) pole of the first cell is linked to the (+) pole of the following cell, and so on. The diagram is not to scale: in practice they are very close and the three patterns generally take a space of less than 1 mm. Thus, they appear to the naked eye as a single line of separation between the cells.

This technique is very useful as it does away with the necessity of physically cutting the cells to assemble them later. All that is needed is a suitable tool, such as, in this case, a laser (or rather a laser guided by a fibre optic), which is then used to 'scribe' the desired format for the module with an adequate number of cells in series. In this way, it is simpler to create all types of operational voltages with thin-film cells than with crystalline silicon. This technique can also be employed, with some adjustments, on polycrystalline CIS or CdTe thin films (see Sections 3.3 and 3.4).

This series integration is especially well suited to 'interior use' amorphous cells of the type used in calculators, the output voltages of which must adapt to these circuits that they have to feed. They can be produced for almost any voltage by adapting their number of stripes.[21]

How many stripes in series are needed to make amorphous silicon modules with voltages of 12, 24 or 60 V? As we have seen in Section 3.2.2, amorphous silicon has a higher optical gap than crystalline silicon, and therefore its open circuit voltage is higher. While 36 crystalline silicon cells are needed for a 12 V panel, only 28 amorphous silicon cells are needed. With an open circuit voltage of 0.8 V under 1000 W/m^2, the amorphous silicon cell will operate at 0.6–0.7 V, which will provide, for 28 cells, a working voltage of 16.8 V under sunlight, which is the voltage needed for a 12 V panel (particularly since the amorphous silicon is less sensitive to voltage loss when the temperature rises, see Section 3.2.2).

The encapsulation process for an amorphous silicon module is not very different from that of a crystalline module. The stacking is slightly different because the amorphous cell is already on a glass substrate (Figure 3.28). But it is important to protect the edges against corrosion. The best way is to keep all the edges away from the side of the module, a few millimetres sufficing for amorphous silicon to keep the working parts away from the outside world (this is the white border seen in Figure 3.27 between the frame and the cells). The same EVA as for crystalline silicon is used for encapsulation, and a back plate that may be opaque or may be Tedlar plastic film or a sheet of glass for more mechanical protection on large modules (because the front glass panel is not toughened[22]).

Framing will be adapted to the module's particular application. Sometimes small amorphous modules have no frame at all (Figure 3.29), nor do large-dimension laminates designed for use on buildings: this will make their integration into a roofing product or a facade easier. For other uses, metal or plastic frames are used as for crystalline silicon.

As for all PV modules, the electrical output point (junction between the module and its output cable) must be carefully designed since it can allow humidity infiltration. Usually, this part is resin-sealed by the manufacturer for additional protection, and not accessible to the user.

Bypass diodes are unnecessary for a single amorphous silicon module, because the shading of a single cell is very improbable in view of the geometry of the cells in long strips (Figure 3.27). On the other hand, in grid-connected systems with many modules in series, they are absolutely necessary, because a panel or part of a panel can be shaded while the rest of the array remains in sunlight.

3.2.5.2   Certification and life expectancy

Amorphous silicon panels have long suffered from a bad reputation on account of their initial stabilisation, but today it is known how to measure and control it, and it has been shown that it is not a continuous deterioration but only leads to a partial loss of performance, which is then stabilised (see Section 3.2.3). These panels have

---

[21] On these small cells and their experimental uses, see *Cellules Solaires* (details in Bibliography) and Figure 3.19.

[22] Even if the glass was originally toughened, the high temperatures that it undergoes when the thin films are deposited would destroy the toughening effect.

Figure 3.29 Small Solems amorphous silicon module (with overcharge limiter integrated to the output cable)

the same life expectancy at their stabilised performance level today as crystalline panels, and are submitted to the same tests as crystalline silicon panels and also tests of degradation under light. The testing standards are given in detail in Section 3.1.4.2. Amorphous silicon panels, like all thin-film panels, had to be certificated according to standard EN 61646 'Thin-film terrestrial photovoltaic (PV) modules – Design qualification and type approval'.

Although they have not been around as long as crystalline silicon panels, at present most reputable brands of amorphous silicon panels are guaranteed for at least 10 or 20 years. The enemies of all panels, whatever the technology, remain humidity, which causes corrosion, and changes in temperature, which can lead to joint failure. Consequently, encapsulation and cabling must be taken particular care of in all solar panels to obtain maximum life expectancy.

3.2.5.3 Current amorphous modules and their manufacturers

Originally this technology was developed by manufacturers who designed their own production machines, the precursors being Solarex (USA), today part of the BP Solar group, ECD-Unisolar (now Ovonics) and EPV in the United States, Kaneka and Fuji in Japan, and Phototronics (Germany) now Schott Solar, Sontor (Q-cells Group), Solems (France) and some others in Europe.

Today, with the development of the technology, the industry is in a higher gear, and since 2007–08, the number of manufacturers of amorphous silicon modules has increased in the United States, Europe and, above all, in Asia.

This is essentially due to three factors:

- the availability of ready-to-purchase production lines;
- the low cost of this technology compared to crystalline silicon: €1–€2/Wc currently, as against €2–€4/Wc for crystalline silicon (2009 prices);
- the superior annual output of electricity in kWh produced per installed Wc.

The two first points are closely linked, since the manufacturers of these production lines, Applied Materials (USA), Oerlikon (Switzerland) or Jusung (Korea), can guarantee a production cost to their clients (figures confidential!).

These companies, whose original speciality was manufacturing deposition machines for the production of thin-film transistors for flatscreen monitors, adapted this equipment for solar cells and added additional equipment such as lasers and laminators. What they offer is complete production lines for thin-film silicon modules from the sheet of glass to the encapsulated module ready for use, either single or double junction (amorphous/amorphous) or micromorph (amorphous/microcrystalline) with variable production capacities, by 30 MW multiples.

Driven by the increase in size of flat screen monitors in recent years, the sizes of glass handled have been increasing. Today, 1.4 m^2 is quite normal, and the new Applied Materials Sunfab factory can already handle 5.7 m^2 (Figure 3.30), and the sizes on pilot equipment already go as large as 10 m^2.

This tendency towards increasingly large sheets of glass is not necessarily an advantage for the PV modules. If the sheets are not subdivided, very large modules can only be used on ground level power stations or very large buildings, and special equipment is needed to handle them. Additionally, any breakdown will require a large size module to be replaced. For individual houses, modules of 1.4 m^2 are quite big enough and easier to install by technicians.

Many new factories equipped by these manufacturers can now be found, especially in Germany (CGS Ersol Thin Films, Inventux, etc.), in Asia (Sunwell, Bangkok Solar), in United States (XsunX) and soon in Italy in collaboration with Sharp, already one of the world leaders of crystalline silicon, which has announced

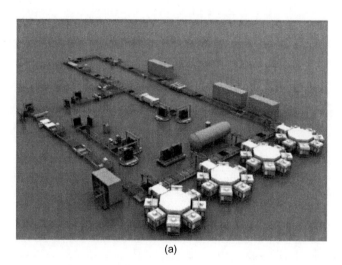

(a)

Figure 3.30   (a) Overview of a SunFab amorphous silicon module production line;
(b) details of a SunFab amorphous silicon module production line
[photo credit Applied Materials]

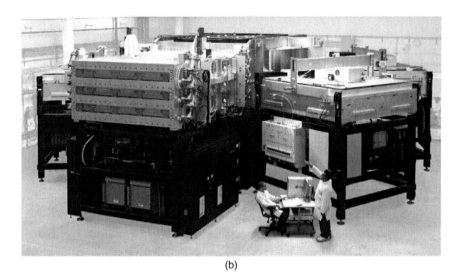

(b)

Figure 3.30   Continues

a massive expansion in amorphous silicon with a predicted capacity of between 500 and 1000 MW in 2010, with suitable production equipment.

Table 3.3 gives details of a selection of thin-film modules, both based on amorphous silicon and on CIS and CdTe, which are described in detail in the following sections.

The amorphous silicon modules include some using tandem and micromorph technology and some with flexible film, with rather disappointing performance. We will return to this in Section 3.5.1.

3.3   CdTe modules

Between crystalline materials and amorphous thin films, there is another family of materials – polycrystalline thin films. This material would be interesting if it could combine the efficiency of crystalline silicon cells under strong illumination with the manufacturing simplicity of thin films and their good performance both in low illumination and high-temperature conditions. But for the time, this material being is still at the laboratory stage.[23]

The other polycrystalline films already on the market are based on other semiconductors such as CdTe and alloys based on copper, indium and selenium (CIS or CIGS).

CdTe is interesting on account of its optical gap of 1.45 eV and its high absorption rate, which enables a film of less than 2 μm thickness to absorb virtually all of the visible spectrum. The panels are in attractive black colour. CdTe is generally p type, coupled with n-type cadmium sulphide (CdS), which serves as the

---

[23] See, for example, the work of the PHASE laboratory of CNRS Strasbourg.

Table 3.3 *Temperature coefficients of thin-film PV modules in standard test conditions [Data supplied by firms]*

	Type	Module dimensions (mm)	U_{oc}	I_{cc}	P_{max}	Module efficiency STC (%)	Temperature coefficient
			Power > 60 Wc				
Kaneka (Japan)	Amorphous silicon single junction	960 × 990	91.8 V	1.19 A	67 Wc	7	N/a
Schott Solar (Germany)	Amorphous silicon single junction	1108 × 1308	23.4–23.6 V	6.60–6.69 A	90–95 Wc	6.2–6.5 stabilised	−0.2%/K
Inventux (Germany)	Amorphous silicon single junction	1100 × 1300	136–139 V	1.09–1.16 A	79–94 Wc	5.5–6.5	−0.21%/K
Inventux (Germany)	Micromorph cells (amorphous/microcrystalline)	1100 × 1300	128–133 V	1.32–1.42 A	105–130 W	7.3–9	−0.25%/K
XsunX (USA)	Amorphous silicon double junction	1000 × 1600	58 V	3 A	127 Wc	7.9	N/a
Sharp (Japan)	Micromorph cells (amorphous/microcrystalline)	1129 × 934	65.2 V	2.11 A	90 Wc	8.5	−0.24%/°C
Unisolar (USA)	Amorphous silicon triple junction flexible	5486 × 394	46.2 V 2.1 V/cell	5.1 A 6 mA/cm²	136 Wc	6.3	−0.21%/K
First Solar (USA – Europe)	CdTe	1200 × 600	88–91 V	1.13–1.15 A	55–65 Wc	7.6–9	−0.25%/°C
Sulfur Cell (Germany)	CIS	1258 × 658	50–52.5 V	1.65–1.7 A	50–60 Wc	6–7.2	−0.3%/°C

front layer, to form a *heterojunction* (junction of two different PV materials). These cells have the advantage of a fairly high efficiency in sunlight, but also under reduced or diffused sunlight. Laboratory results are interesting with, notably, a record efficiency of 18% recorded in England in 2002.[24]

Industrial production, which was first held up by problems of controlling production processes like the p-doping of CdTe and by problems of the stability of modules sensitive to humidity, is today flourishing. The largest manufacturer of this type of panel, First Solar, has factories in the United States and Germany whose annual manufacturing capacity is 210 MW, and is currently building two extra factories in Malaysia for a total annual capacity of 240 MW. This expansion will take the production capacity of the company to 570 MW (some reports say even 1 GW) when all the projects announced towards the end of 2009 are completed. Production is highly automated and based on a single-panel format. The panels have an efficiency of 8–10% and a rather favourable temperature coefficient of –0.25%/°C (Table 3.3). Problems of life expectancy seemed to have been overcome, now reaching the market standard of 25 years, at the price of encapsulation between two sheets of glass to improve weatherproofing: the back cover is glass and not plastic film, see Figure 3.31.

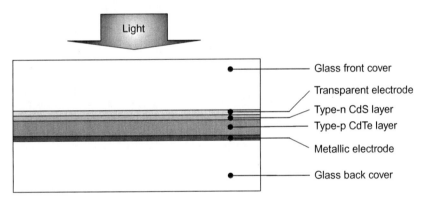

Figure 3.31   Stacking of the layers of a CdTe module [Calyxo]

But the most spectacular aspect of this technology is the low price of the modules, the manufacturing cost having fallen in 2008 below the symbolic level of $1/W. For some years now, First Solar modules have been the cheapest on the market, €1–€1.50/W (as against €2–€3/W for other technologies), which makes them the preferred product of the big energy operators like EDF Energies Nouvelles (France), to the annoyance of European manufacturers such as Conergy (Germany) or Assyce Fotovoltaica (Spain).

Consequently, in certain regions of the world, with prices as low as this, a high annual production because of a sunny climate and sometimes grid electricity prices higher than in France, parity with the grid has been reached. In other words, *the price per kWh of PV electrical production has fallen to the level of the production*

---

[24] Sheffield Hallam University, results unconfirmed.

price of traditional electrical power stations, in California and southern Spain, for example (see Section 4.7). This is a spectacular development that opens the way to the general use of PV for mass electrification.

However, there are several factors that may slow the development of CdTe technology. One factor is the toxicity of cadmium, which was listed by the EU *Restriction of Hazardous Substances* (*RoHS*) Directive published in 2003, forbidding the use of cadmium, lead and other toxic substances in electrical and electronic products, with exceptions (from July 2006).[25] Some countries, such as the Netherlands and Japan, have completely banned the use of cadmium. It is largely a problem of image, because the risk is not linked to the use of PV modules containing cadmium but rather to the handling of cadmium elements in the factory, and this can be controlled. Another usual counterargument is to say that cadmium is still widely used in other industrial sectors, such as pigments.

To get round this difficulty and reassure its clients, First Solar claims to have set up a complete network for collecting and recycling its products (criticised by some as a potential source of toxicity). The question, however, does not arise in the immediate future, bearing in mind the life expectancy of these modules, which is around 20 years.

The other main producers of this technology are Antec Solar and Calyxo (Germany) (Figure 3.32).

Figure 3.32   500 kW CdTe module power station at Springerville (USA)

---

[25] Original Directive 2002/95/EC can be downloaded at http://eur-lex.europa.eu/LexUriServ/LexUriServ.do?uri=CELEX:32002L0095:EN:HTML

3.4 CIS and CIGS modules

CIS, or more correctly $CuInSe_2$, is another PV material composed of an alloy of copper, indium and selenium. It is coupled, like CdTe, with a window layer of type-n CdS.

Theoretically, this heterojunction can attain an efficiency of 25%. It has excellent absorptive properties, but its gap is somewhat low (1.04 eV). This is why gallium is added to increase it, since the optical gap of $CuGaSe_2$ is 1.65 eV. The alloy called CIGS, for $Cu(In,Ga)Se_2$, is formed by adjusting the concentration of gallium to obtain an optical gap around 1.45 eV. The basic structure of the complete cell is shown in Figure 3.33. It is deposited beginning with the back plate and finishing with the layers exposed to the light, the reverse of the process is seen with an amorphous silicon cell. It is made of a substrate of glass (typically 3 mm) covered in molybdenum (Mo), which acts as the back contact, on which is deposited the active layer of p-doped CIGS, then a layer of CdS to form the heterojunction and finally a transparent layer of ZnO as transparent electrode. The CIGS absorbs most light, the CdS acting as a 'window' because with its high gap of 2.4 eV it admits all visible light.

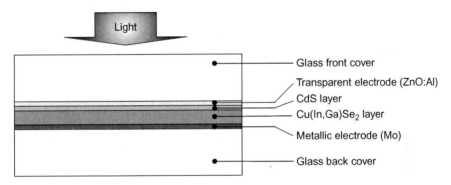

Figure 3.33   Structure of a CIGS module

Many different technologies can be used to produce CIGS. Either it is deposited directly in a single stage by co-evaporation (simultaneous evaporation under vacuum of different materials), or metallic layers of copper and indium deposited by evaporation (or cathodic pulverisation) and subsequently re-fired in an atmosphere of selenium or sulphur.

The company Nanosolar has developed a 'printed' CIGS using a much simpler process: the material is deposited in the form of layers of nanoparticles by screenprinting, then fired. Efficiency is over 10%, but the methodology remains complex and the products have not yet reached the market. It is also possible to deposit using electrochemical techniques: according to the CISEL process (developed by EDF R&D and ENSCP, the École nationale supérieure de chimie de Paris), the electrodeposition is done in one stage followed by firing, without a vacuum stage. It reaches an efficiency of 11.4% at a cost potentially much lower than with other processes. But so far these are still pre-industrial results.

As far as life expectancy in the open is concerned, these CIGS modules, like CdTe modules, appear to have a higher sensitivity to humidity than silicon modules. Double-glass encapsulation and a wide border outside all the layers seem necessary to ensure their long-term stability (20–25 years as other modules on the market).

However, the development of this line is due, above all, to its combination of the advantages of crystalline technology (high efficiency, although lower than the best crystalline silicon cells) and amorphous technology, which is in fact the characteristic of all thin films: the ability to deposit a large surface with series connection integrated rather than applied afterwards. These CIGS panels are some of the cheapest on the market (around €2/Wc, in 2009, for bulk orders) like most thin-film modules, but they are not yet as cheap as CdTe.

But for the reasons mentioned above (toxicity of cadmium), this technology may meet some barriers to its commercialisation, at least in Europe on account of the RoHS Directive. Work is continuing on finding an alternative to the CdS layer (ZnS, for example).

The use of indium creates another difficulty. The cost of this rare element, which is widely used in the booming production of flat screens in the form of ITO (indium tin oxide) electrodes, has soared because of raw material shortages. Indium is only found in minuscule quantities in zinc mines. Yet its consumption is still increasing, and geologists estimate that at the current rate global deposits will be exhausted within a few years, 10 years at the most. Alternative solutions – substitute materials and recycling – are being studied. For the CIGS industry, this is a real problem that could limit its long-term development.

Among the manufacturers of this technology are Avancis, joint-venture Shell/ Saint–Gobain (Germany), Ascent Solar (USA) that produces CIGS on flexible plastic film, Wurth Solar (Germany), Sulfurcell (Germany) and Honda Soltec (Japan) (Figure 3.34). The 2009 market price for power station quantities is around €1.5–€2/Wc (panels only).

Figure 3.34   The Honda Soltec building with its façade of CIG modules

3.5   Special modules

Some PV modules or elements have particular characteristics, often to meet the precise needs of certain applications, and we detail some of these below.

3.5.1   Flexible modules

Truly flexible panels require thin-film technology because crystalline silicon cells are not flexible by nature and can only support a slight curvature without breaking. At best they can be inserted in a 'curve' module, for example, for nautical applications or demonstration racing cars.

A truly flexible module is produced on a sheet of plastic or metal, the latter being much more economical. An amorphous silicon injunction is deposited at 150–200 °C, so any plastic material must be of the high-temperature type (polycarbonate or fluoropolymer (polyimide, for example)), then surface-treated to allow adherence. This method is already used on small and medium-sized formats, but it remains expensive (Figure 3.35).

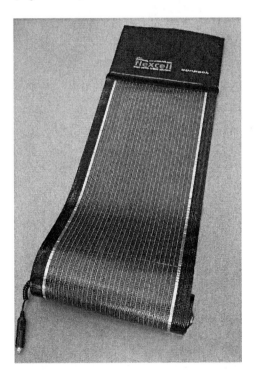

Figure 3.35   Amorphous silicon panel on flexible plastic (Flexcell, Switzerland)

Using stainless steel, the American company Ovonics (Unisolar) produces small cells of a few watts, highly flexible, which are then assembled into more powerful panels. The substrate has the disadvantage of being a conductor. So there is no problem for making a single cell, but when they are connected in series, the classic technique of patterns on amorphous silicon is no longer possible (Figure 3.28). The stainless steel must be insulated and then metallised again or steel cells must be assembled in a ring configuration like crystalline cells, as Ovonics or Global Solar do. The collection of the charge on surfaces of these large cells is then carried out by conducting combs deposited on the front surface, see Figure 3.39.

As can be seen, the manufacturer of flexible modules is not simple. But by using 'roll-to-roll' manufacturing technology, as used in the printing industry, for example, where plastic film revolves at high speed in depositing machines, it is possible to produce them at a lower cost, as has already been announced by Nanosolar (see Section 3.4) (Figure 3.36).

Figure 3.36   CIGS panel produced by 'roll-to-roll' technology (Global Solar, USA)

However, the results of more than 10 years' experience should be available before the life expectancy of these flexible products can be finally determined, on account of uncertainty on the durability of the plastics used and the mechanical handling of flexible PV materials.

Even with new modules, the measured efficiency is not always up to the values announced by the manufacturers. And yet there is much demand for these products because they have many applications. In outdoor activities: tents, boat sails, clothing, sports bags, etc.; military applications: clothing and field chargers, etc.; humanitarian aid: tents for field hospitals, etc.; and also for use on buildings.

3.5.2   Architectural elements

The current tendency for PV use on buildings is to integrate increasingly skilfully the panels into the architecture: this is more aesthetic, and in France it is essential at the moment to benefit from the feed-in tariff of €0.60/kWh (2009). This is known as *building-integrated photovoltaics* (BIPV), where the PV element is both the producer of energy and an element of construction.

This has given rise to a number of products adapted to integration in roofing, particularly solar tiles and flexible panels in rolls to be installed on terrace roofs, providing weatherproofing at the same time. Sometimes it is no more than a single module with a frame integrating a fixing system suitable for mounting on a roof, but sometimes there are far more innovative applications.

3.5.2.1   Roofing products

Flexible solar roof

Based on the requirement for a weatherproof roof covering, essential for terraced roofs on industrial or commercial buildings, the flexible solar roof combines PV production and a waterproof membrane. As it is an element including an architectural function, this solution is considered as an integrated one and remunerated as such through feed-in payments for the electricity generated. It is easily applied on terrace roofs, even for large surfaces. Its lower efficiency is compensated by its use over large surfaces. For this reason it is less attractive for individual dwellings. This technique also has the advantage of being light in weight (less than 5 kg/m^2 as against around 20 kg/m^2 for classic technology), but it does require a slope of at least 5% of the roof for the draining of rainwater. On renovated buildings, a substructure is therefore sometimes necessary, which must be taken into account in the calculation of profitability.

The product shown in Figure 3.37 is based on triple junction Unisolar cells, bonded to a synthetic membrane.[26]

Integration with steel roofing

Another solar roofing solution is Arsolar steel solar roofing, developed in partnership between Tenesol and Arcelor, which incorporates mono- or polycrystalline PV cells on classical steel roofing panels: they are installed in the same way as any other steel roofing, apart, of course, from the electric cabling of the modules (Figure 3.38). The idea is good, but experience with these products has unfortunately revealed problems of overheating and detachment. The materials used do not have the same coefficients of expansion, since the metal expands and contracts under the influence of temperature much more than the panels.

Integration into roofing is more often achieved with panels provided with special fixings, placed over a watertight membrane (see the PVtec system, Figure 4.10, or the Conergy Delta system), which enables a watertight roof to be combined with ventilation to avoid the high temperatures that reduce the performance of the panels.

[26] http://www.urbasolar.com/IMG/pdf/Brochure_SOLAR_ROOF.pdf

Figure 3.37   Waterproof solar membrane for roofing (Solar Roof product by Solar Integrated, USA)

Photovoltaic tiles

Many products come into this category, ranging from large surface modules integrated into roofing 'like tiles', with more or less aesthetic success, to small real tiles onto which a small module has been fixed, by way of imitation slates. Aesthetics plays an important part in this area, especially in France. Local planning regulations must be respected if building permits are to be obtained. The Germans were more lax about this at the beginning, and this did not always result in good results, visually.

Considering colours, the dark red of amorphous silicon blends fairly well with the colour of traditional tiles, whereas the dark blue of crystalline silicon is closer to the colour of slates.

Figure 3.38   Arsolar steel solar roofing

The photographs in Figure 3.39 show that it is possible to have good-looking solar tiles in a variety of dimensions and designs. The cells integrated in the solar slates are virtually invisible.

Figure 3.39   Various solar tiles and slates (Century Solar among others)

Apart from meeting aesthetic criteria, as for all PV modules integrated into roofing, these tiles must meet the following requirements:

- *PV production*: The distribution of the tiles in series and parallel has to be planned for a coherent PV array for both voltage and current so as to be suitable for the inverter.
- *Connections*: The smaller the tiles, the more they look like real tiles, but the more connections are needed.
- *Water resistance and mechanical resistance*: They must comply with weatherproofing standards and mechanical behaviour for PV modules (hail, wind, etc.).
- *Watertightness of the roof*: The design, jointing, compatibility with the slope of the roof had to be thoroughly thought through so that the result is a real roof with all the guarantees of traditional roofing.
- *Cost*: The more sophisticated the tiles, the higher will be the price of the installed kWc.

3.5.2.2   Facade and window elements

Architects interested in PVs have their own requirements, which have pushed the manufacturers of PV elements to innovation, and some very fine buildings have been constructed from PV elements.

Manufacturers of glass, glazed elements or facing blocks such as Schüco, Schott Solar or Saint–Gobain have also entered the PV market with their experience and their commercial networks, and this has contributed to the wider use of PV as an architectural element.

Here are some examples of PV products for this market.

Semi-transparent modules
With mono- or polycrystalline cells, semi-transparency is possible by allowing light to pass between themselves by hiding as far as possible the flat conductors necessary for the connections between themselves. On large panels, this gives a handsome appearance from a distance, as shown in Figure 3.40. And it can be an interesting solution for a glass roof (Figure 3.41).

Semi-transparent thin-film modules are even more beautiful. By changing the characteristics of the layers on the modules, their thicknesses and their optical properties, modules can be produced that allow a proportion of light to pass through their whole surface, either by partial transparency of the stacking of the layers, which gives an orange appearance, or by scribing fine patterns in the material according to the technology developed by Schott, which gives a neutral semi-transparent appearance (Figure 3.42).

Facade elements
These elements are often produced to measure, and can provide pleasing architectural and aesthetic creativity. Manufacturers of facade cladding or glass combine with manufacturers of special solar panels to offer original solution worked out with the architects.

Figure 3.40   Semitransparent crystalline modules used in a facade

Figure 3.41   Station platform at Morges, Switzerland, equipped with semitransparent roofing panels

Figure 3.42   Thin-film semitransparent modules: Schott Solar (top), MSK by Kaneka (bottom)

Here are some examples (Figures 3.43–3.46).

Figure 3.43   Schott Solar modules on the facade of a hotel

Figure 3.44   Buildings of the Ecole Polytechnique de Lausanne, Switzerland

Figure 3.44   Continues

Figure 3.45   Public administration building (Carhaix-Plouguer, France)

Figure 3.46   The Sanyo Solar Ark (Gifu Prefecture, Japan)

And finally, the beautiful Sanyo Solar Ark, 350 m long, 37 m high, 5000 PV panels, 630 installed kWc.

Chapter 4
Grid-connected photovoltaic installations

A grid-connected PV system is made up of an array of panels mounted on rack-type supports or integrated into a building. These panels are connected in series or parallel to achieve optimal voltage and current, and feed into an inverter transforming direct current into alternating current at a phase and at the same voltage as the grid. The typical operating voltage of an array of panels is around 150–400 V DC for small systems (1–3 kW) and 400–700 V DC for inverters of 10–500 kW. Maximum voltage is generally limited on the one hand by problems of insulating panels to avoid any current leakage, and on the other hand by the maximum voltage accepted by the inverter. The inverter will be equipped with a maximum power point tracking (MPPT) system that constantly adjusts the entry voltage to the characteristics of the PV modules, which vary according to temperature and solar radiation.

As the system is linked to the grid, the rules and standards to be followed are those of small producers of electrical energy not controlled by the electricity company, and the safety measures and precautions to be taken during installation and operation are more numerous than for a stand-alone installation. Here, the grid replaces the battery of the stand-alone system and offers the great advantage of accepting all energy produced (like a battery of infinite capacity) and being able to return, if necessary, more energy than has been fed into it. At first glance, a grid-connected system would seem easier to size because there is no need for a battery or charge receivers. However, to achieve optimal performance, careful preparation and sizing are necessary.

Experience feedback: The first European solar power station connected to the grid, with a power of 10 kWc, was installed in May 1982, and is still functioning well after more than 20 years. The system is mounted on a terrace of the LEEE-TISO laboratory in Lugano, Italian-speaking Switzerland. This laboratory, which specialises in alternative energy, has constantly monitored performance of the power station and published a complete report after 20 years of operation.

The report's main findings are as follows:

- While the aspect of the panels is not always perfect, the plant is still functioning correctly and the average loss of power of the panels has been −3.2% in 20 years.
- It is estimated that the plant should continue to function for at least another 10–15 years.

- The main cause of performance degradation has come from hotspots appearing on 24% of the modules (see Section 3.1.4).
- The solar panels manufactured by Arco (leading US and world manufacturer of the 1980s) were encapsulated in PVB (polyvinyl butyril), a material no longer in use today because it is not sufficiently stable and suffers from actinic deterioration over the years. The most affected panels have a short-circuit current between 10% and 13% weaker than the original. The other main defect of PVB is its propensity to absorb water, which has caused de-lamination (detachment through loss of adhesion between the plastic and the cell) on 92% of the modules.
- After 20 years, three modules out of 252 (1.2%) had a completely de-laminated cell at the location of a hotspot. In 1997, one module was replaced and the two other modules still operating produce, respectively, -20.2% and -14.8% of their nominal outputs.
- The performance of this plant enables us to predict an operating life of over 30 years for panels using modern and more stable encapsulation.

4.1   Grid-connected PV systems: feed-in principles and tariffs

Most European countries have a policy of buying in renewable energy dictated by the EU engagement in 2008 to reduce its greenhouse gas emissions by 20% by the year 2020, to reduce energy consumption by 20% through improved energy efficiency and to increase the share of renewables to 20% of the total energy consumption of the EU. These political undertakings are translated by new legislation and directives supporting the expansion of PV energy.

The French Environment and Energy Management Agency (ADEME)[1] has calculated that measures taken by the French Environment Ministry in 2008 will allow 218,000 jobs to be created by 2012 with 66,000 in renewables and 152,000 in improving energy efficiency. These measures should also result in economies of 12 MTOE of fossil fuels (7 MTOE of renewables and 5 MTOE of energy efficiency improvements).

In 2008, the renewable energy sector in Austria, France, Germany, Netherlands, Poland, Spain and Slovenia represented 400,000 jobs and a turnover of €45 billion.

France has also introduced attractive conditions for PV feed-in supported by an ambitious plan for 5400 MW of installed power by 2020. The feed-in tariff for grid-connected systems below 3 kW is maintained along with tax breaks (no income tax or VAT). Until the end of 2009, there was also a tax credit allowing 50% of material costs to be recovered (not including installation costs), up to a maximum of €8000 per person (€16,000 for a couple). More information on the latest incentives is available from ADEME.

---

[1] http://www2.ademe.fr/servlet/getDoc?id=38480&m=3&cid=96

4.1.1   2009 Tariffs

Most European countries try to reward the production of PV electricity by offering a feed-in tariff to cover the costs of PV generation, and sometimes make a profit from it.

Table 4.1 shows the terms on offer in early 2009 in selected European countries. The table is a summary of the main conditions, as most countries have fixed quotas of installed power, which, if exceeded during the year, result in lower tariffs.

The tariffs shown in the second column apply when the PV panels are integrated into a new building, while those in the third column refer to panels added to existing buildings. France awards the highest premium for integrated systems, following strict criteria.[2]

Switzerland launched a feed-in policy for renewable energy in May 2008 in which PV is allocated 5% of the funds available, originating from a tax on electricity sold over the grid. During the first week of May 2008, 4000 requests for PV installation were registered, which saturated the system, and at the start of 2009 there were still 3500 clients on the waiting list. In some areas, the local electricity company is supportive, but this depends strongly on the political orientation of the cantonal governments. For example, Geneva operates an attractive green policy that makes up for the shortcomings at the federal level, whereas the neighbouring canton Vaud offers nothing. Each canton makes its own choices in the absence of a political consensus on national direction in energy matters.

4.2   Components for grid-connected systems

4.2.1   PV panels for the grid

The grid-connected PV system uses traditional panels like those used in stand-alone systems, the only difference being that the number of cells is no longer tied to multiples of 36, the usual number used for the recharging of lead batteries. But panels for grid-connected systems are limited by the sizes made available by the manufacturers. The tendency is for this size to increase; today it is often limited to 1.7 m^2 for a typical output power of 200–300 W. This size corresponds to an approximate weight of 25 kg, which enables two people to carry out the installation. Larger (double-surface) panels are available, but their installation is more difficult without a crane. The size of crystalline panels today is also linked to the size of the wafer, which is mainly either 125 or 156 mm^2. Monocrystalline circles have a cut-off corner, which shows that they come from a circular crystal. These dimensions come from the production equipment for semiconductors, which is designed to deal with diameters of 150 or 200 mm. In future, there is likely to be equipment to deal with 300 mm, but for the time being no solar manufacturer uses wafers with this dimension, which would allow the production of a cell of 25 cm^2 generating around 19 A. Manufacturers are currently more interested in automating production to allow the

---

[2] http://www.industrie.gouv.fr/energie/electric/pdf/guide-integration.pdf

Table 4.1 Feed-in tariffs for PV electricity generation in Europe (2009)

Country	Tariff (€ct/kWh)			Duration (years)	Remarks
	Integrated	Roof	On the ground		
Germany	43 < 30 kW 40.9 > 30 kW 39.6 > 100 kW 33 > 1 MW	43 < 30 kW 40.9 > 30 kW 39.6 > 100 kW 33 > 1 MW	31.94	20	8–10% reduction/year
Italy	49: 1–3 kW 46: 3–20 kW 44 > 20 kW	44: 1–3 kW 42: 3–20 kW 40 > 20 kW	40: 1–3 kW 38: 3–20 kW 36 > 20 kW	20	2% reduction/year
Portugal	65 < 3.65 kW	65 < 3.65 kW	65 < 3.65 kW		5% reduction/year
Spain	34 < 20 kW	34 < 20 kW	32 < 10 MW	25	Tariffs and ceilings revised quarterly
France, Mainland, Corsica, DOM	60.2 60.2	32.8 43.7	32.8 43.7	20	Indexed according to inflation, max. production equivalent to 1500 kWh/kWc in Mainland France, 1800 in Corsica and DOM-TOM
Switzerland	60 < 10 kW	50 < 10 kW	50 < 10 kW		System saturated. Several thousand clients on waiting list

manufacture of cells less than 0.2 mm thick, which would enable them to economise on raw material (see Chapter 3 for details concerning the panels).

The most widely used crystalline panel for grid connection is available in two sizes: either around 1.3 m^2 for powers ranging from 150 to 230 W, or around 1.7 m^2 for 190–330 W. The full range of power available goes from bottom of the range polycrystalline to high-tech monocrystalline Sunpower cells. The same module is generally available with the power of $\pm 10\%$. The leading manufacturers carefully grade their cells and offer, for example, a 220 W range available from 205 to 240 W by 5 W steps. This type of offer is much more attractive than a module sold at 220 W $\pm$ 5%, where the purchaser knows perfectly well that they will receive a batch of cells ranging from 210 to 220 W and that the maximum power of the panels will be badly graded. A good supplier will send with the order the list of module output data, which will theoretically allow them to be sorted and paired for installation. But this task is difficult in the case of a large system where hundreds of modules are received packed in batches of 30 on pallets. This graded power criterion should also apply to thin films, but amorphous silicon panels do not have their definitive power set at the time of installation (because of stabilisation; see Section 3.2.3), and for this technology, pairing is virtually impossible. For other thin films, the principle of sorting by power is the same as for crystalline.

Table 4.2 shows a list of typical modules available today.

The table shows some characteristic parameters that enable the different technologies to be compared: thin-film modules on glass substrate operate at high voltage (80–116 cells in series) and low current (<2.3 A), which implies more parallel cabling. The form factor increases with efficiency, with the ratio between V_m and V_{oc} (maximum voltage and open voltage, respectively) facilitating the choice of inverter, this ratio varying between 1.18 and 1.37. Finally, the temperature dependence of maximum power is not always lower for thin films; the single-junction Kaneka module shows poor performance in this area. The gap in temperature loss between the best crystalline (Sunpower − 0.3) and a good thin film (Unisolar − 0.27) is only 10%.

4.2.1.1   Criteria for choosing grid-connected panels

Beyond performance criteria, most constraints are linked to installation and to the dimensions of modules when these are to be integrated into a roof or facade.

- Reliability and reputation of the manufacturer. You are buying a system that should last more than 20 years and it is essential that quality should be good to avoid service problems. The supplier will generally only guarantee his product for 1 or 2 years and offer a performance guarantee up to a maximum of 25 years. But the costs of exchange in case of breakdown will not be included, and to change the module in the middle of a large PV roof is a costly and sometimes very complicated operation.
- Good price.
- Closely power-matched modules.
- Good mechanical quality, well-designed frame and easily installed panel.

Table 4.2 Typical panels of the different technologies available

Type	Manufacturer	No. of cells in series	Tech	V_{oc} (V)	I_{sc} (A)	V_m (V)	I_m (A)	P_c (W)	FF (%)	V_{oc}/V_m	$P(T)$ (%/°C)
P200-60	Solarwatt	60	p-c-Si	36.1	7.8	28.0	7.2	200.8	71.6	1.29	−0.42
M230-96	Solarwatt	96	m-c-Si	59.4	5.3	47.7	4.8	230.4	73.5	1.25	−0.39
HIP-270	Sanyo	96	m-c + a-Si	66.7	5.6	53.8	5.0	270.1	73.0	1.24	−0.33
SPR-315	Sunpower	96	m-c-Si	64.7	6.1	54.7	5.8	315.1	79.3	1.18	−0.30
TEA 108	Kaneka	100	a-Si	85.0	2.3	62.0	1.7	107.9	55.7	1.37	−0.47
PVL-136	Unisolar	22	a-Si	46.2	5.1	33.0	4.1	135.3	57.4	1.40	−0.27
FS-275	First Solar	116	Cd-Te	92.0	1.2	69.4	1.1	75.0	67.9	1.33	−0.27
SCG-60-HV	Sulfurcell	80	CIGS	52.5	1.7	41.5	1.5	60.2	67.4	1.27	−0.32

- Good quality connections.
- Cooled anti-return diodes, with a junction box designed to dissipate their heat in case of a hotspot.

4.2.2   Mechanical installation and cabling of panels

4.2.2.1   Supporting structures

Since PV modules are always (or nearly always) installed outside, the supports must be resistant to corrosion, and therefore structures and fixings must preferably be in stainless steel, or aluminium, if the frames of the modules are themselves made of that material. Corrosion normally appears at the junction of two metallic materials of different electrochemical potential, and progressively destroys the material of which the potential is the weaker. For example, an aluminium solar panel frame should not be placed in contact with copper roofing material, or it will slowly wear away, with the aluminium being deposited on the copper. Different metals must therefore not be associated without protection, especially near the sea or close to roads that are salted in winter, since a saline atmosphere is an electrolyte that speeds up corrosion. Electrolytic potential is higher for a noble metal that behaves like a positive electrode attracting the current ions from the less noble metal, a process accelerated in the presence of an electrolyte.

Painting the support structure or the use of wooden supports (treated against rot and insects) is the most economical solution sometimes employed even in power stations in developed countries (Figure 4.1).

The sizing of the support structures needs to be carried out by a good mechanic according to the weight of the modules, wind resistance, and possibly the weight of snow in the mountains. In Switzerland, the standard SIA 160 indicates the typical values to take into account for the weight of snow and wind force for installations (not only PV), according to altitude and the height of the construction. Wind force is also dependant on the height above the ground: 70 kg/m^2 between 0 and 5 m increases to 100 kg/m^2 between 15 and 40 m.

4.2.2.2   Mounting

The information given in Chapter 3 on the assembly of PV panels is useful here, but needs to be adapted to the constraints necessary for grid-connected systems. The often high operating voltage of the PV generator calls for particular precautions, which we detail below, and the requirements of the grid companies impose other special rules (earthing, lightning conducting, insulation, etc.).

Safety during mounting

This is one of the paramount safety considerations for PV power stations. Recommended procedures are as follows:

- Install panels supporting the maximum open voltage of the generator.
- If possible, use modules equipped with insulated cables and floating plugs to avoid the need to access junction boxes on the site. Most manufacturers supply their modules already pre-cabled with sufficient connections for series

*Figure 4.1   Partial wooden support (1 MW power station at Verbois, Geneva)
[photo M. Villoz]*

connection. For the connection between the end of the series and the junction box, cables must be used fitted with the same plugs.

- Make all high-voltage connections between insulated plugs, with the ends fixed to already cabled screw or claw terminals.
- When making any modification to terminals on the voltage, work very carefully to avoid any electrical arc that could seriously damage the connections or cause a fire. If necessary, cover the panel array or work at night-time.
- Ensure that all the material used can support a maximum direct current voltage: for plant operating at around 600 V, the material used by trams and trolley-buses is often suitable.
- Closely respect the prescriptions relating to voltage and lightning protection (earth points, cable insulation, etc.).
- Use cabling resistant to outside conditions, particularly with stability to ultraviolet.
- Respect all safety regulations for workers working on roofs and facades and inform the installers of the specific dangers of direct current.

Types of mounting
There are five main types of mechanical installation:

- superimposed mounting on the roof or facade,
- integration into a building,

- mounting on a frame,
- mounting on a pole,
- mounting on a box.

The first two on the list are adapted to buildings already built and apply essentially to grid-connected installations, while the two last ones are reserved for small stand-alone systems (see examples in Chapter 5).

Figure 4.2 shows the surface use coefficient and efficiency of different mountings on buildings, the optimum arrangement being a system with no shade at 30° roof tilt, facing south at a latitude of around 45°.

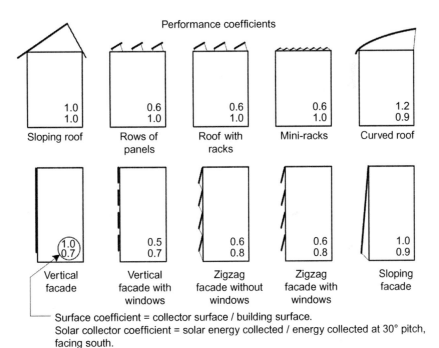

Figure 4.2 shows.

Surface coefficient = collector surface / building surface.
Solar collector coefficient = solar energy collected / energy collected at 30° pitch, facing south.

Figure 4.2   Available surface for a photovoltaic building

Superimposed mounting on flat roofs
Superimposed mounting on flat roofs is mainly used in countries where integration is no longer financially attractive. There is a large range of supports that are generally weighted to avoid affecting the waterproofing of the building. The supporting materials are sometimes a plastic trough ballasted with gravel, and aluminium structure fixed on anchoring of recycled plastic ballasted with gravel, or simply concrete blocks onto which the panels are bolted. Figure 4.3 shows some examples of terrace mounting.

Figure 4.3   Flat roof mountings [photo M. Villoz]

Mounting on a frame

Frame mounting is a classic form of installation for PV modules and is widely used in hot countries where roofs are flat, and on the ground for many technical applications where the surface of the PV generator is 5 m^2 or more. Mounting on poles is normally reserved for smaller surfaces. Frames are also indispensable on uneven terrain.

Module manufacturers sell frames for two, four or six modules, and to measure for larger systems. They are often made up of mountings in stainless steel or aluminium to assemble the modules and adjustable slope props to adjust for different angles. Usually they are fixed to the ground, but with concrete blocks as ballast they can also be held in place by their own weight, which gets round problems of waterproofing on flat roofs.

When the system is bigger, a supporting structure will be made to measure all the panels, often with local partners. A civil engineering company is often necessary for these works, which can call for excavation and the laying of a concrete slab.

In tropical countries and southern Europe, it may be worthwhile installing a *tracker*, a motorised mechanical support following the Sun's trajectory. It can either work in two axes, the PV array always remaining perpendicular to the Sun's rays, or in one axis, either vertical or inclined. Large power stations using trackers of more than 100 m^2 have been installed in Spain and Portugal in the last few years, with an annual production of over 2000 kWh/kWc. Their long-term behaviour is not known, or how they age or their real cost compared to fixed systems. Some tracker sizes with estimated typical production data for systems in the south or north of France are given in Section 4.4.3.

Integrated roof mounting

Integrated roof mounting is described in Section 4.3.2.

4.2.2.3 Cabling and protection against lightning

All cables, mechanical devices, fixings and electrical components must be installed according to International Electrotechnical Commission (IEC) standards and appropriate local regulations. In France, conducting cables must comply with regulation NF C-15100 to reduce voltage losses, which should be below 1% of the nominal voltage for a solar radiation of 800 W/m². On the AC side, the falling voltage between the inverter and the circuit breaker at the limit of the property should be below 1% of the nominal voltage at maximum power.

The installer should refer to 'Protection guide against the effects of lightning in installations using renewable sources' prepared by the ADEME. Figure 4.24 shows a diagram of a classic grid-connected system equipped with a three-phase inverter.

A string of panels should be connected in series without making loops (U-shaped cabling) to avoid any induction effect likely to attract lightning. The series connection should preferably be Z-shaped and the connecting cables should be firmly linked to earthed metal conduits. Depending on any shading in the vicinity, it may be better to cable vertically (if the shadow of a mast or pole crosses the array) or horizontally if there is any horizontal shading (panels on racks, for example). If the shadow touches a panel, its effect will not be much greater if it touches other panels in the same string, the current already being reduced by the shaded panel. For roofs in the North, which may be covered in snow in the winter, horizontal cabling is more suitable, the upper part of the roof in general losing the snow cover more quickly and possibly allowing one string to produce current while the rest of the panels are still covered in snow.

Most panels have a frame of anodised aluminium that is not a surface conductor (insulating aluminium oxide). To earth the array, fixings through the oxide may be used or an earth point on the frame of the panel that has been protected during the anodising of the aluminium could be used. Suitable earth contacts (Solklip by Tyco) can be screwed to the panel frame with a spring-loaded contact for rapid connection to the earth cable.

4.2.3 Grid inverters

As with a stand-alone inverter in a remote location, the principle is to transform direct current into alternating current at a frequency and voltage equivalent to that of the grid. The essential differences are that it must be a sine wave frequency and the AC voltage to be fed into the grid must be in phase with it and comply with a number of regulations and safety requirements, which are more demanding than for a stand-alone system. The other difference is that the entry voltage is not the stable voltage of a battery but the fluctuating power arriving directly from the generator.

All inverters connected to the grid incorporate an MPPT — explanations in Section 5.12 — and a number of common characteristics such as automatic disconnection in the case of absence of the grid, and minimal production of harmonics and a high frequency precision. Today's devices incorporate a processor ensuring the standards applied in most countries. At start-up, the first step is to set the language and country to adapt thresholds and operational limits to local standards.

4.2.3.1   Grid inverter designs

'Module' inverter

The smallest models (100–200 W) are fixed behind the solar panel, which then will directly produce 230 V AC. Its advantages are reduced cabling, only in AC, simplified connection to the building and reduced sensitivity to shading, since one shadow does not usually affect all the panels of the system.

'String' inverter

This is a sort of more powerful module inverter, connected to each string of solar panels in a series. Its advantages are an economy in cabling and DC protection. Its design is similar to the preceding one with the advantage of working at a higher power and voltage and so more efficiently.

Central inverter

Intermediate-sized models (1–5 kW) are generally single phase and intended for individual houses or small buildings. The solar power station type models are generally three phase for power in excess of several hundred kilowatts. The advantages are clear separation of the DC and AC parts and simplified maintenance; the disadvantages are more complex cabling and an increased sensitivity to shading.

4.2.3.2   Technology

Grid-connected inverters use two techniques to generate alternating current: either the sine wave is produced by the device, which uses the zero point of the grid to synchronise, or the grid is used as signal and synchronisation source.

Some inverters use a transformer to ensure a galvanic separation from the grid, which enables the panels to be insulated from the grid; other inverters dispense with galvanic separation in order to improve efficiency (+2%) and reduce cost. For these latter inverters, the cabling of the panels must be floating as the inverter continuously tests all current leaking in the direction of the earth. In case of leakage, the inverter cuts out to avoid all direct contact between the panel frame and the grid.

All inverters use a certain amount of energy for their internal operation. This can either be supplied by the PV generator (advantage: does not use current at night; disadvantage: no system of continuous measurement possible), or from the grid (advantage: regularity and stability of the feed; disadvantage: continuous consumption). Such continuous consumption by the inverter has a slight influence on the annual efficiency of the system.

Inverters, in general, are equipped with devices to measure basic data. They often also offer an interface allowing the collection of these values by a data logger or a computer.

To ensure the safety of the system, prescriptions relating to the local grid and those relating to the inverter must be distinguished. Electricity companies generally require supervision of the voltage and frequency produced and a very rapid disconnection when the grid is absent. The power limitation of 3.5 kW per phase for a simple installation in some countries implies a limitation of the current produced. The inverter, which saturates at this power, uses its MPPT controlled by a microprocessor to vary the point of maximum power of the generator in the direction of open source voltage, and thus limit the input power instead of dissipating it thermally.

Inverters do not have a constant efficiency: they are generally more efficient at three quarters of maximum power, and less efficient at low power. This efficiency curve of the inverter according to its power output is the main parameter to understand. The weighted European efficiency is calculated according to average solar radiation data and the efficiency at partial charge of the inverter (see detailed definition in 'Efficiency of the inverter' in Section 4.3.2.2). The best values of the current technology for the European efficiency factor are >96% for devices without transformers and >94% for devices with transformers. However, another value is important − the dynamic efficiency of the inverter that in turn depends on the efficiency of the MPPT. When power changes rapidly, for example, during the passing of a cloud, the current drops strongly and the voltage drops slightly, and the MPPT must adapt rapidly, increasing its input impedance so as not to overload the input and over-limit the inverter. In the reverse case, when the cloud disappears, the charge impedance needs to fall rapidly to take advantage of this power increase. This efficiency is difficult to measure and may only become clear when data is available over a period of time. A good static efficiency may well be useless if the dynamic efficiency is poor. An inverter's highest efficiency is often reached at a considerably lower entry voltage than the operating MPPT voltage. This information is useful to know when planning a system, as one will then try to choose a number of panels and series producing a voltage as close as possible to this reduced range of the MPPT. Some inverters have several MPPT circuits that function either separately or in cascade. With separated MPPTs, strings with different voltages can be connected (number of panels with different series), which can be useful when covering the surface of the roof, and the strings of identical voltages cannot be divided. Separate MPPTs are also suitable for arrays of different orientations, each array then being treated separately and able to function at its maximum power. If the strings arrive at a common entry point on the inverter and several MPPTs are operating in cascade, the inverter connects them according to their power level, so that at low radiation a single circuit will operate, which will allow the partial-charge efficiency to be improved. This type of inverter is particularly suitable for the operation of systems at high latitudes where solar radiation variations are wide during the year and where the inverter is usually operating at low power. We will show that this parameter is already clearly visible in the case study of a small system installed in the south or in the north of France (see Section 4.3.1).

4.2.3.3 Criteria for choosing grid inverters

A number of criteria need to be considered before ordering an inverter. The following list gives the essential points to respect in making this choice:

- High European efficiency factor: this criterion is more important than the cost of the device, since at current electricity feed-in prices, a higher efficiency will bring in more revenue after some years of operation. A higher efficiency also guarantees lower thermal dissipation. The durability of electronic components is strongly influenced by temperature.

- Good price.
- Reliability and reputation of manufacturer: these two parameters are essential as the inverter is the component most likely to break down. Some manufacturers offer renewable guarantees (through insurance).
- Guarantee, after sales service: in the case of breakdown, rapid repair is necessary, especially in summer. Some manufacturers offer an insurance of production revenue: the device will be repaired after a maximum of x days, otherwise the client will receive a sum equivalent to the production of the generator that has broken down.

These criteria, in practice, limit the choice to a manufacturer either local or sufficiently well-established to be able to provide an adequate local service.

4.3   Grid-connected systems – sizing of integrated roofs

Even if the theoretical design of a grid-linked system is simpler than a stand-alone installation, it is still necessary to choose and assemble components capable of functioning efficiently together. For the optimum sizing of generators, it is strongly recommended to rely on a simulation made with appropriate software. This will enable the energy produced to be forecast on the basis of local solar radiation statistics and will also provide a reference in case of doubt on the generator's performance. All the calculations given below were produced with PVsyst[3] software developed by André Mermoud of the University of Geneva. This well-designed software is currently the only product capable of evaluating losses due to any nearby shading. Regularly updated, it contains a database of most solar components on the market, which enables it to quickly size a PV system (whether grid-connected or stand-alone).

4.3.1   Sizing of the inverter

We have considered in Section 4.2.3.3 some of the criteria for choosing a grid inverter, and we will now look at how to pair an inverter with the array of panels in order for it to be as efficient as possible. To do this, we take the example of a 3 kW roof system, the favourite of many French installers.

4.3.1.1   Inverter power

Initially, one would be tempted to choose an inverter of equal power to the STC (Standard Test Conditions) power of the PV generator, but on examining in more

---

[3] http://www.pvsyst.com

detail the operating parameters of these two linked components, various differences will be noticed:

- The inverter will lose a few percent of its power for its own operation, the best devices attaining a European efficiency factor of the order of 95–97%.
- When irradiance is at its peak, temperature of the panels is almost always well above 25 °C, the temperature of the nominal power of the generator under STC conditions. And as all the panels undergo a loss in efficiency at temperatures >25 °C, the power of the PV generator will be inferior to the STC power. The NOCT (nominal operating cell temperature) (measured at 800 W/m^2 of irradiance) +10 °C to take account of the radiance of 1000 W/m^2 could be taken as maximum temperature initially. A typical NOCT value is 45 °C, or 20 °C higher than STC conditions at 25 °C; by adding the 10 °C for 1000 W/m^2, we arrive at 55 °C. A normal crystalline panel loses around 0.4% of its power per °C, which at 55 °C corresponds to 12% loss. The inverter and array temperature losses will then amount to 15–17%. For very well-ventilated systems (in open fields or on racks on a flat roof) in temperate Europe, the DC power of the generator can comfortably be oversized by 17% of the inverter AC output.
- For integrated roofs with relatively little back ventilation, an extra 5–10% of temperature losses will have to be assumed.
- In the particular case of amorphous silicon panels, instead of undergoing losses as above, the generator can produce more when it is first commissioned, up to +25% for a single junction, on account of the Staebler–Wronski effect (see Section 3.2.3). For double junctions, +15% can be assumed. In these cases, it must be verified that the inverter saturates when the power of the generator is higher, when the panels are first commissioned.
- For thin-film panels, an additional difficulty is the poor form factor, which provides a bigger gap between the open voltage (open circuit voltage) and the maximum power voltage, but it is sometimes difficult to find an inverter that will accept a high open voltage.

Maximum PV power is often reached in spring during May when the sky is particularly clear after a shower. In summer, the heat prevents the sky from being completely transparent, and maximum insolation is rarely reached. But the density of high irradiance is most frequent in summer, and this factor leads us to again limit the maximum power of the inverter. Oversizing the generator allows one to improve the partial charge efficiency of the inverter, since it will reach a high efficiency more quickly, and this is favourable to the annual output of the system. All these parameters can be simulated, and Figures 4.4–4.8 show the output power of a PV generator inverter for typical solar energy during the year.

Voltage and current
The other criteria for sizing an inverter are to adapt the voltage of the panels to the MPPT of the inverter and to ensure that the maximum open voltage of the generator (at any temperature) is accepted by the inverter. For safety reasons, the maximum

open voltage is taken as greater by a kT factor (clearness index) to STC conditions, with the following values in temperate countries:

- 1.15 for a system at an altitude <800 m;
- 1.20 for a system between 800 and 1500 m altitude;
- 1.25 for a system at over 1500 m altitude.

For example, if 12 panels with an open voltage of 22 V STC are connected in series, the inverter in the high mountains would have to tolerate an open voltage of

$$12 \times 22 \times 1.25 = 330 \text{ V}$$

These maximum open voltage values are important, since problems have been encountered on systems working at 550 V DC nominal at sea level (V_{oc} of 700 V): the junction boxes fixed to these stainless steel frame with stainless steel clamps had considerable current leakages, and if the earth connection was poor, the frames could have carried the panel voltage that would have been very dangerous if touched. All the panels in the system, with an output of over 100 kW, had to be replaced in this case.

Simulated 3 kW system

For the example in Figure 4.4 (simulation carried out with PVsyst), we took a roof in the south of France (using statistics for Marseille), and a typical individual system with 2.9 kW of panels. We also took into account the relatively poor ventilation of integrated modules by increasing the NOCT value by 20 °C, corresponding to 8% of additional thermal losses. The roof in question faces full south without shading and has a slope of 25°. The panels chosen for the simulation are two strings in series of six Sunways SM230 modules of 240 W nominal, and we are assuming their power is guaranteed to +1% of the 240 W nominal (panels perfectly graded). The two strings are connected directly to the inverter (without series diode) with 30 m of 4 mm² section cable. To modify the thermal characteristics, we have reduced the typical dissipation of 25 W/m² (open field situation) to 16 W/m² to obtain a new NOCT value of 65 °C instead of the original 45 °C.

The inverter used for simulation is the one manufactured by Sputnik Engineering, a Solarmax 3000 S, a recent model with a European efficiency factor of 95.5%.

The power distribution graphs show the different energy output by the PV generator during a simulated year. The power output of the generator is divided into a scale of 50 steps of roughly 30 W, and the line shows the energy produced by operating step: thus in Figure 4.4, the PV generator will produce a maximum of energy between 1700 and 1730 W.

The simulation of the system gives an annual energy production of 1370 kWh/kWc, or 1370 operating hours at the nominal power of 2.9 kW corresponding to 3946 kWh/year. This gives a performance ratio (PR) of 78.6%. For a better ventilated system, the PR would exceed 80%. PR is the global system performance measure that takes into account all losses, detailed in the following paragraph.

It will be seen that the generator never reaches its nominal power of 2.9 kW: the few moments during the year when the array generates more than 2.3 kW are rare. Most of the annual production is generated at powers between 1 and 2.2 kW DC.

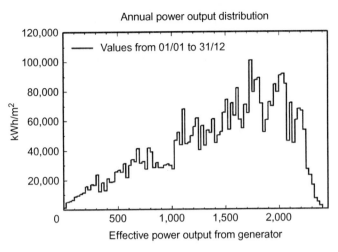

Figure 4.4   Individual 2.9 kW PV system in Marseille: distribution of annual production

We have intentionally chosen an inverter that is a bit overpowered so as to never saturate the system. To improve annual production, the power of the inverter could be reduced to around 2.2 kW AC (improving conversion at low power), but at present there is no inverter with the same performance in this power range.

Figure 4.5 shows a simulation of the same roof, this time in Strasbourg, and we have reduced thermal losses to take account of the lower ambient temperature: we are assuming this time an NOCT of 60 °C. The annual production figures fall to 952 kWh/kWc, which is 952 h at 2.9 kW corresponding to 2743 kWh/year. The PR has slightly increased to 79.6%, reflecting a lower estimated temperature.

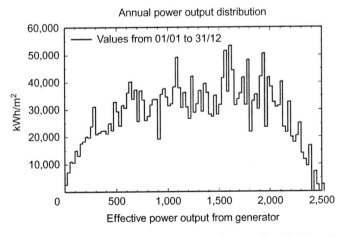

Figure 4.5   Individual 2.9 kW PV system in Strasbourg: distribution of annual production

It will be noticed that the annual power distribution has moved towards the left, because the inverter is more often operating at lower power.

As a final example of the power sizing of an inverter, we have moved our roof to Davos in the Swiss Alps, a town well known for its meetings of high financiers and for its abundant sunshine.

As Davos is situated at 1590 m altitude, we have further reduced losses from the badly ventilated roof to an NOCT value of 55 °C. We have also taken account of the presence of snow from December to March in assuming an albedo effect of 0.8 during these winter months.

The annual production performance increases to 1338 kWh/kWc, which is 1338 h at 2.9 kW corresponding to 3853 kWh/year. The PR has increased to 84.0%, reflecting a lower estimated temperature.

The power distribution shown in Figure 4.6 has completely changed with an obvious saturation of the inverter, which imposes a ceiling on the generator of 2.6 kW. The high production peak of 2600 W shows that for part of the time the system has produced exactly 2.6 kW, because it is unable to produce any more. The loss diagram of the simulation shows a value of −0.3% of inverter overload.

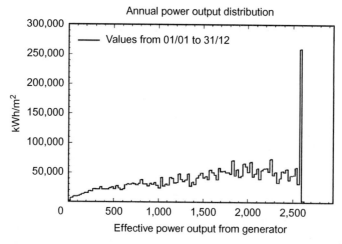

Figure 4.6 Individual 2.9 kW PV system in Davos: distribution of annual production

In this case, it would probably be worthwhile selecting an inverter of slightly higher power.

We will repeat the simulation by choosing an inverter of the same brand so that its efficiency and operation are very close to the last simulation. The Solarmax 4200 S model has a nominal output of 3.8 kW AC with a European efficiency factor of 95.8%. With this inverter, the system produces 3854 kWh/year with a PR of 84.1%. There is an annual improvement of 1 kWh due to the non-saturation of the inverter, but there will have been efficiency losses at low power, which explains the minimum difference.

As seen in Figure 4.7, there is no longer a problem of saturation, but, on the other hand, a few points exceed the power of 2.9 kW, with the graph showing that output rises occasionally to 3 or 3.2 kW. In winter, because of the increase in irradiance caused by the snow and because of the lower temperature, the system reaches and even exceeds its STC power several times during the season. This example shows a particular characteristic of PV systems installed at high altitude. In these areas, and in the Nordic countries, the low temperature and snow cause the generator to operate beyond its STC performance. But care must be taken that the panels are not covered by snow in order to benefit from these high-power moments. When systems are sized at high altitude or in cold countries, these conditions must be taken into account, and then it will be best to mount the panels at a steep angle to take advantage of winter conditions with a favourable albedo effect from the snow.

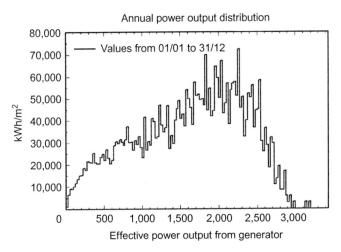

Figure 4.7   Individual 2.9 kW PV system in Davos: distribution of annual production with inverter of higher power than generator

Keeping the same parameters but changing the angle of pitch to 55° (very steep), the high altitude system produces 3914 kWh/year with a PR of 84.3%, +60 kWh or 1.5% more than at 30°.

This final simulation (Figure 4.8) shows that the system can occasionally produce up to 3.45 kW with 2.9 kW of panels (STC). The albedo effect is very clear, and the number of moments when the power exceeds STC power is still more frequent.

4.3.2   Sizing of a complete system

4.3.2.1   Preliminary study

At the pre-project stage, a preliminary study will evaluate what the planned generator should produce and give an idea of the costs and components needed. For the

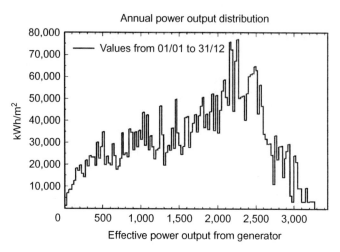

*Figure 4.8   Individual 2.9 kW PV system in Davos: distribution of annual
production with inverter of higher power than generator and roof
pitch of 55°*

evaluation of a solar plant, one normally uses the annual quantity of energy produced in kWh divided by the system's peak power (in kW), called the *energy density* of the system, denominated in hours. This measure corresponds to the time of effective operation at nominal power. It depends on the quality of the system (efficiency, leakages, etc.) and of the solar situation of the location. This figure may be compared to the annual duration of operation of a classic power station using a conventional source of energy. Some countries publish at the end of each year the average energy density of all the systems connected to the grid.

It is useful to know what a PV generator can produce annually in relation to the budget available in order to optimise revenues, taking into account feed-in tariffs. To facilitate this process, we give below an example of a small solar plant with different calculation stages using PVsyst software to estimate energy production for three different types of panel.

4.3.2.2   Integrated PV roof of 450 m²

The owner of the shop with a well-oriented roof (30 × 15 m, sloping at 20° and facing due south) in the region of Limoges, central France, wants to know whether it would be worth his while to cover this surface with different forms of solar panels. We give below this case study with estimated costs accurate in early 2009; it is probable that these would reduce as the range of panels on the market increases and the costs of manufacture come down.

We start with the assumption that the whole south slope of the roof will be covered in panels and that the total generator could slightly exceed the size of the roof if it was too big. If this happened, special roofing sheets would have to be estimated for.

All the figures given in this example are derived from the PVsyst software already described.

Choice of panels and inverter

Three different technologies with different costs and performances are compared to offer a fairly wide choice to this potential client. These comprise a 61 W thin-film CIGS technology array from Sulfurcell, standard polycrystalline 130 W panels from SolarFabrik and a top of the range 350 W monocrystalline array from Sunpower.

Figure 4.9   Solrif support system in process of being installed: the panels are arranged like tiles with their upper edge sliding under the row above [photo http://www.dynatex.ch]

Integration supports

The two first types of panels are available with a frame of the Solrif type (Figure 4.9), which results in perfect integration with the panels being laid like tiles. The width of the panel mounted in landscape format is then that of the laminate (module without frame) expanded by 29 mm. The height is also the height of the laminate, the top of the frame being fitted just under the panel row above. For the Sunpower panels, an aluminium support is fitted, which means that the panels are separated by 1 cm laterally to allow room for fixing bolts. Waterproofing is achieved by means of a synthetic lining. Figure 4.10 shows this system of fixing being installed with its plastic protecting layer (the cells in the photo are standard monocrystalline with a metallisation grill on the face, and not Sunpower cells). Fixing on the roof is completed with protective sheeting fabricated to measure by a roofing contractor.

*Figure 4.10   PV generator support system in process of being installed: the panels
are fixed to aluminium rails and rendered watertight by a plastic
underlay [photo www.pvluberon.com]*

Operating voltage of the PV generator

The panels connected in series produce a generator maximum voltage (V_m) that
must always be within the operating range of the inverter MPPT so that

- at minimum value, at 60 °C, the V_m voltage is still within the range of the
  inverter MPPT;
- at minimum value, at −10 °C, the generator open voltage (V_{oc}) should be less
  than the insulation voltage of the panels (typical value of 700–1000 V) and
  below the maximum voltage supported by the inverter (typical value around
  100–200 V above the MPPT range).

These two parameters are valid for systems in temperate climates where, in
summer, panels will often reach 60 °C, whereas in winter, if the inverter has not
started, temperatures of −10 °C can be experienced with the panels having open voltage. In hot countries, these two criteria can be increased by 10 °C. In cold climates on
the other hand, very low temperatures must be allowed for, as must the power behaviour of the generator in the presence of snow albedo.

Insulation voltage and grounding

The insulation voltage criterion is important to guarantee the safety of personnel. If
the frame is not properly earthed and leakage current gives a direct current voltage
of several hundred volts, the installation becomes very dangerous.

These insulation problems are today well managed by most manufacturers. With the first power stations connected to the grid, a major Japanese manufacturer had to exchange 100 kW of panels mounted on a motorway barrier in central Switzerland. High leakage currents had been detected, and as a precaution, all the panels were changed.

If the Sunpower panels are chosen, the inverter must incorporate an insulation transformer. Sunpower cells have a different form of construction to traditional cells: to avoid any shading by the front grille, the cells are manufactured with back contacts with an interlacing of n and p zones, producing electric fields separating the positive and negative charges generated by the light photons (see 'Back-contact cells' in Section 3.1.1). In a generator of several hundred volts using such panels, a very small leakage current escapes from the surface of the cells to the earth and the frame: when this happens, negatively charged electrons remain on the surface of the cells, which tend to attract positive charges (holes) generated by the photons, preventing these from passing via the back contact and lowering the performance of the cells. To avoid this effect, the positive pole of the generator is connected to the earth, and the whole system is then polarised with negative voltage in relation to the earth. The necessity of grounding the system imposes the choice of an inverter with a transformer, a characteristic that slightly lowers the total efficiency. Inverters without transformers have a slightly higher efficiency, but they must use a floating potential: during their operation, to ensure safely, all leakage of current from the generator to the ground is permanently monitored.

At the time of writing (Spring 2009), it was not permitted in France to ground one of the poles of a PV generator, which in principle prevents the use of Sunpower panels or of thin-film panels with transparent conducting oxide (TCO) electrodes on a glass substrate. For amorphous silicon panels or CIS or CdTe panels using such contacts, corrosion[4] from sodium ions (present in the glass), in the presence of water, can cause de-lamination of the tin oxide deposited on the glass and forming the front electrode. To avoid this electrolytic corrosion, the negative pole of the generator can be grounded, which is currently not permitted in France. These two phenomena are a serious constraint on the choice of panels that can be used in France. It is to be hoped that these standards will be changed and that the practice of polarising the generator will be authorised in the future. It is well known that EDF (Electricité de France) had invested considerable sums in an American manufacturer of thin-film modules using this type of electrode. It would be surprising if these modules were to be used in large-scale systems without protective polarisation. For thin-film or crystalline cell panels, problems of corrosion can also arise from the quality of the seal of the laminate, which needs to be very well manufactured to guarantee the durability of the contacts close to the edge. Many problems linked to corrosion arise from the poor quality of encapsulation of the cells and of their cleanness during this stage of manufacture. The critical parts are the

[4] C.R. Osterwald, T.J. McMohan, J.A. del Cueto, J. Adelstein, J. Pruett, 'Accelerated stress testing of thin-film modules with SNO2:F transparent conductors', Presented at the National Center for Photovoltaics and Solar Program Review Meeting Denver, Colorado, 24–26 March 2003

edges of the panels and the areas where the connecting cables pass through, where the quality of the seal must be particularly high.

Efficiency of the inverter

The other criteria for choosing an inverter are first the highest efficiency possible to generate the maximum energy: this criterion is also a good indication of the intrinsic quality of the electronics, which should be better if it produces fewer losses, as losses generate heat. The efficiency curve is also important: it is best to have equipment functioning at high efficiency even at the lowest power, since the generator will often be operating at partial power in cloudy weather or at the beginning or end of the day.

Summary of inverter characteristics

Figure 4.11 shows the summary characteristics of an inverter. One can see at once the operation range of the NPPT (400–800 V), the maximum voltage permissible (900 V) and the power characteristics (45–52 kW in DC and 35–38 kW in AC). An important consideration is the European efficiency factor (95.5%), which takes into account the average operation of the inverter in temperate latitudes under average irradiance encountered. This efficiency factor is the weighted average of annual operation: if E80 is 80% of the nominal power, the European efficiency factor is

$$\text{EuroEff} = 0.03 \times \text{E5} + 0.06 \times \text{E10} + 0.13 \times \text{E20} + 0.1 \times \text{E30} + 0.48$$
$$\times \text{E50} + 0.2 \times \text{E100}$$

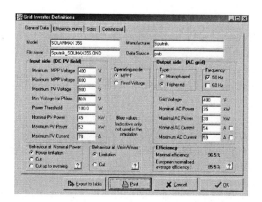

Figure 4.11   Characteristics of an inverter

Simulation of a complete PV installation

With the data characteristics of the selected panels, a calculation sheet is prepared summarising the possible alternatives for covering the solar roof. Table 4.3 shows these dimensions.

With the Sunpower modules, the width of the roof is slightly exceeded. For the three panel types chosen, the dimensions take into account the size of the supporting structure.

Table 4.3   *Fitting solar panels to a roof*

Solar panel	Width (m)	Height (m)	Number in width		Number in height	
			Max	Rounded	Max	Rounded
Sulfurcell SCG60HV	1.296	0.656	23.1	23	22.9	23
Solarfabrik SF130/4	1.487	0.663	20.2	20	22.6	22
Sunpower SPR 315	1.076	1.559	27.9	28	9.6	9

Table 4.4 summarises the characteristics of the three PV generators. The next step is to choose an inverter that will correctly process the output of this number of panels.

Table 4.4   *Characteristics of the PV generators*

Solar panel	Power (W)	Number total	Power (kW)	Efficiency total (%)
Sulfurcell SCG60HV	61	529	32.3	7.17
Solarfabrik SF130/4	130	440	57.2	12.71
Sunpower SPR 315	315	252	79.4	17.64

The efficiency shown is the STC efficiency (1000 W/m^2 and 25 °C) of the complete generator including the supports. It should be noted that the lower the cost of the panels, the more the ancillary costs of installation, supports and cabling become significant, which is a handicap for the low-efficiency panels as these costs become a more important factor.

In the case described here for panels mounted on a roof of given dimensions, with high-efficiency modules the ancillary costs per watt will be considerably lower:

- the installation infrastructure (scaffolding, protection, installation staff) will be the same;
- the cabling costs will be lower (fewer panels, less cabling);
- the mounting costs will be lower (fewer modules to handle, any grading easier).

Looking at the difference in efficiency between SunPower and Sulfurcell in Table 4.2 and taking into account the differences in ancillary costs, thin-film panels need to be significantly cheaper to be financially worthwhile. This is the case for large generators in the open country, and we will calculate the price of the electricity produced in Section 4.7.

Figure 4.12 shows the characteristics of the first Sulfurcell PV panel.

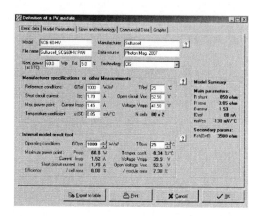

Figure 4.12   Characteristics of the Sulfurcell PV panel

The parameters shown are those of the manufacturer measured in STC conditions as well as those used by the software for the simulation. When choosing a thin-film module, it must be remembered that their low power increases the number of modules to be installed for a given power output and of the low operating current. This panel generates its power at a relatively high voltage (52.5 V) and small current (1.52 A), which will increase the number of strings of panels in parallel. Today's crystalline panels normally supply currents of 5 to 8 A, which limits the number of strings in parallel and lowers the cost of cabling.

Figure 4.13 shows the parameters of the generator of Sulfurcell panels coupled to a 30 kW Solutronic inverter; 528 panels were successfully mounted out of a possible 529. The system is made up of 44 strings of 12 panels in series.

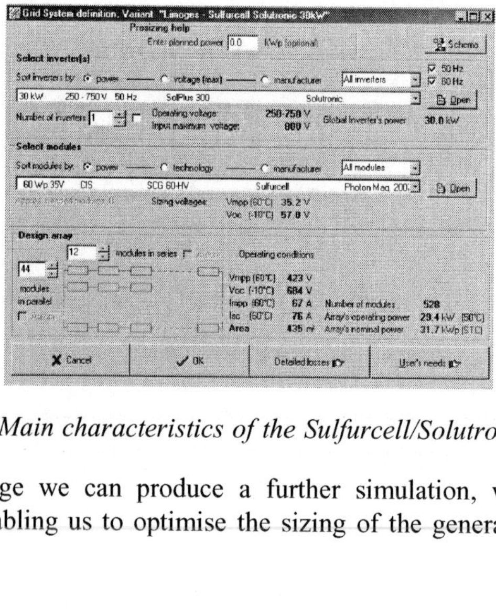

Figure 4.13   Main characteristics of the Sulfurcell/Solutronic PV system

After this stage we can produce a further simulation, which generates a detailed report enabling us to optimise the sizing of the generator. The first step may be to change the inverter to try to find one with better performance. In a second stage, we will look in more detail at the losses of the panel array (see Section 4.3.2.3).

Figure 4.14 shows the results of this simulation.

PVSYST V4.3		25/01/09	Page 1/2

Grid-connected system: principal results

Project: Grid-connected project at Limoges
Simulation model: Limoges – Sulfurcell Solutronic 30 kW

Principal system parameters	Type of system	Grid-connected		
Orientation of collectors	pitch:	20°	azimuth:	0°
PV modules	model:	SCG 60-HV	nominal power:	60 Wp
Panel array	number of modules:	528	total nominal power:	32 kWp
Inverter	model:	SolPlus 300	nominal power:	30 kW AC
Consumer needs	unlimited (grid)			

Main results of the simulation
System output   Energy produced: 36.5 MWh/year   specific: 1151 kWh/kWc/year
                                Performance ratio (PR): 83.2%

Standardised output (per installed kWp):
nominal power: 32 kWp

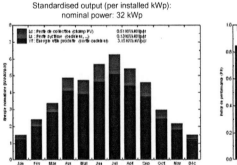

Performance ratio (PR)

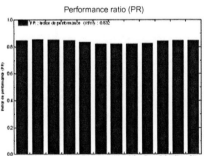

Figure 4.14   Main results of the simulation with thin-film modules

Three essential parameters emerge from this report:

- The annual energy produced (36.5 MWh) simulated from irradiance data for Limoges and the characteristics of the system mounted on the roof at an angle of 20° and facing full south.

- The same energy but specific, in other words divided by the installed power. This value (1151 kWh/kWc/year) corresponds to the number of hours annually when the system is operating at its nominal power. Typically, the same system in the north of France would produce between 900 and 1000 h and in the south of France from 1300 to 1400 h.

- The PR of 83.2% is a measure of the general quality of the system and shows the total efficiency in relation to the installed power. This value is generally higher for thin-film modules that usually performed better under low and diffuse irradiance (see Chapter 2 on their sensitivity to the blue end of the spectrum). For a given installed power, these modules can easily produce 5–8% more energy than crystalline panels. This difference is accentuated in northern countries where the share of diffuse energy is greater than the direct part. In general, when the system has a PR over 80%, it may be assumed that the sizing is correct.

Figure 4.15 shows a diagram of losses to the system.

Diagram of losses over the whole year

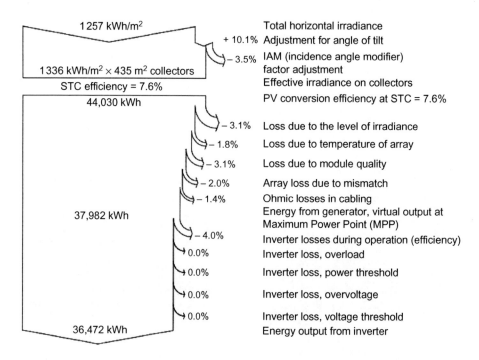

1 257 kWh/m²	Total horizontal irradiance
+ 10.1%	Adjustment for angle of tilt
– 3.5%	IAM (incidence angle modifier) factor adjustment
1 336 kWh/m² × 435 m² collectors	Effective irradiance on collectors
STC efficiency = 7.6%	PV conversion efficiency at STC = 7.6%
44,030 kWh	
– 3.1%	Loss due to the level of irradiance
– 1.8%	Loss due to temperature of array
– 3.1%	Loss due to module quality
– 2.0%	Array loss due to mismatch
– 1.4%	Ohmic losses in cabling
37,982 kWh	Energy from generator, virtual output at Maximum Power Point (MPP)
– 4.0%	Inverter losses during operation (efficiency)
0.0%	Inverter loss, overload
0.0%	Inverter loss, power threshold
0.0%	Inverter loss, overvoltage
0.0%	Inverter loss, voltage threshold
36,472 kWh	Energy output from inverter

Figure 4.15   Diagram of annualised losses (thin-film panels)

The different loss-causing components of the system are detailed thus:

- The +10.1% adjustment corresponds to the gain from the 20° slope of the roof in relation to horizontal irradiance (the optimum is 30° at this latitude).
- The −3.5% IAM (incidence angle modifier) factor is a measure of the losses due to light being reflected when it does not arrive perpendicular to the panel.
- The −3.1% loss due to the level of irradiance corresponds to the system not having started when the light is too low. This loss arises from the parallel (shunt) losses of the cells. The other losses of the PV generator considered in this simulation are average losses and will be modified as soon as more precise data are available.
- Finally, the losses of the inverter allow us to verify the accuracy of the sizing: it can be seen here that the only loss mentioned is that due to the effective efficiency of the device.

Table 4.5 summarises the characteristics of three possible solar roofs, according to panel technology, with their first simulations.

Table 4.5   Solar roof – three possible choices

Panels				
Supplier	Sulfurcell	Solarfabrik	Sunpower	Unit
Type	SCG60HV	SF130/4	SPR 315	
Power	60	130	315	W
Module V_{mp} (STC)	41.5	17.72	54.7	V
Module I_{mp} (STC)	1.45	7.34	5.76	A
Number in series	12	22	12	
Generator V_{mp} (60 °C)	423	340	582	V
Generator V_{oc} (–10 °C)	684	541	838	V
Parallel strings	44	20	21	
Quantity	528	440	252	
Generator power	31.7	57.2	79.4	kW
Inverter				
Supplier	Solutronic	Sunway	Sunway	
Type	SolPlus 300	TG 75-ES	TG82/800	
Power	30.0	50.0	63.0	kW
European efficiency factor	96.0	95.8	95.5	%
Simulation				
Performance ratio	83.2	79.9	79.8	%
Annual energy produced	36.5	63.2	87.7	MWh/year
Specific energy	1151	1105	1104	kWh/kWc

4.3.2.3   Optimisation of the final system

Assuming that the owner of the shop can finance the purchase of traditional crystalline panels, we shall look more closely at the losses and attempt to improve the forecast production. Figure 4.16 details these losses.

There are three modifiable losses.

- *Module quality loss*
  If the supplier can guarantee the power of the modules delivered, this loss may be reduced to 0.

- *Array 'mismatch' loss*
  This arises from variations in panel characteristics in terms of currents and voltages different from their maximum power values. If flash tests are available, the panels can be better paired to connect panels with identical currents in series in an attempt to achieve a uniform V_{mp} voltage between the different strings. This loss can be reasonably assumed to be reduced to less than 1%.

Diagram of losses over the whole year

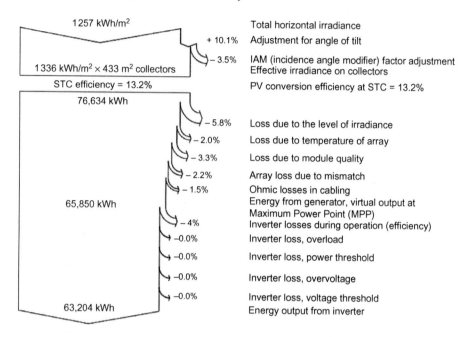

1257 kWh/m²	Total horizontal irradiance
+ 10.1%	Adjustment for angle of tilt
– 3.5%	IAM (incidence angle modifier) factor adjustment
1336 kWh/m² × 433 m² collectors	Effective irradiance on collectors
STC efficiency = 13.2%	PV conversion efficiency at STC = 13.2%
76,634 kWh	
– 5.8%	Loss due to the level of irradiance
– 2.0%	Loss due to temperature of array
– 3.3%	Loss due to module quality
– 2.2%	Array loss due to mismatch
– 1.5%	Ohmic losses in cabling
	Energy from generator, virtual output at Maximum Power Point (MPP)
– 4%	Inverter losses during operation (efficiency)
65,850 kWh	
–0.0%	Inverter loss, overload
–0.0%	Inverter loss, power threshold
–0.0%	Inverter loss, overvoltage
–0.0%	Inverter loss, voltage threshold
63,204 kWh	Energy output from inverter

Figure 4.16   Diagram of annualised losses (crystalline panels)

- *Ohmic cabling losses*

These losses are easily manageable by installing adequate diameter cabling and doing away with the anti-return series diodes often installed in strings of panels. We saw in Section 3.1.4 that the panels are protected against hotspots by parallel diodes cabled every 15–20 cells. These diodes are sized to support directly the current equivalent to 1.5 or 2 times that of the panel. But if several strings are connected in parallel, when a hotspot occurs the reverse current supplied to the shaded zone can come from all the strings and exceed the panel current, which can burn out a protective diode causing the destruction of unprotected cells. To avoid this problem, the diode is sometimes installed in series with another string of panels, causing a loss of voltage. The alternative solution suggested by the manufacturers of inverters is to install a fuse with a value higher than the maximum direct current of each string, so that in the case of a hotspot, the reverse current would blow the fuse before reaching the maximum value of the parallel protective diodes. Figure 4.17 shows a multi-string connection box being installed: each string is protected by a fuse and a needle galvanometer to show the current; it can immediately be seen if one string is supplying less current than the next one. The assembly is also monitored remotely through a modem by the supplier of the inverter, who will warn the manager of the generator if one string is supplying less power than its neighbours.

Figure 4.17   Panel string connection box with fuses and current monitoring by modem (Solarmax system)

This string connection box has components typical of this equipment (Figure 4.17).

- On the upper left (1), two variable resistances are fitted as lightning conductors (plus and minus poles earthed).
- On the upper right (2), the main circuit includes the string galvanometers and fuses.
- Below left (3), the main output cable box connecting to the inverter.
- To its right (4), the general DC switch controlling the input to the inverter.
- Next (5), the negative terminal strips.
- Finally on the right (6), the circuit containing the current sensors with computer processing of the data for remote monitoring.

Looking again at the simulation, we will enter a loss of 0% for minimal efficiency and 1% for mismatch. For the ohmic losses, we would choose the fused connection box model and estimate the average lengths of the cabling.

- Each panel uses a cable of 4 mm² section and 2 m length, making 44 m for the 22 modules in the series, and the average length to the connection boxes 26 m, or 70 m for each string. The cable chosen is of the same type as used by the panels, double-insulated halogen-free Radox cable with a 4 mm² section. The software calculates that the resistance of 20 parallel strings corresponds to 19.3 mΩ.

- Cables connecting the outputs of the two connection boxes to the inverter are 35 mm^2 section (STC current of 73.4 A for 10 strings in parallel), which corresponds to an additional 3.14 mΩ.
- The total ohmic losses amount to 0.8%.

We are then ready to redo the simulation, which is shown in Table 4.6. The improvement in performance is 5.1%, a significant amount when the feed-in price for electricity is high.

Table 4.6   Optimisation of losses of the PV generator

Losses	Standard	After optimisation	
Performance ratio	79.9	84.0	%
Annual energy produced	63.2	66.5	MWh/year
Specific energy	1105	1162	kWh/kWc

4.4   PV generator on a terrace roof or in open country

This type of installation is used for the biggest PV generators in the open country (Figure 4.19) or on flat roofs.

4.4.1   Installation on racks

When a large number of panels are connected, their currents and voltages must be balanced if the maximum power output is to be achieved. On the basis of the measured characteristics of the panels, panels of the same nominal current are linked in series, taking care to equalise the total voltages at nominal power of each string. The panels must have the same orientation to avoid one panel receiving less irradiance, which would then limit the current of the whole series. Figure 4.18 shows the parameters of the dimensions useful for calculating the shading of panels on racks. The objective of the calculation is to find the optimum between loss of shading and the gain in pitch for the chosen panel density: if to begin with it is

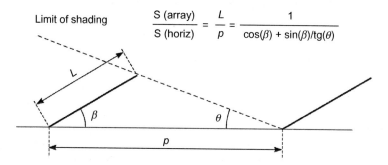

$$\frac{S\,(array)}{S\,(horiz)} = \frac{L}{p} = \frac{1}{\cos(\beta) + \sin(\beta)/tg(\theta)}$$

Figure 4.18   PV racks: shading limit angle and horizontal surface occupied

decided to tilt the modules at 30° (European optimum), the maximum installed power will be defined with shading losses of a few percent. The compromise will be a choice between maximum installed power and shading losses acceptable to the client.

We have given in Section 4.2.2 a number of rules for the installation of panels in a PV generator. Those principles remain valid here with a supplementary rule governing the distribution of panels on the surface available. In the case of systems mounted on racks on flat roofs or in the open, the exploitation of the surface available will depend on the losses allowed when the Sun is low, and casts a shadow from one rack onto another. To limit the effect of this shading, it is best as far as possible to connect the panels in horizontal series so that one partially shaded panel does not reduce the current of a panel completely exposed to the Sun. It is recommended to add a diode or, better, a fuse in series with each string of panels so that one string shaded at the beginning or end of the day does not affect the output of the remaining panels in full sunlight.

To optimise annual energy output in relation to the area of land or roof available, the angle that the panels are pitched at and the distance between the racks will be adjusted to take into account local climatic conditions and irradiance when the Sun is low on the horizon. By running a simulation, the benefits of various arrangements can be quickly calculated, and the best one is chosen. Obviously, if the surface area is limited, the best option will be to reduce the pitch and put up with a few percent of loss in relation to the maximum attainable without shading or limitation of surface area. The reduction of pitch has other advantages:

- the supports can be smaller and therefore cheaper;
- sensitivity to the wind is lower;
- installed power is much higher;
- efficiency under diffused light is better, a characteristic that is more suitable for thin-film panels having better sensitivity to blue light;
- the architectural impact is diminished and therefore is more acceptable.

Table 4.7 shows an example of the exploitation of a given surface area for various places at different latitudes.

The coverage ratio is the ratio between the surface of the collectors and the surface available. The first four examples have an optimal pitch of 30%, which is the normal reference used. For Bombay, the optimum of 20° is taken as the reference. The optimisation of the system will subsequently depend on the cost of the supports and the ease of installation. For generators in cold countries, it is usual to leave a clear space at the foot of the panels where snow can accumulate without shading the last row of panels.

The losses indicated are total for array racks of infinite length and do not take into account the geometry of the cabling. If one reckons on finite racks, the losses at the ends are smaller, and if the site lends itself to it, with the lengths of racks corresponding to a whole multiple of a string of panels, then the losses from shading can be sharply reduced by cabling the strings by horizontal rows. This is the option chosen for a recently installed 1 MW power station at Verbois near Geneva.

Table 4.7   Coverage ratio of generators on racks

	Pitch (°)	30	20	10	5
	Coverage ratio	45.5	54.0	69.1	81.4
	Relative power	100	119	152	179
Place, latitude					
Hamburg (D),	Losses (%)	0	1.4	4.5	7.0
53.30°	Final energy (%)	100	117	145	166
Bourges (F),	Losses (%)	0	1.2	4.2	6.5
47.04°	Final energy (%)	100	118	146	167
Barcelona (E),	Losses (%)	0	1.2	4.9	7.3
41.32°	Final energy (%)	100	118	145	166
Algiers (Al),	Losses (%)	0	0.7	4.0	6.2
36.34°	Final energy (%)	100	118	146	168
Mumbai (India),	Losses (%)		0	1.3	3.4
19.17°	Final energy (%)		100	133	161

Figure 4.19 shows a detail of the arrays that comprise four rows of modules. Note, in particular, the supporting structure that is made up of concrete anchor points in the ground with panel supports of Douglas fir sourced from local forests.

Figure 4.19   1 MW power station, detail of the arrays [photo M. Villoz]

4.4.2   Solar trajectory and shading

Figure 4.20 shows the curves of annual solar radiation and shading of PV arrays at Bourges (central France), with panels at 20° pitch and an active surface area of 54%. It will be seen that the shading in December is between 0% and 20% from 9 AM to 3 PM and subsequently increases at either end of the day. If each rack is

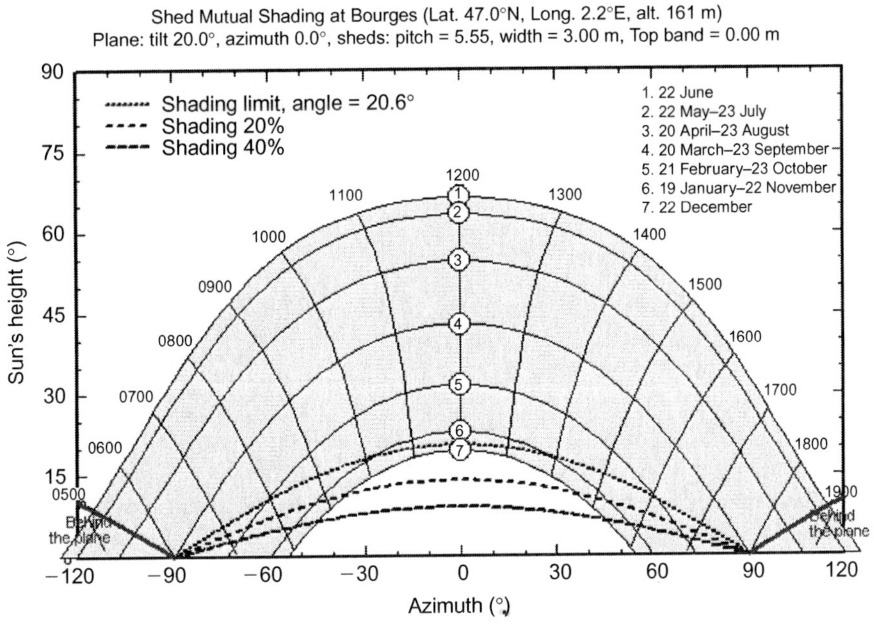

Shed Mutual Shading at Bourges (Lat. 47.0°N, Long. 2.2°E, alt. 161 m)
Plane: tilt 20.0°, azimuth 0.0°, sheds: pitch = 5.55, width = 3.00 m, Top band = 0.00 m

Figure 4.20 Curves of sunshine and shading on PV arrays at Bourges

made up of four horizontal rows of panels, the loss will only affect a quarter of the system for the greater part of the winter and will disappear completely between March and September.

The pitch of the racks is also influenced by the type of energy received. At low altitudes in regions subject to fall or high cloud cover, the annual diffuse energy is greater than the direct energy, and in this case a flatter angle of pitch will result in more diffuse light being collected, and the reduction in performance compared to the optimal is slight; at high altitude on the other hand, direct radiation exceeds diffuse radiation, and when snow is present, the higher angle of pitch will allow more energy to be collected in winter and will reduce problems of snow covering the panels.

4.4.3 Trackers

A tracker is an array of panels mounted on a movable surface, which follows the trajectory of the Sun on one or two axes. With only one axis, the panels can be mounted in one plane pitched at a fixed angle on a vertical post, which will point the PV array in the direction of the Sun throughout the day. The single axis can also be in the plane of the inclined panels that will then swing from east to west following the Sun. A dual-axis tracker, more complex mechanically, will keep the plane of the panels perpendicular to the Sun whatever its position in the sky. The orientation of the plane perpendicular to the Sun's rays is achieved by measuring the current of four photosensors arranged on either side of a shading partition

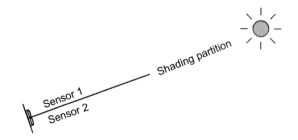

Figure 4.21   Photosensors for the orientation of a tracker

separating them (Figure 4.21). The motors or hydraulic systems governing the orientation are activated by the differences in current of each sensor: as soon as one of them is slightly shaded, the system will move to compensate for this loss and reorient precisely to the axis of the Sun.

Naturally such systems are more attractive when there is much direct sunlight, in less cloudy climates. In recent years, trackers have been mainly used in Spain and Portugal where the climate is more suitable.

When the Sun is hidden by clouds, the systems can be repositioned to the horizontal to capture the maximum of diffuse energy. But often when it is cloudy, the system follows the Sun according to a pre-programmed pattern, the pitch and orientation being defined by the hour and day of operation. The trackers are sometimes lowered to the horizontal when strong winds are expected, to reduce mechanical strain.

For the sizing of such a system, it is advisable to use software that takes account of shading: this will quickly determine optimum performance in relation to the investment budget and the area of land available. Again, the following calculations have been obtained from PVsyst software that provides for this type of calculation.

4.4.3.1   1 ha available

Let us assume that a farmer in the south of France (solar radiation statistics of Nice) has a small parcel of land of 1 ha lying fallow because the soil is too stony and poor for crops. However, the field is close to a line of electricity supply and could easily be the site of PV panels connected to the grid. We compare two solutions for the installation: a generating plant with panels on racks at a fixed pitch, and one with trackers of around 90–100 m^2 of panel surface. At present, the French feed-in tariff is limited to 1500 peak hours/year, but this limit may be removed and the tracking system would then become more attractive. To compare these two installations, we've chosen the same panels and inverters, and the rack mounted generator will be an assemblage of several small systems.

The plot of land is 100 m^2 facing due south. We assume that PV arrays are both linked to a local inverter mounted below the panels in a watertight box. Thus, the whole surface of the plot of land is utilised. Only a transformer for connection to the grid is placed at the north of the plot behind the panels.

Table 4.8 Basic system, either fixed or with a dual-axis tracker

Description	Systems			Gains and losses				Simulation		
	No.	Power (kW)	Density (W/m²)	Inclin. (%)	Shading (%)	Tot. (%)	PR (%)	kWh/kWp	Energy Rel. (%)	MWh/year
System with 30° pitch	1	13.3	110.8	13.4	0.0	13.4	84.1	1402	100.0	18.7
Dual-axis tracker	1	13.3	78.7	50.3	0.0	50.3	86.1	1904	135.8	25.4

Basic system

The plot of land will be covered with multiple systems of a basic tracker. We have chosen for this simulation, polycrystalline panels manufactured by Atersa supplying 222 W for a surface of 1645×900 mm. These are mounted on trackers in landscape format 12 wide and 5 high, giving a surface of 11 m $\times$ 8.25 m, including the fixing bolts between the panels. The power of this array of 60 panels will be 13.32 kW for a surface of 91 m². It is linked to the grid through a Danfoss TLX 12.5 K inverter, a recently developed device with a European efficiency factor of 97%.

Table 4.8 shows the results of the simulation of the system first at a fixed pitch of 30° and second on a dual-axis tracker accurately following the Sun's trajectory.

The data presented will provide a reference when more systems are mounted close together and will be affected by shading. In detail, the data show

- in column 3, the total installed power;
- in 4, PV power by plot surface;
- in 5, 6 and 7, the optical gains and losses due to inclination and shading (here zero for a single system);
- in 8, the performance ratio for the ohmic losses and those due to pairing of panels;
- in 9, the annual energy density;
- in 10, the energy variation in relation to the reference surface (here fixed at 30°);
- in 11, the annual energy production.

It will be noted that the dual-axis tracker gain is excellent, reaching 35.8% at this latitude and climate (see column 10). The optical gain from the pitch of the PV array is 13.4% for 30° fixed pitch due south and 50.3% when the tracker keeps the panels perpendicular to the Sun's rays.

Use of space available with panels on racks

The basic arrays of 60 panels (the same as above) are this time mounted in portrait format 15 modules wide and 4 high. Seven systems can therefore be mounted on a rack with a total width of 95 m, which leaves a gap of 2.5 m free around the PV array. Each rack thus has a power of $7 \times 13.32 = 93.24$ kWc. Figure 4.22 shows the appearance of the complete system with shade cast at 8 AM on 21 December for a variant of six racks at 30° pitch and spaced 18 m apart.

Table 4.9 shows simulation results for four variants of use for the 1 ha field available. We have calculated successively the energy produced for six to nine racks and then sought the optimal angle producing the maximum energy for the final simulation.

It is arranged in the same way as Table 4.8, the example given without shading. We have kept for comparison with a smaller system of 560 kW (six racks of 93.24 kW) at optimum pitch of 30° and producing the smallest shading losses. It will be noted that the increased density of racks does not produce too many losses, the optimum performance of the biggest system with nine racks (20° pitch) being only

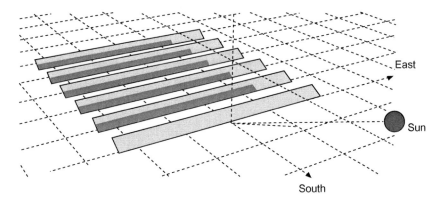

Figure 4.22 PV generator on racks – shading in winter

3.1% less efficient than our reference example, while the installed power has increased by 50%.

Optimum performance is achieved when the racks are well spaced and produce little shading, but the financial optimum is probably for a high density and better use of the available space, as the costs of preparing the ground and connecting to the grid as well as engineering and other infrastructure expenses will not change very much according to the number of racks.

Use of space available with trackers

The problems of shading are much more difficult to resolve when the trackers increase their tilt towards the horizon when the Sun is low. Shading appears at the bottom of the arrays or on the sides, and the cabling of the strings should be optimised according to the simulation, which enables the impact of this shading to be seen directly. The simulation will also be valuable for the financial optimisation of the investment.

Figure 4.23 shows the example of the 36-tracker variant of the generator and the impact of shading at 8 AM on 21 December.

Table 4.10 shows the simulation results systems with 25, 30 and 36 trackers on the 1 ha project plot.

It will be seen that the gain relative to our basic six-rack system varies between 30.6% and 25.2% when the density of the trackers is increased. If the cost increase due to the trackers is of the order of 12–15% of the total price, then the trackers are useful addition at this latitude. But the use of this equipment calls for a closer supervision of the system, and the costs related to the ageing of the equipment are not really known.

4.5   Typical 12.6 kW system in different countries

To have an idea of performance as affected by latitude, we have included tables giving the production in different parts of the world of the same typical 12.6 kW

Table 4.9 *Systems mounted on racks over 1 ha*

Description	Systems			Gains and losses				Simulation			
	No.	Power (kW)	Density (W/m²)	Inclin. (%)	Shading (%)	Tot. (%)	PR (%)	kWh/kWp	Energy		
									Rel. (%)	MWh/year	
6 racks at 30°	42	559.4	55.9	13.4	−2.0	11.4	82.5	1376	100.0	770.0	
7 racks at 30°	49	652.7	65.3	13.4	−3.5	9.9	81.4	1357	98.6	886.0	
8 racks at 25°	56	745.9	74.6	12.7	−3.3	9.4	81.4	1349	98.0	1006.0	
9 racks at 15°	63	839.2	83.9	9.4	−1.7	7.7	82.4	1326	96.4	1112.0	
9 racks at 20°	63	839.2	83.9	11.4	−3.2	8.2	81.3	1333	96.9	1118.0	
9 racks at 25°	63	839.2	83.9	12.7	−4.9	7.8	80.0	1326	96.4	1113.0	

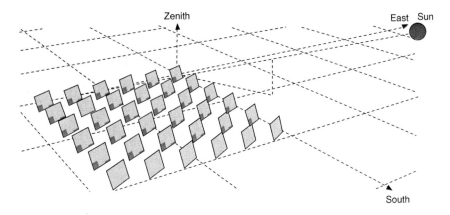

Figure 4.23 PV generator showing shading of trackers in winter

system of panels feeding an inverter of 10 kW nominal. Tables 4.11–4.13 give the peak hours of operation calculated with PVsyst software, as well as the optimal angle for panels without shading and facing south (north in the southern hemisphere). The tables show the energy produced at the output of the inverter and take account of all losses arising from cabling, the inverter and the pairing of the panels.

The specifications of the system are as follows:

Inverter

- Minimum operating point of MPPT: 450 V
- Maximum operating point of MPPT: 800 V
- Maximum permitted open voltage: 900 V
- Nominal power: 10 kW
- Output voltage: 400 V, three phase
- Operating threshold: 100 W
- Maximum efficiency: 94%
- Dimensions and weight: 555 × 554 × 1170 mm, 150 kg

PV generator
This is composed of 68 modules in four parallel strings of 17 panels in series with the following individual characteristics:

- STC power: 185 W
- Power guarantee: 5%
- Insulation voltage: 1000 V
- MPP voltage at 60 °C: 30.3 V
- Open voltage at −10 °C: 49.8 V

Table 4.10 Systems mounted on trackers over 1 ha

Description	Systems			Gains and losses				Simulation			
	No.	Power (kW)	Density (W/m²)	Inclin. (%)	Shading (%)	Tot. (%)	PR (%)	kWh/kWp	Energy		MWh/year
									Rel. (%)		
Dual-axis trackers	25	333.0	33.3	50.3	−5.6	44.7	81.3	1797	130.6		598.0
Dual-axis trackers	30	399.6	40.0	50.3	−7.1	43.2	81.5	1768	128.5		706.0
Dual-axis trackers	36	479.5	48.0	50.3	−9.4	40.9	78.0	1723	125.2		826.0

- Number of cells: 72
- Dimensions and weight: 1575 × 826 × 46 mm, 17.0 kg

The characteristics of the generator are therefore

- MPP voltage at 60 °C: 515 V
- Open voltage at −10 °C: 847 V
- MPP current at 60 °C: 21 A
- Power at NOCT of 50 °C: 11.3 kW
- STC nominal power: 12.6 kW

Table 4.11   Typical 12.6 kW system in Europe and Africa

European and African countries	Place	Latitude (°)	Optimal angle of pitch	Annual production (kWh)	Energy density (kWh/kW) (h)
Iceland	Reykjavik	64.25	40	9,507	755
Finland	Helsinki	59.57	39	11,114	882
Germany	Berlin	52.25	32	11,094	880
England	London	51.36	33	10,244	813
France	Paris	49.12	30	11,402	905
Switzerland	Geneva	46.12	30	13,062	1,037
France	Carpentras	44.02	36	17,125	1,359
France	Nice	43.40	33	15,964	1,267
France	Ajaccio	41.55	33	17,176	1,363
Italy	Naples	41.01	32	19,004	1,508
Spain	Madrid	40.39	30	18,516	1,470
Portugal	Lisbon	38.48	30	18,471	1,466
Spain	Seville	37.12	30	19,184	1,523
Tunisia	Tunis	36.48	30	19,847	1,575
Lebanon	Beirut	33.55	27	18,407	1,461
Morocco	Casablanca	33.29	27	19,173	1,522
Jordan	Amman	31.46	26	22,524	1,788
Egypt	Cairo	30.12	25	22,045	1,750
Namibia	Windhoek	−22.23	25	23,316	1,850
Senegal	Dakar	14.48	15	22,156	1,758
Mali	Bamako	12.45	14	20,522	1,629
Ethiopia	Addis-Ababa	9.12	13	19,812	1,572
Kenya	Nairobi	−1.20	0	18,238	1,447

At first sight, the energy produced increases as the latitude approaches the equator; however, local climatic variations such as the presence of mist or more frequent precipitation can modify this rule — see the differences between Helsinki, Berlin and London (Tables 4.11–4.13). Although at similar latitudes, Geneva and Quebec show considerable differences, with 24% more energy being generated in Canada, which has much higher irradiance in summer. The highest energies are generated in desert regions or at high altitude, as is shown by the examples of

Table 4.12   Typical 12.6 kW system in America

Countries of the Americas	Place	Latitude (°)	Optimal angle	Annual production (kWh)	Energy density (kWh/kW) (h)
Canada	Edmonton	53.25	44	16,193	1,285
Canada	Quebec	46.49	36	16,175	1,284
Canada	Montreal	45.44	36	15,869	1,259
USA	Salt Lake City	40.49	33	19,656	1,560
Argentine	Buenos Aires	−34.45	28	17,252	1,369
Uruguay	Montevideo	−34.45	27	15,362	1,219
Chile	Santiago	−33.23	20	13,294	1,055
Cuba	Havana	23.01	20	16,292	1,293
Brazil	Rio de Janeiro	−22.50	22	15,971	1,268
Mexico	Mexico City	19.19	20	17,916	1,422
Puerto Rico	San Juan	18.43	17	17,738	1,408
France	Guadeloupe	16.26	16	21,879	1,736
Bolivia	La Paz	−16.24	19	24,837	1,971
Guatemala	Guatemala	14.41	15	18,221	1,446
Nicaragua	Managua	12.19	12	16,419	1,303
Peru	Lima	−12.10	15	20,083	1,594
Venezuela	Caracas	10.40	11	14,968	1,188
Guyana	Georgetown	6.44	7	15,395	1,222
Colombia	Bogota	4.23	4	16,638	1,320
Brazil	Belem	−1.18	6	16,281	1,292
Ecuador	Quito	−0.11	2	17,421	1,383

Table 4.13   Typical 12.6 kW system in Asia and Oceania

Countries of Asia and Oceania	Place	Latitude (°)	Optimum angle	Annual production (kWh)	Energy density (kWh/kW) (h)
New Zealand	Auckland	−37.22	31	15,864	1,259
Japan	Kyoto	34.56	23	11,617	922
Australia	Adelaide	−34.52	28	17,803	1,413
Australia	Perth	−32.07	28	19,529	1,550
India	Delhi	28.32	28	19,635	1,558
Pakistan	Karachi	25.03	24	17,624	1,399
India	Calcutta	22.16	22	16,652	1,322
France	La Réunion	−20.52	20	19,481	1,546
India	Mumbai	19.17	20	18,261	1,449
Madagascar	Tananarive	−18.49	18	18,679	1,624
Indonesia	Jakarta	−6.11	8	14,473	1,149

Jordan, Namibia and Bolivia. The systems installed near the equator have the benefit of regular higher solar irradiance with a few variations between winter and summer, but most of these tropical countries have higher rainfall during the monsoon, which limits the duration of irradiance during that period. The tables show the optimum pitch for panels to maximise the annual energy produced, but a higher pitch value can be chosen to prioritise energy received in winter or if the site is at high altitude and receives heavy snowfall.

Influence of pitch

At high altitude, the increased pitch of the modules also serves to encourage any snow to slip off the panels. Table 4.14 shows the albedo effect (reflections from the ground) with different angles of pitch for a system situated at Davos in the Swiss Alps at 1560 m altitude with regular snowfall from the end of November to the end of April. The units for each month show the daily energy density in kWh/kWc/day. This unit is useful because it gives a good idea of the maximum daily energy that can be produced in a given place: typically in hot countries, the daily irradiance received on the horizontal varies between 4 and 7 kWh/m^2 during the year, and by tilting the panels these values can be increased slightly. In cold or temperate climates, during winter, the values are often below 1 kWh/m^2, and in June, maximal values can rise to 6.5 kWh/m^2. It will be noted that a pitch angle of $\pm10°$ from the optimum does not hugely change annual energy output for a due south orientation. For an azimuth other than south, the loss at $\pm30°$ is below 5.2%. For panels mounted vertically on a facade, the loss is 35% for southerly orientation (low altitude without snow), but it increases for less favourable orientations. The simulation gives the results for an Alpine climate where winter sunshine is considerably higher than the nearby location in the plain ($>2\times$). It will be seen that the loss for a vertical system is less than in the plain (22%) on account of this particular climate and the significant albedo effect in winter.

Table 4.14   Daily production density for an Alpine site

	Daily production density (kWh/kWp)							
	Albedo constant 0.2				With snow albedo in winter			
Pitch (°)	20	30	40	50	40	50	60	90
January	1.9	2.4	2.7	2.8	2.7	3.0	3.1	3.1
February	3.0	3.6	3.8	4.0	4.0	4.2	4.3	4.2
March	4.2	4.6	4.7	4.8	4.9	5.1	5.1	4.7
April	4.9	5.1	5.0	4.9	5.3	5.3	5.2	4.5
May	4.9	4.8	4.6	4.4	4.6	4.4	4.0	2.4
June	4.8	4.6	4.4	4.1	4.4	4.1	3.7	2.1
July	4.8	4.7	4.5	4.2	4.5	4.2	3.8	2.1
August	4.4	4.4	4.3	4.2	4.3	4.2	3.9	2.4
September	3.7	4.0	4.1	4.0	4.1	4.0	3.9	2.9
October	2.9	3.3	3.5	3.6	3.5	3.6	3.6	3.0
November	2.1	2.6	2.9	3.0	2.9	3.1	3.2	3.1
December	1.6	2.0	2.2	2.4	2.3	2.5	2.6	2.7
Annual	1276	1365	1381	1364	1405	1401	1366	1095
% of optimum	−7.6	−1.2	0.0	−1.2	0.0	−0.3	−2.7	−22.1

4.6   Grid company regulations

All electricity companies construct their grid network following precise standards, which may vary from one country to another. For PV producers, a series of standards is used by most companies, but before planning and connecting a system to the grid, all this information must be obtained and applied, and local safety regulations must be respected. We give in the next section an extract from the specifications required for a PV generator connected to the EDF grid in France. This example is not exhaustive but the main regulations are those applied by the majority of grid operators.

4.6.1   General considerations

The main standards to be followed are those of the PV industry and low-voltage installations.

* Standard NF C 15-100 regulating low-voltage electrical installations (May 1991)
* UTE C 57-300 (May 1987): descriptive parameters of a photovoltaic system
* UTE C 57-310 (October 1988): direct transformation of solar energy into electrical energy
* UTE C 18510 (November 1988, revised in 1991): general electrical safety instructions
* C 18530 (May 1990): collection of electrical safety regulations intended for skilled personnel

ADEME has published a technical paper, reference 6257, *Guide de rédaction du cahier des charges techniques des générateurs photovoltaïques connectés au réseau* (November 2007), which gives details of the standards to be followed.

4.6.2   PV array

The modules must be resistant to the following climatic conditions:

* temperature between −40 and + 85 °C,
* relative humidity up to 100%,
* wind speed up to 190 km/h (gusts),
* resistance to rain and hail (hailstones < 25 mm).

The other main specifications are as follows:

* The modules must conform to standard IEC 61215 for crystalline panels and ICE 61646 for thin-film panels (see Section 3.1.4).
* The modules must be identical and interchangeable, and fitted with frames in stainless steel or anodised aluminium.
* The power of the modules must not vary by more than 5% nominal value to avoid mismatch losses.

The installation of the modules in series or parallel must be done respecting the operating range of the selected inverter (open voltage, maximum current, maximum power voltage). The connections between cables must be made in flexible cable, double-insulated and resistant to ultraviolet radiation.

4.6.3   Grid-connected inverter

The following main specifications must be included or respected:

- Synchronisation with the grid, frequency of 50 Hz $\pm$ 1%
- Automatic disconnection in case of fault or absence of grid; for example, fluctuation of grid voltage (<0.85 Un and >1.1 U_n) or frequency (>0.2 Hz)
- Automatic connection and disconnection of the installation
- Protection against overload and short circuits
- Possible cutting off of the direct current source from the modules
- A low rate of harmonic distortion ($<4\%$ THD)
- No electromagnetic disturbance
- High reliability
- High efficiency: $>95\%$ at nominal power; $>90\%$ from 10% of the nominal entry power
- No-load consumption $<1\%$ of nominal power and 0.1% in standby

Several inverters should, if possible, be mounted in parallel to provide three balanced phases (3P + N).

A very important safety specification is the shutdown of the inverter in the absence of the grid to prevent lines being fed into when they have been disconnected from the grid for maintenance, for example.

Finally, the list of certifications to be satisfied:

- CEM: DIN50081 part 1
- EN 55014; EN 60555 part 2
- EN 55011 group 1, class B
- DIN EN 50082 part 1
- Grid conformity: DIN EN 60555
- Voltage regulation: DIN EN 50178 (VDE 0160)
- DIN EN 60146 part 1-1 (VDE 0558 part 1)
- Disconnection protection: DIN VDE 0126

It will be noticed that the inverter is the component that has to satisfy the largest number standards before being accepted by the electricity company.

4.6.4   Protective devices and control box

Each cable coming from the PV generator must be equipped with a disconnect or circuit breaker enabling the DC input to be cut off. The characteristics of these disconnecting switches must be appropriate for the DC voltage and the current level so as to be able to disconnect the DC input at any moment. The main components required to ensure the safety of the system are summarised:

- disconnecting switches for the PV generator,
- overload protection devices,
- overcharge and short-circuit protection devices,

- grid AC disconnect switch (visible disconnection protected by a padlock and only accessible to authorised personnel and emergency services).

Equipment for the AC connections will be installed in a lockable control box. The interface between the inverter and the grid is made up of the control box (or e-panel) and a customer circuit breaker (Figure 4.24). The control box includes

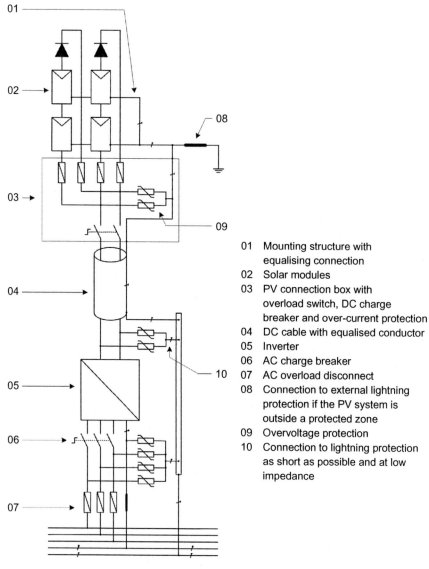

01 Mounting structure with equalising connection
02 Solar modules
03 PV connection box with overload switch, DC charge breaker and over-current protection
04 DC cable with equalised conductor
05 Inverter
06 AC charge breaker
07 AC overload disconnect
08 Connection to external lightning protection if the PV system is outside a protected zone
09 Overvoltage protection
10 Connection to lightning protection as short as possible and at low impedance

Figure 4.24   Typical electrical diagram of grid-connected system

the safety elements mentioned above as well as an energy meter measuring the energy injected into the grid.

4.7   Cost analysis

The dramatic development of solar PV is still strongly influenced by the feed-in tariffs offered by different countries. Germany and Japan were the first countries to offer these incentives from the late 1990s, followed by Spain and the rest of Europe, and the growth in the market that followed has enabled the best PV systems (thin-film panel generators in the south) today to generate electricity at practically the same price as other generating methods.

4.7.1   Cost and revenue analysis of 12 kW solar panel generators

Tables 4.15 and 4.16 give two examples of systems installed at Clermont-Ferrand in France and Barcelona in Spain. We have taken the typical system with 10 kW inverter described in Section 4.5 with varied levels of installed PV power. The financial analysis uses the local feed-in tariffs of these two countries (Table 4.1), that is, €0.602/kWh for France (integrated system) and €0.34/kWh for Spain (overtaxation system). To take into account the poorer thermal dissipation of the integrated system, we have reduced its production by 5%. Calculations are made using an amortisation of the system over 20 years in France and 25 years in Spain to take account of prevailing conditions.

The cost including taxes of the systems in France is estimated at €5.50, €5.40 and €5.30/W, falling with the increase in power, and €0.50/W cheaper in Spain where competition is fierce in a more developed market.

- Fixed amortisation of the installation in 20 (France) or 25 years (Spain)
- In France, indexation of €0.01/year
- Reduction in panel power: -0.35%/year
- Additional costs not taken into account (or taxes)
- Financing costs: 0%, 3% or 5%/year

Table 4.15   Analysis of system at Clermont-Ferrand, France

Installed power (W)	Annual production (kWh)	Total cost (€)	Feed-in tariff (€/kWh)	Revenue per year (€)	Revenue over 20 years (€)		
					Rate 5%	Rate 3%	Rate 0%
10,500	9,762	57,750	0.602	5,877	42,806	54,933	130,874
12,200	11,390	65,880	0.602	6,857	50,372	64,463	152,699
14,000	13,081	74,200	0.602	7,874	62,215	77,797	175,370

It can be seen that the system is clearly attractive, even if installed in the centre of France where solar radiation conditions are less favourable than in the south. These feed-in tariffs will enable France to expand its PV industry.

Table 4.16   Analysis of system at Barcelona, Spain

Installed power (W)	Annual production (kWh)	Total cost (€)	Feed-in tariff (€/kWh)	Revenue per year (€)	Revenue over 20 years (€)		
					Rate 5%	Rate 3%	Rate 0%
10,500	12,835	52,500	0.34	4,364	17,522	31,172	104,147
12,200	14,942	59,780	0.34	5,080	22,609	38,151	121,246
14,000	17,049	67,200	0.34	5,797	27,459	44,931	143,794

For Spain, the situation today is less favourable than a few years ago as the feed-in tariff has fallen. Sunshine in Barcelona is similar to that in Provence, which makes the system economic, but installation costs need to fall considerably to make the financial return attractive.

4.7.2   Cost of PV electricity

The arrival on the market of thin-film modules that are considerably cheaper than crystalline modules brings the price of PV electricity close to the cost of fossil fuel energy. In Tables 4.17–4.19, we look at the price of electricity generated by a large generator of at least 1 MW on open ground at different latitudes and with different technologies. We have calculated costs including the financing of bank loans at rates between 5% and 1%, the 1% rate being one that could be made available by political decision in a region or country anxious to invest in clean energy. We have assumed that the initial investment is paid off in 20 years and that the generator has a life expectancy of 30 years. The cost of electricity given is the total cost of the generator including financing and maintenance divided by production over 30 years.

We have also included the following cost estimates for each generator:

Crystalline system (Table 4.17)

- Cost of €5/W, including all maintenance expenses
- Orientation due south and optimal tilt
- Panels graded
- Ohmic losses below 0.5%
- Inverters at European efficiency of 95.5%
- No losses from shading
- 40 kW subsystems

Dual-axis tracker crystalline system (Table 4.18)

- Cost of €5.80/W, including all maintenance expenses
- Other technical specifications the same as fixed systems
- Shading losses of 5%

Thin-film system (Table 4.19)

- Cost of €3.80/W, including all maintenance expenses
- Other technical specifications the same as fixed systems

Table 4.17   Price of electricity for a fixed crystalline system

Fixed crystalline system		Cost of electricity (€/kWh)		
Place	Energy density (kWh/kWc)	Capital costs		
		5%	3%	1%
Guadeloupe (F)	1950	0.130	0.112	0.094
Seville (E)	1600	0.159	0.137	0.115
Marseille (F)	1420	0.179	0.154	0.130
Caen (F)	1040	0.244	0.211	0.177
Uccle (Be)	860	0.296	0.255	0.214

Table 4.17 shows that, even for a crystalline panel system, parity with other energy sources is almost attained, provided financing costs are low and sunshine is abundant.

Table 4.18   Price of electricity for a tracker crystalline system

Dual-axis tracker crystalline system		Cost of electricity (€/kWh)		
Place	Energy density (kWh/kWc)	Capital costs		
		5%	3%	1%
Guadeloupe (F)	2400	0.123	0.106	0.089
Seville (E)	2100	0.140	0.121	0.102
Marseille (F)	1850	0.159	0.137	0.115
Caen (F)	1220	0.242	0.208	0.175
Uccle (Be)	1010	0.292	0.252	0.212

With panels mounted on trackers (Table 4.18), the financial incentive depends strongly on the location and climatic conditions: in Guadeloupe, despite particu-larly favourable sunshine data, the advantage of the tracker is only 5.4% while in Seville or Marseille, it is 11.2%. In the North, it is no surprise to see that the tracker advantage is only around 1.1%, since the solar radiation is mainly diffuse, and the complicated tracker mechanism is probably not worth installing. The example is perhaps slightly unfavourable to trackers that require a huge surface of exploitation if shading is to be avoided. This is the reason of our choice for shading losses of 5% if they are mounted on surfaces comparable to fixed arrays. The example of the limited gain in Guadeloupe is explained by the prevailing climate: humid tropical zones have a more variable solar radiation that is less favourable for trackers. The additional cost of the trackers of €0.80/Wc compared to a fixed array is the value current at the end of 2008 if the equipment is regularly maintained. On the other hand, it is difficult to estimate the long-term cost of systems of this kind.

Table 4.19   Price of electricity for a thin-film system

Fixed thin-film system		Cost of electricity (€/kWh)		
Place	Energy density (kWh/kWc)	Capital costs		
		5%	3%	1%
Guadeloupe (F)	2040	0.095	0.082	0.069
Seville (E)	1660	0.116	0.100	0.084
Marseille (F)	1485	0.130	0.112	0.094
Caen (F)	1095	0.176	0.152	0.128
Uccle (Be)	900	0.215	0.185	0.156

With thin-film technology modules, the annual production benefits from a better spectral response and an improved efficiency to low light levels compared to crystalline panels: the gain is around 4–5% in these examples. As with the crystalline panels, we have not considered shading, which implies a much bigger installed surface, some 2–2.5 times the crystalline array. The estimated cost does not take into account the cost of the land, which should be fallow, desert or unusable in some way. If all these conditions are met, it can be seen that the price of electricity generated in the south is close to that supplied by the grid. It may even be cheaper: in the United States, electricity costs more in the middle of the day when demand is very high because of air conditioning.

These cost estimates show that parity with non-renewable energy is almost achieved. If these systems last 10 years longer, the average price will be further reduced by 33% and the costs of maintenance or of renewing inverters can be easily covered and made economic.

4.8   Examples of installed systems

We give below examples of typical individual or industrial systems that are good illustrations of the current grid-linked PV market.

4.8.1　3 kW villa

This system was installed on a new building exceeding the Swiss Minergie[5] standard for a low-energy dwelling. This standard imposes high insulation criteria (typically more than 20 cm of quality insulation), excellent windows and minimal consumption of energy. In this villa, solar thermal collectors produce most of the domestic hot water requirements and give a boost to the wood pellet burning central heating system. The building is in the Swiss Canton of Valais, in a very sunny environment with little mist.

The PV system is made up of twenty-four 130 W Solarfabrik modules coupled to a Sputnik Solarmax 3000 S 2.5 kW nominal inverter. The integrated roof uses Solrif supports (Figure 4.9) that provide an excellent finish, with the modules replacing tiles. The roof has a pitch of 25° and faces 10° E.

The system was one of the lucky ones approved by Swissgrid, an operator buying in the electricity produced at approximately €0.60/kWh for a duration of 25 years, which enables the cost of the investment to be fully amortised.

For the simulation, we have taken account of shading from the horizon, as the house is near high mountains to the south-west. On the eastern side, the horizon is not entirely open as the house is low down in the Rhône valley. Figure 4.25 gives a simplified horizon curve.

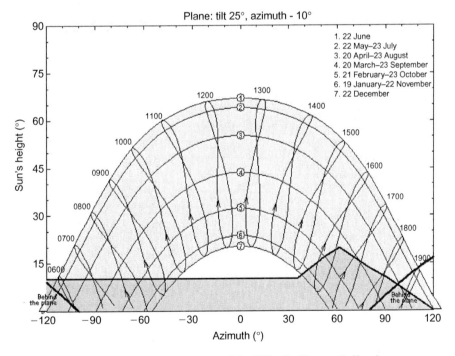

Figure 4.25　Horizon curve of the Villa Boillat at Collombey

[5] http://www.minergie.ch

The losses represented by the shading from the mountains are estimated at 5.2% in the simulation, which assumes an annual production of 3015 kWh, corresponding to an energy density of 966 kWh/kWc.

Table 4.20 gives the main data of this system.

Table 4.20   Characteristics of 3 kW system

Panels	Power (W)	Simulated energy		Feed-in tariff (€/kWh)	Revenue (€/year)
		kWh/kWc	kWh/year		
24 × Solarfabrik 130 W	3120	966	3015	0.6	1809

As the system costs a little over €21,000, it can be seen that with the anticipated production, the financing is covered and will enable other long-term maintenance expenses to be met.

Figure 4.26 shows the villa just completed in the middle of the winter, the garden still unplanted.

Figure 4.26   Villa Boillat with 3 kW PV system [photo M. Villoz]

4.8.2   110 kW solar farm

The system described here was commissioned in November 2005. The installation completely covers the south roof of an agricultural shed belonging to the Aeberhard family in the canton of Fribourg, French-speaking Switzerland. Installation was carried out by the Solstis company of Lausanne. We give a simulation of the system calculated with PVsyst software as well as the results of the first years of operation. The basic characteristics of this are summarised:

- STC PV power installed: 110,160 W
- Solar panels: Nine-hundred and eighteen 120 W Kyocera modules mounted on Solrif-type supports
- Inverter: Sputnik Solarmax 80 C, 80 kW nominal
- Solar roof surface: 960 m^2
- Panel tilt: 20°
- Azimuth: 24° E

The solar panel array replaces the standard ventilated roof. The panels are slotted in place like tiles (Figure 4.27), with the under roof of wood being closed and covered with a watertight membrane. In winter, around 100 head of cattle produce enough heat to melt the snow on the roof rapidly. Even during the winter 2008–09, there was only one day in December when no electricity was produced.

Figure 4.27   Farm with a 110 kW system [photo M. Villoz]

The investment made by the family is around €600,000 for the solar generator to which €13,000 of costs and tax from the electricity company had to be added to increase the connection power rating, in total around €613,000. To cover these costs, the owner negotiated with the local electricity company a feed-in contract rate over 15 years, which will enable him to amortise the system over this period; after 15 years, the production sold at the market price will produce a small profit or complete the amortisation. The financial balance is a delicate one in that interest rates of 3% at the time of the installation could rise and make the total cost more expensive. The company has accepted to buy the whole production within the range of ±30% of the calculated simulation, which is sufficient to absorb annual sunshine variations.

4.8.2.1   Simulation

The system characteristics, number of panels in series and parallel, tilt, azimuth, module type, inverter type as well as the statistical data of solar radiation of the region enable an accurate simulation of expected performance to be made. Table 4.21 shows the performance of the generator in the first years of exploitation compared to the simulated results.

Table 4.21   Simulation of solar generation and first years of exploitation of 110 kW solar farm

	Data (kWh/m²/ month)	Simul. (kWh)	2006 (kWh)	Diff. (%)	2007 (kWh)	Diff. (%)	2008 (kWh)	Diff. (%)
January	33.2	3,814	4,960	30.0	4,495	17.9	5,383	41.1
February	48.6	5,398	5,093	−5.7	6,642	23.0	8,776	62.6
March	87.0	8,845	9,091	2.8	11,476	29.7	9,768	10.4
April	113.0	10,573	11,714	10.8	17,371	64.3	10,110	−4.4
May	148.0	13,144	12,977	−1.3	14,883	13.2	15,716	19.6
June	158.0	13,605	17,678	29.9	14,948	9.9	15,023	10.4
July	178.0	15,289	18,091	18.3	15,617	2.1	16,740	9.5
August	151.0	13,689	11,586	−15.4	13,755	0.5	14,672	7.2
September	106.0	10,223	11,061	8.2	13,255	29.7	10,823	5.9
October	62.4	6,419	8,668	35.0	8,461	31.8	7,136	11.2
November	32.8	3,519	6,265	78.0	4,539	29.0	4,725	34.3
December	26.1	2,980	3,971	33.3	2,754	−7.6	3,011	1.0
Total	**1142.2**	**107,498**	**121,155**	**12.7**	**128,196**	**19.3**	**121,883**	**13.4**

These three years (2006–2008) show generation considerably higher than the simulated forecasts. The simulation relied on sunshine statistics from 1960 to 1990, but since 2000, sunshine has been higher every year than the period, possibly due to global warming. The most astonishing difference is the mildness of the winters: during the winter of 2006–07, there was an increase over the simulations of 78% of energy produced in November and 64.3% in April. During the months of April 2007, the system produced more energy than for an 'average' month of July.

4.8.3   167 kW agricultural shed

The example of the large shed shown here is typical of recent French installations. Commissioned in spring 2009, it is situated in Provence and belongs to a fruit grower, Philippe Manassero. Its location is in one of the sunniest climates of southern France. The installation was carried out by the Groupement Photo-voltaique du Luberon.[6]

---

[6] http://www.pvluberon.com/

The corrugated iron roof on the south slope of an existing large agricultural shed with an area of around 1250 m² was completely replaced by PV panels fixed on supports of the type Mecosun,[7] specially developed for this type of installation. These supports are attached to the shed's metal beams. The advantages are that the supports accept standard panels with aluminium frames of all dimensions and that the installation is done by working laterally: a vertical strip of the old covering is removed to the top of the roof, the first rails fitted and then the panels, and so on, with only small sections of the old roofing being removed at a time, which enables the work to be carried out even in light rain. Figures 4.28–4.30 show the roof installation in progress and a striking change in appearance between the old and the new covering.

The specifications of this installation are summarised:

- STC PV power installed: 167.6 kW
- Solar panels: Solarwatt M230, 4 × 182 modules
- Inverters: 4 × Sputnik Solarmax 35S, 35 kW nominal
- Solar roof surface: 1200 m²
- Panel tilt: 17°
- Azimuth: south

*Figure 4.28   Installation of 167 kW generator on agricultural shed
        [photo M. Villoz]*

---

[7] http://www.mecosun.fr

Figure 4.29   167 kW generator on agricultural shed – old roof covering [photo M. Villoz]

Figure 4.30   Installation of 167 kW generator on agricultural shed – new roof [photo M. Villoz]

*Figure 4.31   The underside of the roof panels of the agricultural shed
shown in Figure 4.30*

- The hooks linking the Mecosun rail to the structure of the building can be seen: these allow the fixing to be done without drilling the sections.
- A small piece of bent corrugated iron placed under the horizontal joints of the panels provides a channel to collect any water penetrating during heavy rainfall and takes it under the main rail where a gutter takes it to the ground.
- The cabling is fixed vertically within the supporting rail and then in perforated channels first horizontal and then vertical to the junction boxes.
- The cabling and the structure allow easy access to any panel showing a problem: the panels are accessible and can be dismantled from above and can be put in place or removed by large glaziers' suction cups. The new roof can be carefully walked on without a problem.
- The new roof structure binding the vertical rails to the frame improves the stability of the shed by making the whole building more rigid.

Table 4.22   167 kW agricultural shed: data and simulated performance

		Unit		Total	Note
Panels	Solarwatt M230-96 GET AK	230 W	728 pieces	167.6 kW	STC
Inverters	Sputnik Solarmax 35S	35 kW	4 pieces	140 kW	Nominal
Cabling	56 strings of 13 modules	V_{mp}/I_{mp}	553 V	272 A	at 50 °C
Simulated production	1401 kWh/kWc	235	MWh/year		
Estimated revenue		143,000	€/year		0.61 €/KWh

Table 4.22 shows the main characteristics of the system with the results of the simulation.

Figure 4.32 shows the inverter room.

Figure 4.33 shows the shed completely covered with its new PV roof.

Figure 4.32   Inverter room of 167 kW agricultural shed [photo M. Villoz]

Figure 4.33   167 kW agricultural shed: PV roof completed [photo M. Villoz]

Chapter 5
Stand-alone photovoltaic generators

Stand-alone PV generators are the oldest historically. They first appeared in the 1970s as solar panels mounted on satellites. These were soon followed by applications for rural pumping and electrification in developing countries, then domestic and technical stand-alone applications in industrialised countries.

The reader will perhaps be surprised by the size of this chapter, which is disproportionate to the share of stand-alone in PV systems today. The length is justified by the fact that the stand-alone solar system is not just a PV generator but also includes energy storage, regulators and appliances specially developed for this purpose. Designing a stand-alone system often requires reviewing energy consumption and optimising it so as to only produce the essential energy necessary for the function to be supplied. In addition, the applications described in this chapter are highly diversified: stand-alone energy generation is attractive in a wide variety of circumstances, which is best to identify and deal with on a case-by-case basis.

5.1 Components of a stand-alone system

As we have seen in Chapter 1, for true autonomy without any other energy supply, all the energy must be produced during the day by the PV array, and storage is indispensable if there is to be any consumption outside daylight hours, as opposed to a grid-connected PV system, which can take energy from the grid at night.

5.1.1 Storage of energy

The storage of energy in stand-alone PV systems is generally provided by batteries or accumulators.

Only some applications using energy directly from the Sun, such as, for example, pumping or ventilation, can manage without storing energy. Therefore, a thorough understanding of batteries is essential for the success of stand-alone systems.

In such systems, energy storage represents around 20–30% of the initial investment, but over a period of 20 years of exploitation of the system, this cost can reach 70% of the total. It is, therefore, important to try to reduce the price by increasing the life expectancy of the storage elements, which is always less than that of the panels. Batteries, therefore, have to be replaced several times during the life of the system (every 2, 5 or 10 years depending on the type). High temperatures can shorten life, because corrosion is speeded up when the batteries are hot.

The batteries used in stand-alone server systems are generally of the lead-acid type (Pb). Nickel cadmium batteries (NiCd) are only rarely used as their price is much higher and they contain cadmium (toxic). They have been superseded by nickel-metal-hydride batteries (NiMH), which are more attractive; we will give some typical parameters for them, but their use is more frequent in top of the range professional applications or very small applications (<2 Ah). We will also discuss lithium batteries (Li-ion, for example), since certain models can be useful for solar, provided certain precautions are taken, because they are compact, but still expensive. Other batteries are being developed, particularly by electric vehicle manufacturers. We may also mention the development using compressed air as energy storage; the potential interest of this process is the long expected life and the absence of chemical components to be recycled.

5.1.1.1   Lead batteries

The lead battery was developed during the nineteenth century, and its operation is well understood. Two electrodes of lead and lead oxide (PbO_2) are inserted into an electrolyte made up of dilute sulphuric acid. When the two electrodes are connected to an external appliance consuming current, they are converted to lead sulphate ($PbSO_4$) and the acid is diluted, a property that enables the state of charge of the battery to be measured by checking the specific gravity of the electrolyte. When a reverse current is applied to the system, the acid re-concentrates and the two electrodes return to their initial states (Figure 5.1).

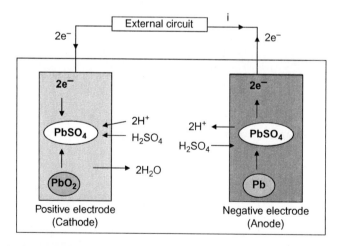

Figure 5.1   Diagram of a lead-acid battery and chemical reactions (here during discharge)

Positive electrode	Acid	Negative electrode	Positive electrode	Water	Negative electrode
Charge ←					
PbO_2	$2H_2SO_4$	Pb	$PbSO_4$	$2H_2O$	$PbSO_4$
→ Discharge					

At the positive electrode:

$$PbO_2 + H_2SO_4 + 2H^+ + 2e^- \leftrightarrow PbSO_4 + 2H_2O \qquad (5.1)$$

At the negative electrode:

$$Pb + H_2SO_4 \leftrightarrow PbSO_4 + 2H^+ + 2e^- \qquad (5.2)$$

The nominal equilibrium potential is the sum of the potentials at the two electrodes:

$$E + (PbSP_4/PbO_2) = 1.7 \text{ V} \quad \text{and}$$

$$E - (PbSO_4/Pb) = -0.3 \text{ V, that is, } E = 2.0 \text{ V} \qquad (5.3)$$

We will not go further here into the details of the electrochemical reactions of electrodes and the origin of the 2 V potential, but will give some more general and practical information; the reader wishing to get more deeply into these questions can refer to the excellent article on lead-acid batteries by Dr D. Berndt.[1]

The positive electrode is made of brown lead dioxide and the negative electrode of spongiform grey lead. During charging, lead dioxide forms on the anode, while the cathode is transformed into pure lead and the sulphuric acid becomes more concentrated. During discharging, a part of the electrolyte binds to the lead and transforms it into $PbSO_4$; this reaction produces water, which reduces the density of the electrolyte. This variation in density can be calculated as the ratio of the charge/discharge: for each Ah of discharge, 3.654 g of acid binds to lead and 0.672 g of water is produced, with the charge producing the same values in reverse. This variation in specific gravity provides an easy way of checking the state of charge of open batteries.

The variation in specific gravity of the electrolyte has another important effect that limits the use of batteries in low temperatures, because the freezing point rises as the battery discharges.

[1] D. Berndt, 'Valve-regulated lead-acid batteries', *Journal of Power Sources*, 2001;**100**:29–46.

Each cell of a lead battery supplies an average voltage of 2 V, and the appropriate number of elements is assembled in series or parallel to produce the desired voltage and current. Small capacities are often met by using batteries of 6 or 12 V (three or six cells in series), while larger requirements use multiples of 2 V elements connected in series and in parallel, and may reach several thousand Ah.

Construction

The first lead battery developed by the French physicist Gaston Planté in 1859 used sheets of solid lead. The oxide was formed on one sheet during charging. This Planté type is still used for special applications today. The disadvantage of these batteries is that they are slow to manufacture (and therefore expensive), and their solid structure provides little contact between the mass of lead and the electrolyte, which increases the internal resistance R_{Bi} formed by the resistances of the electrolyte and contact between active material and grids. Furthermore, the low proportion of active lead makes the battery very heavy with a poor energy density.

To get round these problems, Sellon invented in 1881 an electrode in two parts: a perforated lead plate filled with a paste of PbO_2 active material on the anode and solid lead on the cathode.

These grids serve as a solid support and conductor of current. The necessary chemical reactions only involve the active filling substances. This separation of functions makes for easier and cheaper manufacture. Also, the porous structure of the electrodes gives quicker access to the current and reduces the internal resistance. This method of construction is still widely used today. Porous plastic separators are placed between the electrodes allowing the acid to pass but preventing the electrodes from short-circuiting.

The active material is made of granules separated by numerous cavities or pores. The chemical reactions take place on the surface of these granules and this requires the diffusion of the electrolyte through the pores. This movement is slowed by the small size of the pores, which makes the acid concentration non-uniform through the battery plates.

Uniformity only appears when the battery has been at rest, without charging or discharging, for several hours. At this point the open circuit voltage V_{Bi} can be empirically determined by the formula:

$$V_{Bi} = \rho_e + 0.84 \tag{5.4}$$

where ρ_e is the density of the electrolyte in g/cm^3.

For fully charged batteries, ρ_e is between 1.20 and 1.28 g/cm^3, which gives values between 2.04 and 2.12 V for V_{Bi}.

$PbSO_4$ has a volume 1.5 times greater than the oxide PbO_2 and 3 times greater than lead. Consequently, the active material of the electrodes swells during discharge, preventing the diffusion of the electrolyte in the pores and producing mechanical constraints in the plates. These constraints may dislodge the active material from the grids, especially from the positive electrode. The dislodged material is useless and collects at the bottom of the battery, reducing its capacity, and in the end possibly short-circuiting the plates, rendering the battery useless.

Electrolyte

The electrolyte, consisting of sulphuric acid diluted with distilled water, is an excellent transporter of ions. As it is transformed during charging or discharging, its specific gravity varies.

In *open batteries*, at the end of the charge, a small part of the water is electrolysed and hydrogen and oxygen escape. This loss must be regularly compensated by topping up with distilled water.

In *sealed or valve-regulated batteries* (valve-regulated lead acid, or VRLA), the electrolyte is maintained by a sodium silicate gel or absorbed in a glass fibre separator (absorbent glass mat, or AGM). The nature of this electrolyte has a direct impact on the life expectancy of sealed batteries – the gel electrolyte is the most durable. For this reason, this type of electrolyte is often preferred for stand-alone PV systems to prolong the life of the battery. This type of battery does not have an electrolyte top-up facility, and it is important to avoid any overcharge that would electrolyse the water. However, at the end of the charge, a small amount of hydrogen may escape through the valves and oxygen may diffuse from the positive to the negative electrode where it chemically recombines.

Negative electrode

To reduce large variations in volume of the negative electrode, dilating materials, such as barium sulphate, are added to the active material of pure lead, and carbon is added to improve its conductivity. With the composition of the plates, the important thing is to conduct current and resist corrosion while keeping contact with the active material. The plate is normally cast, and antimony is added to lead to improve the fluidity of the active material, and thereby the solidity of the alloy. Formerly, the proportion of antimony was as much as 5% or 10%, but these batteries used a lot of water and suffered from a considerable self-discharge. To reduce these effects, the antimony added was reduced to less than 1%, and as the alloy is very difficult to cast, other materials are added, such as arsenic, tin, copper, sulphur or selenium, to make casting easier. Finally, to produce a low maintenance battery that hardly uses any water, antimony is replaced by calcium, aluminium and tin to obtain a stronger alloy, resistant to corrosion and yet ductile enough to be cast. For the grid, it is usual nowadays to use an expanded metal containing much less lead to reduce its weight. This is only possible for the negative grid that contains an active material of pure lead to favour conductivity.

Spiral electrodes are another possibility that enable the lead mass to be lightened, and so to be used pure, which virtually eliminates self-discharge, and to increase the surface of the electrode/electrolyte exchange.

Positive electrode

There are three types of positive electrode: flat plates, tubular electrodes and the Planté type.

The *first Planté-type electrodes* were plates of pure cast lead with ribs and channels to increase the surface area. The plate is oxidised with sulphuric acid and an anion (nitrate or perchlorate) that creates a soluble Pb^{2+} salt and prevents the total passivisation of the plate. The final plate of 6–12 mm thickness is formed of a thick layer of spongy oxide bound to pure lead, a very solid and durable mixture but with a low proportion of active mass. The typical energy density of a Planté battery is around 7–12 Wh/kg.

For the *flat plates* used in ordinary economical batteries, the electrode is made up of a lead grid filled with spongy PbO_2. To improve solidity and avoid a loss of active material, a porous cover of fibre glass or plastic is sometimes added to retain the oxide within the grid.

In *tubular batteries*, the positive electrode is made up of a series of tubes aligned like the teeth of a comb. The lead teeth of the grid are surrounded with spongy PbO_2 held in place by porous fibreglass tubes and protected by a sheath of polyester or polyvinyl chloride (PVC). This type of plate is very solid and allows a large number of charge/discharge cycles.

Separators
The separators used to prevent contact between the positive and negative plates must be

- resistant to the electrolyte,
- solid with a precise and uniform thickness,
- permeable to the electrolyte and to gases but electrically insulating.

The separators must resist the expansion of the active material during discharge while allowing the electrolyte to circulate. They are usually made of plastic sheets with an indented surface to mechanically guarantee the distance between the plates, the permeable part being in the ridges between the indentations.

In batteries designed to be frequently cycled, porous covers in fibreglass or plastic are added to the active material to prevent it from falling out. In tubular batteries, these covers are made of a series of tubes.

Charge characteristics

Note
To simplify matters, we speak of energy stored or released by the battery both in Wh and in Ah, units that relate to a system of nominal voltage V_B. In this case, the real energy is equal to $V_B \times$ Ah.

Figure 5.2 shows the charge curb at constant current of a lead battery at different temperatures. During the charge, current enters the battery by the anode, which results in the production of acid increasing the specific gravity of the electrolyte. Consequently, the ionic density also increases, which reduces the series resistance R_{Bi} and increases the voltage V_{Bi}. Towards the end of the charge, the active material density falls and a part of the current is no longer absorbed. This current then tends to electrolyse the water, producing oxygen on the positive

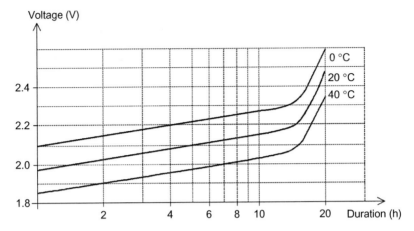

Figure 5.2 Charge characteristics of a lead battery (at constant current)

electrode and hydrogen on the negative electrode. This phenomenon, called gasification, has both advantages and disadvantages.

- The disadvantages include the loss of water and the corrosion of the positive plate by oxygen. If the overcharge is very intensive, active material can be dislodged by the gas bubbles.
- The advantages are mainly that the bubbles circulate the electrolyte and make it more homogeneous, avoiding the *stratification of the electrolyte*: if it is not circulated (as in the battery of a moving vehicle), the electrolyte tends to concentrate at the bottom of the battery, which leads to faster corrosion of the bottom of the plates.

When overcharged, the battery loses water that must be replaced. If part of the plates is not covered by electrolyte, irreversible damage can occur.

During gasification, internal resistance increases considerably along with a sharp increase in voltage (see the curves in Figure 5.2); this facilitates the design of charge controllers, since the end of the charge is easy to measure.

On the other hand, the temperature curves remain in parallel, and it is possible to determine a linear dependence of the maximum voltage on temperature, which equals, by 2 V step:

$$\frac{\mathrm{d}V}{\mathrm{d}T} = -55 \text{ mV/K} \tag{5.5}$$

For PV installations, the current input is often very weak and it may be assumed that the battery remains at the ambient temperature (the internal dissipation due to R_{Bi} is negligible). The charge controllers can thus simply measure the ambient temperature to apply the charge conditions.

When the ambient temperature increases, the electrolyte expands, which lowers the acid concentration. Consequently, the specific gravity and the battery voltage fall. But the temperature increase makes the ions more mobile and this effect overcomes dilatation, and in the end lowers R_{Bi} at high temperature.

The proportion of active material by stored capacity is 11.97 g/Ah. As the voltage of each element is of the order of 2 V, the maximum theoretical energy density of the battery is around 170 Wh/kg. In practice, this value is much lower because the lead plates, the case and the electrolyte are also heavy, and density is reduced generally to between 20 and 40 Wh/kg.

Discharge characteristics

Figure 5.3 shows the discharge curves for different currents. The electrical discharge equation is

$$V_B = V_{Bi} - I \times R_{Bi} \tag{5.6}$$

The discharge process is the transformation of PbO_2 and of Pb, the active materials, into $PbSO_4$, accompanied by the absorption of the electrolyte acid. This absorption of acid increases R_{Bi}, which contributes to lowering the V_B. This characteristic varies with the age of the battery: the curves shown below would be lower with an old battery as the internal resistance would have increased. When the battery is deeply discharged, the active material is swollen and blocks the pores, which prevents the electrolyte from reaching all of the active material. At this point, the voltage falls abruptly, and this phenomenon also enables one to easily size a regulator to disconnect consuming devices. However, if the discharge is very deep and the battery remains in this state for long, the sulphate on the plates tends to form large permanent crystals that prevent current from flowing. If it is partial, this phenomenon results in a lowering of the battery capacity, but if it is major, it can completely block the flow of current, a process called sulphation.

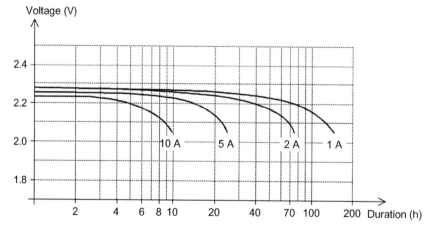

Figure 5.3   Typical discharge characteristics of a lead battery of 100 Ah

Capacity

The nominal capacity C_B of a battery is the quantity of Ah that can be extracted in a given time (see units of measurement in Appendix 1). Typically, the end of nominal discharge is a voltage V_B of 1.85 V/element. The discharge equation (5.6) shows that if the current of discharge increases, V_{Bi} will be higher when V_B reaches 1.85 V. Therefore, capacity falls as current increases. Figure 5.4 shows the detail of this phenomenon.

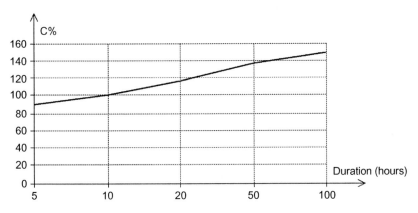

Figure 5.4   Capacity as a function of the duration of discharge. C (%) is a relative capacity equal to 100 at C_{10} (see explanations in the text)

Capacity is strongly dependent on temperature (see Figure 5.5).

State of charge

The state of charge of a battery E_{CH} is the quantity of electricity still available in Ah divided by the nominal capacity of the battery: if $E_{CH} = 1$, the battery is full, and if $E_{CH} = 0$, the battery is empty.

The depth of discharge P_D is the complement of the state of charge: $P_D = 1 - E_{CH}$. To compare two batteries, their capacities must be known at the same discharge current or least at the same speed of discharge. This is how batteries work. The capacity of professional batteries (for example, tubular batteries) is often given for a discharge in 10 h, whereas for small PV systems the capacity value for a discharge of 20 h is more practical: this is often the current level of battery operation. It is usual to describe capacity values as C_{10}, C/10 or 0.1C (for 10 h), and C_{20}, C/20 or 0.05C (for 20 h).

Note

By extension, it is also possible to speak of charge (or discharge) at C/10 or at C/20. For example, charging a 20 Ah battery at C/10 is to apply a charge current of 2 A.

Temperature effect

The ambient temperature has a direct influence on capacity, which reduces as the temperature falls (Figure 5.5). It can be seen that below 0 °C, capacity reduces rapidly. For operation at this temperature, the use of the battery must be limited or its capacity must be considerably increased to avoid destruction by frost. In this case, a voltage regulator with adjustable disconnection should be used, which allows the appliances to be cut-off above the freezing point of the electrolyte. For systems working below 0 °C, a regulator should be used that compensates the maximum charge voltage by −5 mV/°C. Figure 5.6 shows the minimum charge state to be respected for a battery of 100 A/10 h according to temperature to avoid the freezing of the electrolyte.

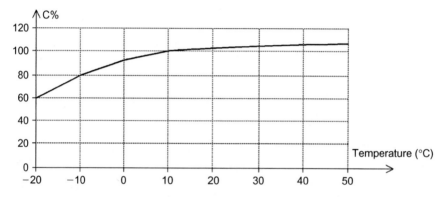

Figure 5.5   Typical capacity variation of a lead battery according to temperature (see explanations in the text)

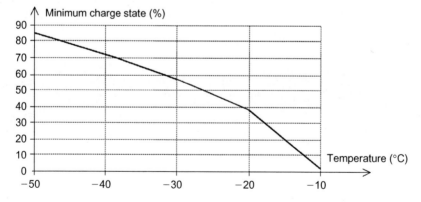

Figure 5.6   Minimum charge state before freezing

At high temperatures (>25 °C), thermal compensation must be allowed to avoid evaporation of the electrolyte. At these temperatures, a less concentrated electrolyte is often used (ρ_e between 1.20 and 1.22) to limit internal corrosion of the battery. The additional circulation caused by the temperature compensates for the lower specific gravity, and R_{Bi} does not vary too much.

Efficiency

Efficiency at small charge/discharge currents is virtually constant: for a new battery we can assume a value of 0.83 in Wh or 0.9 in Ah. The efficiency is strongly dependent on the state of charge of the battery: for an 'average' charge state, it is high, but falls rapidly at the end of charge and when the current is no longer absorbed by the active material but begins to electrolyse the water.

Cycles and life expectancy

The maximum number of cycles and the life expectancy of batteries are strongly dependent on the technology of manufacture and conditions of use. For example, when used in very hot countries (ambient temperature at >35 °C), if one assumes that corrosion phenomena will be the first to limit the life expectancy of the battery and the maximum number of cycles will not be reached, there will be a tendency to limit the capacity and so the material investment at the start, and daily cycles of charge/discharge of the order of 80% will be chosen. On the other hand, in temperate climates, using a quality charge controller, corrosion phenomena can be limited and the choice of the sizing of the battery and the depth of discharge will depend on the number of criteria such as

- degree of autonomy desired taking into account variations in sunshine;
- the replacement cost of batteries in relation to transport, ease of access to the site and labour costs;
- investment capacity at start-up and financing costs;
- environmental aspects and local facilities for recycling batteries.

Initial estimates will suggest that the number of cycles will be inversely proportional to the depth of discharge: for example, a battery able to provide 300 cycles per 100% discharge should be able to provide 600 cycles per 50% discharge with good regulation.

Similarly, a lead battery of any technology can be expected to have its life shortened by a factor of two for every 10 °C, on account of corrosion.

Thus, if the life expectancy of the battery is 10 years at 25 °C, it will only be 5 years at 35 °C (if this is the permanent average temperature).

Table 5.1 gives some typical figures for cycles and lead battery life expectancy along with investment costs and the energy stored in kWh, without taking into account the charge/discharge efficiency or the financing costs (cost of borrowing). These costs are the before tax values that a final consumer could expect to obtain in France in 2008.

There are some other variations in the technologies described, often manufactured in relatively small quantities for professional applications. For example, there are sealed tubular gel batteries and Planté batteries.

Modern batteries are mainly of the low self-discharge type, in other words, they lose less than 3% of their capacity per month at 20 °C. However, this loss triples at 30 °C and when managing a stock of batteries or when assembling large

Table 5.1   Lead battery technologies

Technology	Cycles (80% depth of discharge)	Life expectancy (years)	Investment (€/kWh)	Energy cost (€/kWh)
Car	100	7	90	1.1
Solar panel	250	7	100	0.5
Sealed AGM	250	5	170	0.85
Sealed gel	400	8	200	0.6
OpZs tubular	550	12	150	0.4
Tubular bloc	1200	15	200	0.2

PV systems in hot countries, this must be taken into account, with top-up charges being carried out if stock remains unused for some time.

Grey energy and recycling

The energy needed to manufacture, install and maintain a battery includes the energy for each component (plates, separators, case and acid), and ranges from raw material extraction to the final recycling when the battery is dead. These figures are taken from the publication of IEA Photovoltaic Power Systems Programme Task 3: Lead-acid battery guide for stand-alone photovoltaic systems.[2]

The energy necessary for manufacture and use are calculated for standard technology batteries and include recycling rates. These rates today are 70% in the United States, 60% in Europe, 50% in Africa and 30% in Asia. These cover all types of batteries, and often for renewable energies in Africa, the costs of returning goods to the factory are too high and the rate falls to 20%.

Table 5.2 shows the total grey or embodied energy as well as the energy needed for storing energy, with the average energy density of open solar batteries being 40 Wh/kg.

Table 5.2   Grey energy of lead batteries

Type/energy	Total kWh/kg	Recycling	Energy/storage Wh Cons/Wh stored
New battery	8.33		208
Battery 100% recycled	4.76		119
Battery for PV in Africa	7.62	20%	191
Battery in Europe	6.20	60%	155

In conclusion, 190 charging cycles would be needed in Africa to recover the grey energy expended.

---

[2] Can be downloaded from http://www.iea-pvps.org

Comparing the figures with those in Table 5.1, showing the average life expectancy of lead batteries, it will be noticed that the car-type battery used in solar home systems (SHS) will not be able to store as much energy as was used for its manufacture and installation. With the best lead batteries available, one can hope at best to store five or six times the amount of grey energy throughout the whole life of the battery. This figure is considerably lower in solar systems because 20–30% of energy is lost before completing the charge or because the battery is full in summer.

Grouping of batteries

Solar batteries that are identical and of the same age can be connected in series or parallel. However, connecting in parallel is not recommended: it should be reserved for installations where the provision of large elements is not possible. In such cases, care must be taken to balance the currents by symmetrical cabling. For each string of batteries cabled in series, include a fuse in the series cabling. For small systems, it is always better to oversize the battery from the start, since the parallel connection of batteries of different ages is to be avoided since the oldest battery will cause premature ageing of newer batteries.

A large battery of the type used in emergency installations (several MWh) is often made up of more than hundred 2 V elements in series. In these systems, the battery is maintained *floating* and occasionally partially discharged. In a PV system of the same dimensions, the battery undergoes daily cycles, regularly completes a charge and if necessary must undergo a deep discharge to respond to demand. If the regulation only controls the global voltage, as each charge is completed, the weaker elements of the series will be overcharged and lose more electrolyte; they will age more quickly and lose more capacity compared with the other elements. In case of deep discharge, one of the elements may be completely discharged while the other elements may still be supplying current. In this case, if the discharge continues, the voltage will reverse and increase rapidly because its internal resistance is high, the last stage before the destruction of that element which could cause a fire. For a high-voltage battery undergoing regular cycles, the intermediate voltages of the elements should be measured and regulated accordingly. The Fraunhofer Institute[3] in Germany has developed a special regulator that monitors each element separately and transfers the necessary charges between the cells to balance the battery.

5.1.1.2   Nickel batteries

We now briefly review the main parameters of the NiMH battery. This battery, which has virtually replaced the NiCd battery, was widely used in portable devices before the Li-ion battery. A few manufacturers are still producing NiCd batteries for industrial applications (telecom networks, transport applications), but with the problems of recycling cadmium, it is unlikely that these batteries will develop further. However, top of the range NiCd batteries, expensive and with long life expectancy (15–20 years), are still used in some special PV systems, when access to the site is very complicated (high mountains and desert). We will therefore only

---

[3] http://www.ise.fhg.de/

concern ourselves here with NiMH, which is less polluting and has other advantages over NiCd, such as higher energy density (around +40%).

Most NiMH batteries are of low capacity, stick or button cells ranging from a few mAh to a few Ah, and are consequently reserved for low-capacity applications (interior photo sensors and modules <5 Wp).

Charge characteristics

Figures 5.7 and 5.8 show typical charge curves at constant currents for an NiMH cell at three different temperatures. It is seen that these charge curves vary according to temperature and current level, and the completion of charge also depends on temperature. It is to be noticed that there is no charge curve at temperatures below freezing, because NiMH batteries are incapable of receiving a charge below 0 °C, although discharging can take place at temperatures below 0 °C.

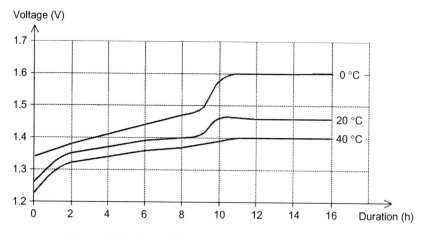

Figure 5.7   Slow charge of an NiMH battery at 0.1 °C

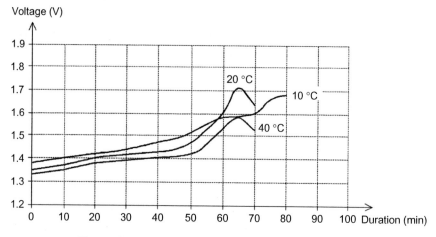

Figure 5.8   Fast charge of an NiMH battery at 1 °C

These effects make the design of a solar charge controller difficult since the current from the panel varies with solar irradiance, and the variations in temperature and current level make it necessary to measure the voltage, voltage variations over time, and the temperature and to include the energy input. The rapid chargers of NiMH batteries working on the mains are generally equipped with a processor monitoring all these values.

Discharge characteristics
The discharge characteristics of NiMH cells vary according to their technology, the temperature and the current level. We will not show here any particular curve since every supplier has different characteristics. One important point to note is that the NiMH can be completely discharged, which makes the presence of the discharge regulator unnecessary if the appliances accept very low voltages.

Table 5.3 shows typical capacity values according to the discharge current and temperature. It will be seen that these values are given for relatively high current levels because these batteries are normally intended to power portable appliances were autonomy is rarely more than a few hours.

Table 5.3   Relative capacities (%) of an NiMH battery according to discharge current and temperature

Temperature/current	0.1C	0.2C	1C	2C	5C
20 °C	106	100	87	81	63
0 °C	100	94	80	82	58
−20 °C	87	80	58	40	

We have seen that the charge curves (Figures 5.7 and 5.8) are strongly influenced by temperature. Capacity is also influenced by temperature and current level. One of the advantages of nickel batteries are that they do not use a water-based electrolyte and do not freeze. So nickel batteries are attractive for professional applications in low ambient temperatures (except for small battery formats that are to be charged above 0 °C).

Self-discharge
A major disadvantage of NiMH for solar applications is its rate of internal loss that is much higher than with lead batteries. Table 5.4 shows the rate of capacity losses according to temperature and duration of storage.

It will be seen that for a battery left uncharged for 5 days at 30 °C, the solar panels need to be oversized by 20% to take into account the self-discharge rate. In hot countries, it is essential to install batteries in a shaded and well-ventilated place or better still in a cellar. Never install a compact device containing batteries in a box under the solar panel. Experience has shown that a compact lamp used in Africa containing a solar panel, a fluorescent tube and an NiCd battery saw its

Table 5.4   Loss of capacity (%) of NiMH according to temperature

Time/temperature	20 °C	30 °C	40 °C	50 °C
5 days	15	20	28	35
10 days	22	28	36	50
30 days	36	48	60	83

useful life reduced to a few minutes per day instead of the 3–4 h expected, as the battery temperature reached 75 °C at midday.

Efficiency

The charge/discharge efficiency of NiMH is generally lower than for lead: calculated in Ah, it is generally 80–85%, and 65–70% calculated in Wh. Also, it falls with a lower charge current, often <50% in Ah below C_{50}. The solar panel to be associated with it will need to be sized to charge between C_{30} and C_{10} in order to have a good charge efficiency (it falls subsequently for lower currents).

Cycling and life

This is the parameter where NiMH batteries are at their most favourable: they generally supply 500–700 cycles at 80% discharge, and they tolerate being stored discharged 3–5 years. Their life is dependent on temperature: they can easily last 10 years at 20 °C but can lose 20% for each 10 °C increase.

Connection in series/parallel

Here also, NiMH is more demanding. When several 1.2 V cells are installed in series, if the system is not equipped with the regulator cutting off appliances below around 1 V/cell, all the battery elements must be measured and their capacities heard before connecting them in series. If the elements are not identical, in the case of total discharge, the element with the lowest capacity will have to support an inversion of its voltage while the other cells are still supplying current: this effect rapidly reduces the life expectancy of this element. For installation in parallel, the same precautions as for lead batteries need to be observed.

Price

NiMH batteries are significantly more expensive than lead batteries. Compared with sealed lead batteries (VRLA), their cost at equivalent capacity is generally four to five times higher. For short duration use, the cost per kWh will be around twice that of lead.

5.1.1.3   Lithium batteries

Lithium batteries are reserved for the time being for portable devices for which their high energy density (around six times better than sealed lead) is the main advantage. Their current price is around €1000/kWh, and the price of the corresponding stored energy is €0.80/kWh without guarantee, their cycling not having been proved. They are making inroads currently at the expense of nickel batteries, as their energy density and efficiency are superior.

Technology

The main advantages of lithium are its high energy density, its storage efficiency in excess of 93% and a potential reduction in costs that could make it competitive with lead. Battery life and number of cycles (6 years and 1300 cycles at 100% discharge depth) are estimates for the time being, as the technology is too young to have reliable data. Their main disadvantages are fragility and the danger of violent destruction when operated outside strict conditions of temperature and voltage, which necessitates an electronic monitor on each 3 V battery cell with charge transfer between the elements to balance them in a large battery. The manufacturers generally offer blocs containing internal protection (thermal cut-off at very low voltage <1.5 V), but this protection can be destroyed by a defective charger or static electricity and transform itself into a simple shunt: in this case, the element becomes very dangerous if the external circuits do not strictly limit the voltage range. This complicated technology may mean limited development prospects for renewable energies, since reliability and the long-term resistance of associated electronics need to be very high to prevent a fire or a major problem. In the short term at least, it would be dangerous to use this technology in hot, dry countries.

Safety

Lithium batteries today power the majority of portable telephones and computers, and this major market is encouraging unscrupulous manufacturers to offer counterfeit products that do not have the indispensable safety features of this technology. Imitation mobile phone batteries that look exactly like those of major manufacturers are in circulation, but if they are too deeply discharged, their internal temperature will exceed 100 °C and the telephone can catch fire.

5.1.1.4   Future trends

In a PV system, batteries represent around 15% of the initial investment, but over an operational period of 20 years the cost can exceed 50% of the total cost of the system. It is therefore very important to improve this component to reduce the cost of the energy produced. In 2000, the market in batteries for renewable energy was around €130 million/year, and with the development of individual SHS and large rural electrification projects in tropical countries, the market could reach €820 million/year by 2010. In 2001, a European project, Investire,[4] began collecting the experience and knowledge of 35 companies and research institutes to review and assess existing storage technologies in the context of renewable energy applications and to propose appropriate research and development to lower the cost of storing renewable energy.

The technologies compared were

- lead-acid batteries,
- nickel batteries (nickel-zinc, nickel-cadmium, nickel-metal-hydride),

---

[4] http://www.itpower.co.uk/investire/summary_project.htm

- lithium batteries (Li-ion, lithium metal and lithium polymer),
- oxide reduction systems,
- supercapacitors,
- flywheels,
- compressed air,
- zinc-air batteries.

Drawing on this work, we will summarise the main recent developments in lead battery technology as these batteries will remain an essential component for decades to come.

Open lead-acid battery

The developments of open or vented batteries are mainly aimed at improving the number of cycles and the life of traditional batteries that can be manufactured on the production lines of car batteries. A recent European project proposed modifying the composition of the electrolyte by the addition of additives to stabilise the specific gravity of the acid, which would avoid stratification problems: the idea is to be able to manufacture the battery on a classic production line and only modify the formulation of the acid, which would not involve any modification to the manufacturing process. The first results of the 'crystal' battery are encouraging[5]: phosphoric acid is added to the electrolyte, which reduces the formation of large crystals of $PbSO_4$ on the positive electrode, and a silicon colloid that stabilises the electrolyte. These batteries should offer twice as many cycles and last longer than a similar battery without electrolyte modification.

Sealed lead battery

Several recent developments have improved the life of sealed batteries that offer certain advantages over vented batteries:

- the electrolyte maintained by gel or AGM avoids stratification;
- transporting the batteries is made easier, as an electrolyte leak is virtually impossible;
- the electrolyte is supplied by the manufacturer and its quality is therefore controlled.

Their disadvantages are, on the other hand, as follows:

- slightly higher cost of manufacture;
- the danger of drying out if not regularly tested and maintained.

To ensure the maximum life of these batteries, overcharging must be absolutely avoided: when sealed batteries are overcharged, excess hydrogen and oxygen are generated and corrosion increases exponentially. The excess hydrogen is expelled through the valves but the excess oxygen remains, corroding the positive

---

[5] L. Torcheux, P. Laillier, 'A new electrolyte formulation for low cost cycling lead acid batteries', *Journal of Power Sources*, 2001;**95**:248–254.

electrode. Additionally, the loss of hydrogen leads to a loss of water and a higher concentration of the acid, which loses volume. As the batteries are sealed, the loss of liquid cannot be compensated, and if this phenomenon is repeated, the battery will dry out and lose its capacity. If, however, the battery remains constantly under-charged, the risks of sulphation and softening of the active material increase, so it is advisable to carry out a compensation charge regularly. A recent study[6] outlines the advantage of keeping sealed batteries in an intermediate state of charge between 20% and 80% of their nominal state of charge (SoC) so as they remain in the zone of maximum efficiency (no loss due to the end of the charge), and suggests a system of management that applies a complete charge based on the preceding cycles.

For example, this system was tested on a gel battery operating between 40% and 70% SoC for 5500 cycles. An equalisation charge was carried out every 84 partial cycles, and at the end of the study, the capacity was still 95% of the nominal value at the start. This corresponds to around 1650 cycles at 100% deep discharge. The total charge/discharge efficiency (in Ah) was over 99% during the same period. Another trial gave similar results with more than 6000 cycles between 60% and 90% SoC with a compensation charge every hundred cycles, which corresponds to 1800 cycles at 100% deep discharge. This type of operation has another advantage for large systems of rural electrification where the compensation charge is provided by a diesel gen-erator: when the diesel operates every day, the battery is at maximum efficiency, accepting all the current produced, and less frequent compensation yields economies of typically 30% of the use of the generator. These phenomena of ageing by sul-phation are due to the increase in size of the $PbSO_4$ crystals on the positive electrode when the battery is discharged. These larger crystals allow less current from the contact grid to pass, and they can also fall off, reducing capacity, and in the case of an open battery fall to the bottom of the case. To avoid this, it would be useful to find a means of compelling the active material to remain in contact with the grid. Mechanical means applying pressure are not possible with AGM separators made of fragile glass fibre or with gels. An American company, Daramic, has developed a new acid jellying separator (AJS) that allows the application of mechanical pressure on the plate group without any deformation of the separator. A recent publication[7] describes the use of these separators and the increase in performance obtained when different pressures are applied on the plate group of the battery:

- 250 cycles with an AGM separator and compression of 30 kPa;
- 530 cycles with an AJS separator and compression of 30 kPa;
- more than 1500 cycles with an AJS separator and compression of 80 kPa.

This test was carried out at a discharge of C_5 (current = 1/5 of capacity) and at 100% discharge; the battery is considered 'old' when its capacity has fallen below

[6] R.H. Newnham, W.G.A. Baldsing, 'Benefits of partial-state-of-charge operation in remote-area power-supply systems', *Journal of Power Sources*, 2002;**107**:273–279.

[7] M. Perrin, H. Döring, K. Ihmels, A. Weiss, E. Vogel and R. Wagner, 'Extending cycles life of lead-acid batteries: a new separation system allows the application of pressure on the plate group', *Journal of Power Sources*, 2002;**105**:114–119.

80% of its nominal capacity. The test of the highly compressed battery was not finished at the time of the publication of the article.

Nickel batteries
Three different materials are used with nickel: cadmium, metallic hydrides and zinc.

- The future of NiCd batteries, faced with the probable outlawing of cadmium, is very uncertain.
- NiMH batteries are replacing NiCd batteries for environmental reasons.
- Few manufacturers still make NiZn batteries, and we do not have enough data to assess their usefulness in this review.

Nickel batteries will probably be never widely used in PV systems, but they will remain useful for some low temperature and portable applications.

Lithium batteries
Lithium batteries use a wide variety of different lithium materials and electrolytes. Among them are lithium/metal (Li-metal/titanium sulphide or iron), lithium/ion (carbon/Li_xCoO_2) and lithium metal polymer (Li-metal/V_6O_{13}). All these technologies have advantages and disadvantages.

- Li-metal batteries (Li/MoS_2) have been abandoned as they are dangerous and explosive in the event of internal short circuit. There are developments on the way to make them more stable.
- Li-ion batteries are much more stable in behaviour. The cells are around 3 or 4 V and the energy density is higher today for a longer-life battery. The electrolyte is based on organic solvent, and there is therefore no consumption of water through electrolysis or loss of liquid at the end of a charge; the battery can be completely watertight and needs no maintenance current.
- Li-polymer batteries use dry technology with the materials in leaf form stacked and rolled. Here the aim is to increase energy density even further for portable applications. Many materials have been studied, and the first models should soon be appearing on the market.

The properties of lithium batteries make them mainly suitable for portable applications where their high energy density and efficiency and low self-discharge rate are great advantages. They are unlikely to be widely used in PV systems of any size.

Supercapacitors and flywheels
These two technologies have properties that make them suitable for the same very short-term and long-life storage applications. They are more useful for their abilities to filter and smooth energy than for real storage. There is very little chance of them being one day used in applications of rural electrification.

Metal-air batteries
Rechargeable metal-air batteries require a complex infrastructure to produce. The low number of cycles can be increased by recycling the electrodes, which must be

regularly removed and replaced. The main interest of this technology is its high energy density, but it is probably limited to portable applications in the short term because self-discharge is very high.

Redox systems

Redox batteries use electrodes submerged in two liquids, which serve to store the energy. Lead batteries also operate by oxide reduction reactions, but the energy is stored in the lead plates and not in the liquids, as here. The size of the electrodes determines the exchange, and thus the power. A selective membrane allows ions to pass but prevents the solutions from mingling. Several different ion couples are possible such as Fe–Cr, Zn–Br and Br–S. The energy storage is carried out by the electrolytes kept in external tanks and which can be maintained for a long time without losses. To exchange the energy, the electrolytes are circulated by pumping, and the exchange occurs in the electrode tank and selective membrane.

Current applications for redox batteries are mainly as sources of peak demand energy for the grid. The large variety of technologies in competition makes it difficult to evaluate their usefulness, and their application for renewable energy storage is not yet competitive.

Compressed air

The final technology described here, compressed air, is similar to the preceding. The input/output of current activate electromechanical transducers: an input compressor, and at output the hydraulic motor coupled to a generator. When electricity is supplied to any user there are inevitable losses relating to the vacuum generator.

The attraction of compressed air is that a large part of the technology is already developed and available anywhere: compressed gas storage and all the necessary components (cylinders, taps, valves, pipes, etc.) can be found in all countries. The only new elements are the adiabatic compressors, which recover the heat produced during the compression.

Compressed air is ideal if mechanical energy is needed, because in this case the discharge efficiency is over 90% with a good hydraulic motor: typical mechanical requirements in a rural setting are agricultural machines for treating cereals (milling, threshing, etc.), milking machines, cold (compressor), pumping and ventilation. Other mechanical applications may shortly be available, such as small transporters that can be quickly recharged (transfer of compressed air in less than a minute). There is also much interest in using compressed air for cars.

Technology

Two main systems are currently being developed: type A where the compression/expansion takes place in the storage tank by displacement of liquid filling up to half the tank, and type B where the compression/expansion takes place in an energy transformer with recovery of heat, which enables the whole storage volume to be used with the compressed air and so reduce the storage volume by a factor of 10. Type A exists already and can be installed with elements widely available (compressor/hydraulic motor, gas cylinders, pressure gauge, DC generator). Type B is more complicated because it needs a new type of compressor/motor incorporating heat exchange, which will enable it to function anywhere at a constant temperature. Type A has a total electrical efficiency of over 73% at 3 Wh/l of storage, and type B should have an efficiency of over 60% for 35 Wh/l at 300 bar. These efficiency rates should rapidly improve in the short term with the development of hydraulic motors and DC motors. Life expectancy is very important and storage cylinders are guaranteed for 100,000 cycles; for calculations, we have assumed 20,000 cycles in 20 years. The self-discharge will depend on the starting method selected: either a flywheel is periodically fed by compressed air (pulse-width modulation, PWM), or an intermediate storage involving supercapacitors, for example, is used (under study at l'École Polytechnique Fédérale de Lausanne [EPFL], Switzerland). With a flywheel and a current DC motor, losses are below 40 W for an output of 1500 W. Some other advantages of these technologies are precise control of storage by a simple pressure gauge, components well-known and available anywhere; the possibility of long-term storage without losses (full-sealed cylinders), increase the volume of storage at any time; the mixing of stocks (ages and sizes) without consequences; and the possibility of using energy directly in mechanical form.

This technology is a real alternative to current lead batteries with major advantages on the environmental level: no heavy metals, no acid, a very long life, no rapid ageing at high ambient temperatures and the lowest current energy cost of storage, estimated at €0.02/kWh. However, it is hard to say if it would ever succeed in ousting traditional batteries that are so widely used everywhere.

5.1.2   Charge controllers

In a stand-alone PV system, the regulator generally represents less than 5% of the total cost of the system, which at first sight may suggest that this component is not important. On the contrary, its function is essential and its quality will deeply influence the final cost of the energy produced. The battery remains the most delicate part of the system, and its maintenance and the quality of its control have an important influence on its life and so for the price of the final kWh generated.

Up to now, relatively little has been done to optimise this component, which is often manufactured in developing countries for small SHS. A recent study[8] comparing 27 regulators on the market has shown that the techniques used to monitor the battery are very diverse and that the typical control parameters vary widely. There is therefore no unanimity today among designers on the best way of regulating a PV system battery.

The charge controller is the central function of the stand-alone PV system since it controls the energy flux. Its function is to protect the battery against overcharging (solar) and deep discharge (consumer). It also has to monitor the safety of the installation (overcharging, alarms, fuses, changes in polarity). In more elaborate systems, it can also activate the recharge from other energy sources (top-up generator, wind power, hydro). In some cases, it can also adapt the impedance (Maximum Power Point Tracker, MPPT).

---

[8] IEA PVPS Task 3, *Management of Batteries Used in Stand-Alone PV Power Supply Systems*, Available from http://www.iea-pvps.org

Additionally, it can supply information on the state of charge of the batteries and the operating parameters of the system.

The charge controllers of stand-alone PV systems fall into three main groups:

- the *series* regulators, which include a switch between the generator and the battery to switch off the charge;
- the *shunt* regulators, which short-circuit these solar generator when the charge is complete;
- the MPPT, which uses a special electronic circuit enabling maximum power to be permanently drawn from the panel array.

A discharge regulator is generally added to all three types of circuit to prevent deep discharge of the battery.

We now describe the two main functions of charge controllers – the control of the charging and discharging batteries. Then we outline the most recent technologies with some recommendations on the most appropriate choice according to the system planned. Finally, we give recommended typical values in the most recent publications.

5.1.2.1 Functions

We limit our descriptions below to controllers used with lead batteries, still the type most used with PV panels.

Charge control

The control of charging is the most critical function affecting the life of a battery. The difficulty of the operation arises from the nature of the energy generated, which is not always available. To guarantee long life for a battery, it would need to be recharged 100% after each discharge. The sporadic nature of sunshine means that it is not always possible to make a complete recharge, and the battery often has to remain several days in a state of 'medium' charge, which can reduce its life in the long term.

There are several possible techniques for regulating a battery, by measuring either the input voltage or current. In fact, measuring voltage is much easier and most charge controllers use this parameter.

The voltage of a battery charged at a constant current increases in a linear fashion almost until it reaches the end of the charge when suddenly it increases much more rapidly because the active material is almost completely transformed and electrolyte begins to give off gases. This gasification arises from the decomposition of the water in the electrolyte to hydrogen and oxygen, caused by electrolysis. If this phenomenon is allowed to continue, the battery will become overcharged, which will accelerate the corrosion of the lead, cause the loss of electrolyte and damage the lead plates. The main function of the regulator is to prevent this overcharge.

Some light gasification is, however, necessary and recommended for open batteries. When this phenomenon begins, the battery is not yet completely charged and a part of the input current will enable the charge to be completed while the rest produces electrolysis, and gases passing through the electrolyte will gently agitate it, allowing the acidity to be homogenised. If the final stage is never reached, the electrolyte will become stratified, with the concentration of the acid being greater at the bottom of the battery, accelerating the corrosion of the plates and also creating sulphation, the transformation of part of the active lead into hard crystals, which reduce the energy storage. These two phenomena will result in a reduction of battery life. This gasification is recommended for open batteries with completely liquid electrolyte, but not for sealed batteries containing gel or an AGM material absorbing the acid, which prevents the agitation of the electrolyte.

Figure 5.9 shows a typical voltage curve for a battery over time. Two characteristic values are used to control the charging process: the end-of-charge voltage (V_{fc}), or upper disconnect, and the recharge voltage (V_{rc}), the value at which the charging process begins again.

The simplest charge controllers work by simple on/off switching and use these two voltages to stop or recommend the charge process. To switch off the current, a relay, a bipolar transistor or a MOSFET (metal-oxide semiconductor field-effect transistor) is used.

The on/off charging with these two switching voltages functions reasonably well for systems with a large storage capacity where the charge current is below C/20 or 1/20 of the storage capacity. When a battery is charged to a higher current, the high current multiplied by the internal resistance of the battery will generate a higher charge voltage, and the switching voltage values will no longer be valid as the charging stops too quickly, which will lead to a battery never reaching its full charge. If the full charge voltage is increased, a lot of gas will be produced with the unfavourable consequences mentioned earlier in this section, and the necessary final charge that calls for lower current will not be realised.

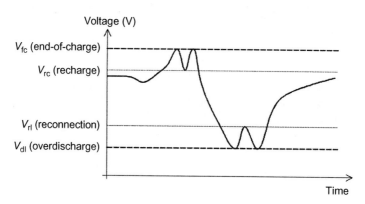

Figure 5.9   Battery voltage during charge cycle

For systems with a high charge current compared to storage capacity ($I > C/20$), it is better to use a regulator with constant voltage. However, the unpredictable nature of solar current requires the installation of what is called a constant voltage regulator. In the first charging phase, the panels provide all they can produce, and when the battery reaches its end-of-charge voltage, this voltage is maintained for a short time to enable the charge to be completed, and subsequently is reduced to maintain the battery at a floating voltage.

This type of regulator is more sophisticated than the two voltage on/off switch and it costs more. Older models dispersed the energy not used by the battery during its constant voltage phase, which required bulky dissipaters. More recent models generally use MOSFETs working with PWM; dissipation is then reduced to switching losses and the ohmic loss produced by the charge current across the transistors. Figure 5.10 shows the flow of current and voltage for a modified constant voltage regulator.

Boost charge
Determining the optimal end of charge voltage is tricky: to get round this problem, modern microprocessor regulators generally use an end-of-charge voltage that is not too high to avoid any corrosion and limit water losses. Subsequently, to guarantee that the acid remains homogenised, they regularly make what is called a boost charge that raises the voltage higher at the end of charge for a limited time for liquid electrolytes only. The frequency of this boost charge depends on the battery manufacturers but, in the literature,[9] a boost charge is recommended every two or three weeks for deep-cycle batteries, whereas once a month is adequate for batteries with float voltage. It is also recommended to carry out a boost charge to completely top up the battery after the overcharge regulator has operated.

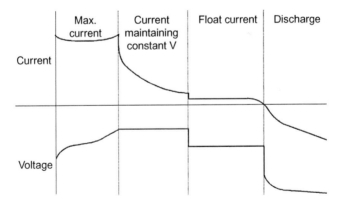

Figure 5.10 Constant voltage regulator

Equalisation
Another phenomenon appears with time: in a deep-cycle battery, differences of capacity between cells, variations in temperature between cells (the internal cells will be warmer than those on the edge of the case) and variations in self-discharge

[9] IEA PVPS Task 3, *Recommended Practices for Charge Controllers*, Available from http://www.iea-pvps.org

create divergences of the state of charge between cells. To compensate for these variations, it is recommended to carry out an equalisation charge that will enable all the cells of the battery to receive a complete charge. Equalisation is a sort of longer boost charge at low current and it allows a cycle to be restarted with a battery where all the cells are fully charged. This effect is increasingly marked when there are a large number of elements (high nominal voltage). This high-voltage charge is not recommended for sealed batteries that cannot lose liquid without at the same time losing some capacity. But it is still possible to equalise a sealed battery system every 100–150 cycles, probably equivalent to one or two equalisation charges per year.

Thermal compensation

The electrochemical activity of a battery is strongly dependent on temperature. At high temperatures, the battery accepts the charge more easily and begins its gasification at a lower voltage. If in a hot country a regulator adjusted for temperate climate (20–25 °C) is used, the high disconnect voltage will accelerate corrosion and lead to a loss of electrolyte. This phenomenon is especially important when the charge current is high and when the ohmic losses of the battery increase its internal temperature. For a battery used in countries that are cold in winter, on the other hand, fixed voltage regulation will prevent the total charging of the battery that will age prematurely because it will never be completely charged. At temperatures below 0 °C, another possible danger is freezing of the electrolyte, which depends on the state of charge of the battery (Table 5.6). Thermal compensation is very important for sealed batteries used in hot countries: overcharging will cause a loss of electrolyte, which cannot be compensated.

To guarantee precise measurement of temperature, it is recommended to attach a temperature probe to the battery case. A temperature probe inside the battery case will often be disturbed by internal heat from power components (diodes and transistors), which would distort the temperature measurement. Temperature probes need to be encapsulated in a material resistant to acid, be accurate within less than 2 °C and be robust and cheap. Should the temperature probe fail (open or short-circuit probe), the regulator should operate with voltages set in the centre of the operating range (25 °C). The typical thermal compensation value is −5 mV/°C, which corresponds to −30 mV/°C for a 12 V battery.

Charge, boost and equalisation voltages

Table 5.5 shows the recommended charge voltages for the most widely used batteries. A well-adjusted regulator does not guarantee that the battery will be well controlled: the measurement of voltage must also be reliable and precise. If the charge current is high, it must be ensured that the fall of voltage between the regulator and the battery is negligible (adequate cable sections). Otherwise a regulator with a separated measurement of battery voltage must be used (current/voltage cabling separated). This problem is important for on/off regulators, which, in case of significant fall of voltage between the regulator and the battery, will never succeed in completely charging the battery. For constant voltage regulators, the problem is less serious because the regulator imposes its own voltage while reducing the current, which allows the final charge to take place at the correct voltage.

Table 5.5   Charge and boost voltages (per battery cell)

Type of regulator	Type of battery			
	Liquid electrolyte		AGM	Gel
	Open		Sealed	
	Sn–Pb	Ca–Pb		
On/off				
V_{fc}	2.40	2.45	2.35	2.35
V_{rc}	2.25	2.30	2.20	2.20
Constant V. reg.				
V_{fbo} (boost)	2.50	2.55	–	–
V_{fc} (charge)	2.35	2.40	2.35	2.40
V_{fc} (float)	2.25	2.25	2.25	2.25
	Equalisation for half a day every 30 days			
V_{feg} (equalisation)	2.55	2.55	–	–

The ratio between V_{fc} and the V_{rc} (end of charge/recharge) is important: if the gap between these voltages is too high in an on/off regulator, the time of the end of charge will be seriously extended with the risk of never charging adequately. If the ratio is closer, the regulator will switch more frequently and end the charge more rapidly (which may shorten the life of the switch if it is a relay).

The choice of regulator values depends on the type of application: for a system near a dwelling house and regularly checked, voltages may be set slightly higher, guaranteeing a complete charge, if the user can top up the electrolyte level. In the case of 'uninhabited' systems such as automatic applications (metrology, tele-communications, etc.) or when the user does not have the skills to manage their own system, it would be better to use the values in Table 5.5. Equalisation voltage depends on the technology of the battery and of the alloy used (lead antimony or lead calcium). If catalysts (recombining the oxygen and hydrogen into water) are installed on an open battery, the equalisation of the battery must be avoided so as not to damage them by saturating them.

Regulator voltages and type of application
The choice of regulator voltages depends on the type of application – floating stand-alone, cycled stand-alone or hybrid.

A system where the battery is almost always fully charged (telecoms repeater, emergency telephone, etc.), mainly autonomous (>2 weeks), or floating voltage requires the lower regulator voltages, with equalisation only once a month or after a complete discharge cycle.

A system where the battery is cycled (for a main or secondary residence example) and with autonomy below 2 weeks should use a regulator with higher voltages and regular equalisation to avoid the stratification of the electrolyte and the divergence of individual cells.

For hybrid applications where another energy producer such as a diesel generator is connected to the system, regulator voltages may be lower because the auxiliary generator will regularly completely charge the battery.

In systems with major energy requirements during the day using an on/off regulator, when a consumer uses a lot of current at the same time as the battery reaches its full charge at V_{fc}, the current from the panels is disconnected and the battery voltage then falls abruptly, and the system may become unstable with rapid variations of voltage, with the current from the panels being switched off and on again too rapidly. In this case, there may be an advantage in lowering V_{rc} to limit this type of oscillation. But it is certainly more advantageous in this type of system to opt for a more sophisticated constant voltage regulator.

Discharge control

Solar regulators not only control the charging of the battery but also monitor the state of the battery when current is withdrawn from it. In order to avoid deep discharge of the battery, which can severely reduce its life, an overload circuit is added that disconnects the appliances when the battery voltage falls below a critical threshold. This threshold is fixed according to several criteria – expected life, ambient temperature and current level. If the point chosen is high enough, the battery will last longer but the unused capacity battery will reduce its total capacity, and for the same level of power supply, the number of batteries will have to be increased. The overload set point also depends on the age of the battery: as it ages, the battery loses some voltage, and if the chosen set point is high, after some years, the useful capacity will have reduced, which can sometimes cause problems. The ideal regulator should also take account of this effect and lower the set points according to the age of the battery. It is a question of finding the best compromise between the cost of the system and battery life. For a system that is not easily accessible, a larger battery would be more sensible insofar as transport and labour costs for its replacement is considered.

Types of control systems

Systems for *professional applications* in telecoms and metrology are designed to operate with high reliability and in general throughout the winter (see Sections 5.3.2 and 5.6.1). In summer, the battery normally remains close to full charge, and at the end of autumn, a large discharge begins that ends in spring. In this type of system, the battery will not age on account of the number of cycles, but through internal corrosion, sulphation and loss of electrolyte. For these applications, overload prevention is not really useful, since the system is always oversized for reasons of service reliability.

In *domestic systems* (see Section 5.3.1), the same reliability and absolute availability of energy are not demanded. In this case, overload prevention is a useful means of facilitating management and preventing a deep discharge resulting from an abusive use of the energy available.

For any system, overload prevention is useful to protect the battery if the generator breaks down, if the panel-regulator connection is broken or if the charge controller does not allow sufficient current to flow to the battery.

Overload parameters

In cold countries, the first parameter that must be respected is the possible freezing of the electrolyte below -7 °C. Table 5.6 gives the limits that should not be exceeded according to temperature and the electrolyte used.

Table 5.6   Maximum discharge according to temperature

Density of the electrolyte according to state of charge (100%/0%)							
E-Ch 0%	1.10	1.12	1.15	1.10	1.12	1.10	1.12
E-Ch 100%	1.30	1.30	1.30	1.25	1.25	1.20	1.20
Temp (°C)	**Maximum depth of discharge permitted (%)**						
-5.0	100	100	100	100	100	100	100
-7.5	100	100	100	100	100	100	100
-10.0	93	100	100	91	100	87	100
-12.5	87	96	100	82	95	73	92
-15.0	81	90	100	74	86	61	77
-17.5	75	83	100	67	77	50	63
-20.0	70	78	93	60	69	40	50
-22.5	65	73	87	54	62	31	38
-25.0	61	68	81	48	55	22	27

The second parameter is battery life. It is generally possible to make a rule of 3 to estimate the number of cycles possible in relation to the depth of discharge. For example, a standard solar battery for which one could expect 200 cycles at 90% discharge and a moderate temperature (20–25 °C) should last 300 cycles at 60% discharge, the number of cycles parameter times depth of discharge being fairly constant. The choice of the depth of discharge in this case is rather linked to the parameters of the cost of access and exchanging the battery. This is valid for temperate countries. In a hot country on the other hand, the life of the battery is generally limited by internal corrosion that is considerably accelerated by the higher temperature; in this case it would be best to strongly cycle the battery to draw on all its capacity during the shorter life expectancy.

The third parameter is the type of battery used. The maximum discharge value recommended by the manufacturer should not be exceeded. The different lead alloys, if sealed or open construction, and the electrolyte additives influence the robustness and the ability of the battery to achieve its full capacity after a deep discharge.

Load-shedding voltages

From the three parameters mentioned above and the type of system, a depth of discharge can be selected that corresponds to a load-shedding voltage. However, this voltage is not easy to determine because it is influenced by the level of discharge current compared to its capacity, by ambient temperature and by the state of the battery (age, sulphation, corrosion). As the discharge current passes through the internal resistance of the battery, it lowers the output voltage; therefore, the higher this current is, the more voltage drops without this effect being related to the real state of charge of the battery. For solar systems with high autonomy, the current is in general below C/30, and in this case the fall in internal voltage can be disregarded and the controller can be regulated at a lower current value, which will improve reliability. The age of the battery strongly influences the output voltage by raising the internal resistance: by regulating the load-shedding voltage of a controller to 11.8 V, which theoretically corresponds to around 70% of discharge depth, for a standard new battery at C/10, a residual capacity of 40% can be measured, whereas for the same battery after ageing, the residual capacity at 11.8 V was still 70%, and there only remained 30% useful capacity.

Also, the type of appliances connected must be taken into account: appliances using motors are inductive, which considerably increases their current when they start. To avoid load shedding due to the motor starting current, the disconnection must be postponed and then kick in after several seconds at a voltage below the chosen set point. As a general rule, it is prudent to choose a load-shedding voltage at a low discharge current and 25 °C ambient temperature, which increases reliability at low temperatures, with the disconnection being made earlier. Table 5.7 gives some values of load-shedding voltages (V_{dl}) according to the depth of discharge. These values are indicative and correspond to those for a new battery at 25 °C.

Table 5.7   Recommended load-shedding voltages

Depth of discharge (%)	Discharge current		
	C/100	**C/20**	**C/10**
10	2.14	2.11	2.08
20	2.12	2.09	2.07
30	2.10	2.07	2.05
40	2.08	2.05	2.04
50	2.05	2.03	2.01
60	2.02	2.00	1.99
70	2.00	1.98	1.96
80	1.96	1.95	1.93
90	1.92	1.91	1.89
100	1.80	1.80	1.80

Reconnection voltages
When the regulator is switched off, the battery has the time to recover its capacity and for the good of the system, the user circuit should only be re-switched when the battery is fully charged. If the appliance that has caused the load shedding is rapidly

reconnected, the disconnection will probably occur again quite quickly, which will cause the system to oscillate at a low charge level and will keep the battery in a state that is likely to accelerate sulphation. It is recommended to only permit the discharge if the battery has been charged for several hours and has recovered a stable state.

Table 5.8 shows the values to apply according to the level of current and the desired capacity for the reconnection voltage (V_{lr}).

Table 5.8   Reconnection voltages after load shedding

State of charge	Discharge level	Charge current			
		C/10	C/20	C/50	C/100
0	100	2.08	2.05	2.01	1.99
10	90	2.09	2.07	2.03	2.02
20	80	2.12	2.1	2.07	2.06
30	70	2.15	2.13	2.1	2.09
40	60	2.19	2.17	2.14	2.13
50	50	2.23	2.21	2.17	2.16
60	40	2.27	2.25	2.21	2.2
70	30	2.34	2.32	2.27	2.26
80	20	2.43	2.43	2.34	2.32
90	10	2.61	2.6	2.47	2.46

To simplify matters, a low charge current can be selected, and if it is higher, which will increase the voltage more rapidly, it may be assumed that it will be able to make a quick connection to the receptors while continuing the charge. For greater reliability, it is recommended to apply to these values the same thermal compensation as that used for battery charging (-5 mV/°C/cell).

5.1.2.2   Regulator technologies

When PV was first introduced, two main families of regulators were used: 'shunt' models and 'series' models. With the more widespread use of microprocessors, new techniques are appearing and each manufacturer puts forward commercial arguments that, in general, are very difficult to verify, because battery testing is reserved for highly specialised laboratories.

'On/off' shunt regulator

The current from the solar panel is sent to a power switch in parallel with the battery when full charge is reached. The principle of the circuit is a simple shunt: all the current from the panel flows normally into the battery, and when the disconnect set point is reached, all the current goes to the power switch. It is essential to add a diode between the power switch and the battery in order not to short-circuit the battery. This diode also serves to block any nocturnal current that may flow from the battery towards the panel. The most sophisticated models use a Schottky-type blocking diode with a fall of voltage of the order of 0.5 V, around twice as low as for an ordinary silicon diode. The power switch used is usually a MOSFET, sometimes a bipolar transistor or even a relay. A MOSFET with low flow resistance (R_{ds} on) is more suitable than a bipolar because it wastes less energy. A relay offers a very low flow resistance (often lower than a MOSFET), but it will need to be delayed so that it doesn't rapidly reach the end of its switching life, which creates problems if the charge current for a given battery capacity is relatively large; in this case, the end of charge will be hard to reach. The disadvantages of shunt regulators are as follows:

- the circuit breaker receives the total voltage of the panel, which means there must be protection against overvoltages,
- the thermal dissipation of the power switch may be considerable at high current values;
- in short-circuiting the panel when the battery is full, the risk of hotspots increases, the reverse voltage flowing to the shaded cell being higher (see Section 3.1.4).

Table 5.9 summarises the advantages and disadvantages of each regulation technology.

Linear shunt regulator

This type of regulator maintains a constant output voltage when the battery reaches its full charge. The advantage is that the total charging of the battery is guaranteed, the disadvantage is that the unused panel power at the end of the charge needs to be dissipated by the parallel transistor, which limits this type of regulator to small outputs. For small systems, sometimes a simple Zener diode is connected in parallel with the battery. In this case, a blocking diode is not needed although it is to be recommended in temperate or cold countries (Figure 5.11).

'On/off'/linear series regulator

This regulator is very widely used and has replaced the shunt regulator (Figure 5.12). Here the power switch is in series with the battery and it opens when the end of charge is reached. The same switches can be used as with the shunts but the relay may be a diverting switch to allow the current to flow to another appliance when the battery is charged. This property is sometimes used in systems where all solar power needs to be recovered. The auxiliary appliance is often ventilation, pumping or possibly heating in cold countries.

In the linear model, an additional source of current is added in parallel with the switch to achieve the equalisation of the battery by floating; the source may be a controlled transistor or a simple resistance in series with a diode. The blocking diode recommended in temperate countries may or may not be part of the regulator; sometimes it is directly mounted on each string of panels. The disadvantage in comparison with the shunt type is that the power switch depending on its flow

Table 5.9   Advantages and disadvantages of different regulator technologies

Type of regulator	Method of charge	Advantages	Disadvantages
Shunt – power switch	On/off	– Low flow resistance between the panel and the battery – Simple, reliable if well sized	– Dissipation of power switch at I_{cc} of panel – Blocking diode indispensable – Higher hotspot voltage – Difficult to reach end of charge – Power switch voltage higher in case of overvoltage
Shunt – linear	Constant voltage	– Optimal end of charge – Low flow resistance between the panels and a battery	– Considerable thermal dissipation – Blocking diode indispensable – Higher hotspot voltage
Series – power switch	On/off	– Dissipation of power switch when $I_{max} < I_{sc}$ – Simple and reliable – Overvoltage on the power switch reduced to one battery voltage	– Difficult to reach end of charge
Series – linear	Constant voltage	– Optimal end of charge – Overvoltage on the power switch reduced to one battery voltage	– Considerable thermal dissipation – Higher flow voltage
Series – PWM	Constant voltage	– Optimal end of charge – Reduced thermal dissipation	– Higher flow voltage – More complex electronics – May cause interference on sensitive equipment nearby
MPPT	Constant voltage	– High efficiency at all temperatures	– Cost May cause interference on sensitive equipment nearby

resistance adds an additional fall in voltage between the panels and the battery. The advantages are as follows:

• lower output voltage at the power switch (reduced from the battery voltage);
• lower hotspot risk (lower reverse voltage passed on from battery voltage).

PWM regulator
This type of regulator attempts to combine the advantages of the two previous technologies by using an active power switch modulated by pulses of variable size

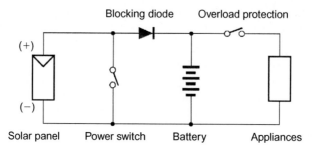

Figure 5.11   Diagram of shunt regulator

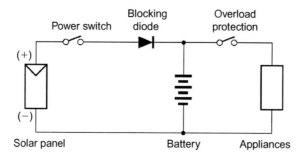

Figure 5.12   Diagram of series regulator

(PWM). The advantages are that a constant temperature can be maintained at the battery input to complete the charge while dissipating through a transistor (generally a MOSFET) only the switching losses and losses due to flow resistance. Either shunt or series technology can be used in this case, but most manufacturers use series. Since the recharging of the battery is a relatively slow chemical process, this technique must be used at fairly low frequency, a maximum of the several hundred hertz in order for the charge to operate, otherwise the current risks escaping in parallel with the battery through the skin effect.

MPPT regulator
In this regulator, a circuit permanently measures the voltage and current of the panel to draw energy at the point of maximum power (MPPT). This enables the maximum of energy to be recovered, irrespective of the temperature and solar radiation. In general, these regulators work either by raising or by reducing the voltage. One circuit adjusts the demand at the maximum power point on all the panels while the second circuit adapts the current and voltage to the type of battery used. The advantage of this type of regulator is that it enables the panels to work in a wide temperature range and so to recover the considerable excess voltage in winter when the maximum power point can exceed 17–18 V in a 12 V system.

This technique is only suitable for systems of several hundred watts where the energy gain compensates for the higher cost of the regulator. Also, before choosing this equipment, the losses associated with MPPT and the DC/DC conversion must be appreciated to ensure that the investment is worthwhile.

5.1.2.3   Regulator accessories and special considerations

The many manufacturers of charge controllers often offer additional functions apart from battery regulation. These accessories or specialities can be very useful and particularly recommended in certain cases depending on the types of solar system and appliances. We give below three categories of accessories according to their degree of usefulness – basic, recommended and sometimes useful.

Basic accessories
Battery state indicators
One needs to know at least whether the battery is in a 'normal' state, that is, between full charge and load shedding (often not indicated, the indicators being reserved for extreme positions), if the battery is full or if the load shedding is activated. These three states are generally located by LEDs – green for a full charge, red for load shedding and sometimes yellow to indicate charging from the panel. More sophisticated models have analogue or digital volt/ampere metres and show whether the regulator is in an equalisation or boost charge state.

Thermal compensation
This is a very important accessory to guarantee long battery life for all systems with wide temperature variations. Systems in countries with a stable average temperature ($\pm$ 5 °C) can dispense with it but it must be ensured that the regulator is adjusted to the local ambient temperature.

Delayed load shedding
This delay is essential to allow inductive appliances to start. Load shedding should only kick in if the battery voltage remains below the disconnect set point for several seconds. More sophisticated regulators will adjust the disconnect voltage according to the level of current to take into account falling voltage due to the internal resistance of the battery (Table 5.7).

Recommended accessories
Adjustable disconnect set points
This function is useful if the system is commissioned by a specialist skilled in regulating set points according to the recommendations of the battery manufacturer and the anticipated discharge characteristics. The set points should be able to be adjusted by small dip switches, calibrated potentiometers or by a regulator with an indicator on the LCD display (regulators with microprocessors).

Separate measurement of battery voltage
This characteristic is useful for systems working with high currents. When the fall in the regulator-battery voltage can exceed some hundreds of millivolts, the

separate measurement of voltage avoids load-shedding oscillation and insufficient battery charge, as the real voltage of the battery is always below the voltage measured at the 'current' terminals of the regulator. But if this function is available, the current cables must be sized with an adequate cross section so as not to lose too much thermal energy during power transport.

Load management
This enables priority to be given to certain energy needs. For example, for the supply of a medical dispensary, the first priority will be given to maintaining the vaccine refrigerator and lighting will be rated as less important. It should be possible to make this type of choice either for priority reasons as indicated above or for other considerations such as taking account of the state of charge of the battery before starting certain equipment (for example, ventilation, which is not indispensable) or according to the time to switch on a pump when the battery is charged or at 1 o'clock in the afternoon.

Sometimes useful accessories
Data logging and modem access
Telephone access function is reserved for large systems controlled remotely. The logging of data enables graphs and operations to be accessed in case of breakdown so that the problem can be better understood and put right. This is very useful when it is necessary to prove that appliances are using more than the solar production and the PV system is not the cause.

5.1.2.4   Criteria of choice of regulators
We list below the essential parameters determining the choice of a regulator according to its use and environment. Table 5.10 summarises these criteria.

Sizing
The first parameter to consider is the power of the regulator or the maximum current that it can control for a given nominal voltage. This value will need to take into account special conditions of irradiation that can generate instant solar values in excess of 1 kW/m^2:

- the albedo value can be quite high at high altitudes with snow or in the presence of reflectors;
- there can be a concentration effect when the Sun comes out in the middle of brilliant cumulus clouds, when the instantaneous radiation can increase considerably (as much as 1350 W/m^2 has been measured in Europe in spring).

A safe value for determining the regulator current is 1.5 times the short-circuit current I_{cc} for a shunt regulator and 1.5 times the nominal current I_m for a series regulator.

For voltage, the regulator will need to support around twice its nominal voltage, the value that is close to the open circuit voltage of the panel (V_{oc}) at low temperature.

Table 5.10   Choice criteria for regulators

Type of controller	On–off/constant voltage – shunt/series
Charge process	Linear/PWM
Ambient temperature	°C
Nominal voltage	V
Nominal current (×1.5)	A
Thermal compensation	Internal/external probe
Losses at no load	As % of solar power
Blocking diode	Schottky/bipolar
Disconnect voltage	V_{fc}
Recharge voltage	V_{rc}
Load-shedding voltage	V_{dl}
Reconnection voltage	V_{rl}
Boost charge voltage	V_{fbo}
Equalisation voltage	V_{feg}
Adjustable set points	V ± x mV
Types of switch	Relays/semiconductors
Load management	Priorities, clock, etc.
Protections	Overcharging/polarity
Installation facilities	Section, type of contacts
Environment	Case, material
Reputation of manufacturer, after-sales service	
Dimensions, weight	
Cost and guarantee	

Internal consumption

The no-load current of regulators varies typically between 1 and 25 mA. It is useful to know the value of this current in order to calculate the losses that this represents over a year, for example. At a latitude of 46° in Europe, some 1000 kWh/m^2 can be collected annually. With a small regulator consuming 2 mA, the no-load current represents an annual loss of 17.5 Ah. If this regulator is used for a small system employing a 10 W panel generating an average of 0.6 A, or 600 Ah/year, the losses from the regulator represent 3% of the energy produced. The no-load losses should not exceed a few percent of the power generated.

The other losses of the regulator come from the flow resistance of the switch (series regulator) and the blocking diode. It is better to choose a regulator with a Schottky diode and sometimes a shunt regulator for small systems.

Setting set points

The disconnect and load-shedding set points of the regulator should remain stable at ±2% of their nominal value during the life of the system. It must also be ensured that the fall in voltage between regulator and battery remains negligible for the accuracy of set points in a two-conductor installation, otherwise it will be necessary to use a regulator with a separate voltage measurement (four conductors).

Protection

All the cables arriving at a regulator need to be protected against transitory overload by adequate components (Zener diodes, lightning protection, etc.). For systems installed in exposed places, in the mountains for example, it is recommended to install surge protection on all conductors before they enter a building (see Section 5.1.4). If possible, use elements with visible indications of operation: some lightning conducting components[10] are equipped with an indicator that changes from green to red when its nominal power is lost after an electric shock.

Protection against reverse polarity is also indispensable.

In the case of absence of the battery, a shunt regulator should be able to dissipate the power from the panels: this is a frequent cause of breakdown if the circuit has not been carefully designed, as the system oscillates around the nominal voltage with a transistor often poorly supplied (insufficient grill voltage), anticipating too much. This problem does not exist for series regulators, which in the absence of a battery can supply the open panel voltage to appliances.

An advanced charge controller should incorporate thermal protection against excessive temperature and disconnect the appliances if the battery is not connected.

In cold and temperate climates, a blocking diode is recommended (if possible of Schottky type). For hot countries, the gains and losses of this component can be evaluated according to the nocturnal current of the panels chosen.

Methods of installation

The connecting terminals should be easily accessible and allow adequate size of cabling. This criterion is often not respected by cheap regulators, making intermediate terminal blocks necessary and increasing the final cost of the regulator. Fixing on the wall makes it easier to achieve the correct ventilation for the regulator dissipaters.

Guarantee and certification

Regulators are generally guaranteed for 1 year or more depending on the manufacturers. The presence of an adequate local after-sales service is also a criterion of choice. For certification, reference may be made to the decentralised rural electrification directives of the EDF (Electricité de France). Specification C7 concerns charge controllers.[11]

5.1.3   Converters

Converters are used for adapting the DC voltage from the panels or the batteries to supply appliances working either on a different DC voltage or an AC voltage. Today, most converters are electronic but one can still find generators supplying 230 V AC powered by DC motors. We describe later in this section some types of DC/DC converters, useful for supplying small appliances or chargers from a

---

[10] http://www.dehn.de

[11] Directives générales pour l'utilisation des EnR dans l'Electrification Rurale Décentralisée (Directives ERD) (Juin 1997).

battery, and also stand-alone DC/AC inverters. We will not discuss the converters directly coupled to solar panels, such as inverters for submerged pumps, because these elements are directly linked to their appliance. Inverters designed for connection to the grid have been dealt with in Section 4.2.3.

5.1.3.1 DC/DC converters

This type of component is used to transform the voltage of the batteries to a different DC voltage to feed a special appliance such as a mobile phone charger, a radio or a laptop computer.

Only described below are active converters with a good efficiency; passive converters that supply an appliance with lower than nominal voltage by dissipating the difference are not described. Two types are possible: 'upward' converters to increase voltage and 'downward' converters to lower the voltage.

Upward converter
The typical need for such an appliance is to convert a 12 V supply for a laptop computer that may call for an exotic voltage such as 19 V. Figure 5.13 is a diagram of the typical components of an upward converter. There also exist converters generating a voltage of 300 V DC to supply appliances at a nominal voltage of 230 V AC, which are equipped with their own primary transformers, such as low-energy lamps, modern television sets, computers and computer peripherals. The advantage of producing direct current for this type of appliance is that the wave produced is not deformed and that by not converting to 50 Hz, fewer harmonics are generated. This type of converter exists in a very compact form (Figure 5.14) with a good performance: an 18 W model supplies from a 12 V source, a compact fluorescent lamp of 15 W with an efficiency of 92%. Another advantage is that, if necessary, several converters can be operated in parallel to supply the larger appliance.

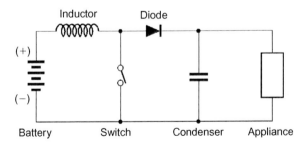

Figure 5.13   Diagram of an upward DC/DC converter

Operating principle
When the switch is closed, the inductor stores the current from the battery. When the switch is opened, the interruption of the current in the inductor creates an overvoltage that is diverted to the condenser and the appliance: the diode prevents

Figure 5.14   On the right, an 18 W–12 V/300 V DC/DC converter, and on the left, a 60 W–12 V/300 V DC/DC converter

any current reversal. The condenser smoothes the output voltage and the switch is controlled by an electronic device that permanently measures the output voltage and current to adjust the frequency and the size of demand of the switch and limits the current to a value safe for the components. There are integrated circuits including practically all these components except the inductor, the condenser and the switch transistor if the current exceeds several amperes. Typical efficiency is generally over 70% and can reach 85–90% for the best converters.

Downward converter

The second type of converter produces a lower voltage than that of the batteries and is normally used for the recharging of NiMH batteries, supplying radios and other small appliances (Figure 5.15).

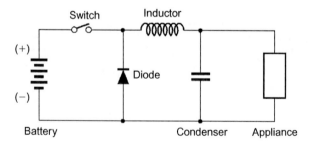

Figure 5.15   Diagram of typical downward DC/DC converter

Operating principle

When the switch is closed, the current flows from the battery to the appliance through the conductor and when the switch is open, the output voltage of the inductor is reversed, which activates the diode that protects the switch. The electronic controls dictate the frequency and duration of operation of the switch according to the desired voltage and the maximum current possible. Typical efficiency is slightly superior to the preceding model and is generally between 80% and 90% for modern models.

These two types of converter operate at between 50 and several hundred kHz, which can cause some interference when used to supply an AM radio (frequent in Africa).

5.1.3.2 DC/AC inverters for stand-alone installations

For systems with a large number of light points or where cabling becomes too cumbersome, it could be attractive to work in 230 V AC, which gives a much wider choice of appliances available on the market; 230 V AC compact fluorescent lamps are available in a huge variety of shapes and powers at lower prices than DC lamps, and some 230 V AC refrigerators have consumption close to the best DC appliances at a much lower cost. For all these reasons, as soon as there is a large number of appliances, it is usually worthwhile to add a good inverter, as savings in the cost of appliances could outweigh the investment for the inverter.

However, care must be taken to avoid sine wave deformations of the alternating current by appliances with cut-off feed, which transform the signal provided into an increasingly square wave: for a system that needs to supply many lamps and an appliance with a motor (for example, a refrigerator), the signal will perhaps become too square to be able to easily start the motor. We have measured on a small sine wave inverter 12 V DC/230 V AC–150 W a no-load distortion of 5% and, when it is loaded with three 15 W CFL lamps, a harmonic distortion of 13% (Figure 5.16).

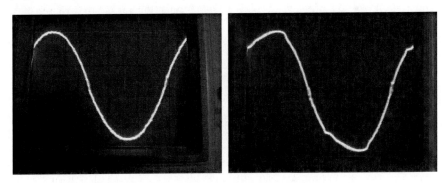

Figure 5.16   Sine wave of 230 V/50 Hz supplied by an inverter with no load (left) and loaded with three compact fluorescent lamps corresponding to 1/3 of its power (right)

Supplies with circuit switching even disturb the signal of the European power grid: we have measured a harmonic distortion of more than 3% (Figure 5.17), and the distortion deforming and flattening the peak of the signal is clearly visible on the oscilloscope.

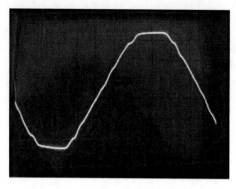

Figure 5.17   Sine wave of 230 V/50 Hz supplied by the grid (harmonic distortion 3.5%)

For the sizing of systems incorporating converters, account needs to be taken of the losses of the inverter in standby and its efficiency.

Modern inverters use microprocessor-based technology to generate a sine wave through pulses of variable width (PWM). These pulses control low-loss MOS power transistors feeding a transformer. At the transformer output, a filter removes any harmonics arising from the digital commands. The technique is widespread and the cost of such devices is falling. For example, reliable sine wave devices generating 150 W from 12 V can be found for around €200.

Historically, inverters can be classed as generators producing either a sine wave or a square wave or what is called a pseudo-sine wave. The choice of inverter will depend on the appliances that will make it work, the choice being based on valid criteria for any wave pattern. We would refer the reader to an interesting publication of the IEA PVPS Task 3 Group,[12] which lists a whole series of breakdowns and malfunctions arising in stand-alone PV systems and often caused by a poor understanding of the operation of inverters.

Criteria for choice
Output voltage accuracy
This figure is given in percent for 230 V AC. It is useful for certain applications if delicate electronic appliances are being powered. The inverter should be stable at all charges and whatever its input voltage.

Resistance to overcharging and reactive current
To succeed in starting certain charges, the inverter must often produce several times its maximum power for a brief period. As examples of charges with a difficult start, we may cite refrigerators (starting P five to ten times nominal P) and motors already charged mechanically (for example, pumps). This criterion is important, and if not respected can be the source of many problems.

---

[12] X. Vallvé, G. Gafas, *Problems related to appliances in stand-alone PV power systems*, Available from http://www.iea-pvps.org/products/pap3_011.htm.

Harmonic distortion

In certain sensitive appliances, presence of harmonics is an audible problem (hi-fi) and can disturb the operation of charging (motors). Non-sine wave inverters not only disturb the electromagnetic environment but are also a source of energy loss in the case of powering motors, for example. The following criterion (efficiency) will thus be strongly influenced for certain appliances by the production of harmonics.

Efficiency

This is undoubtedly the main criterion of choice. It is important to reduce energy loss to the minimum between the batteries and the 230 V AC charge.

The efficiency curve needs to be studied according to the inverter charge: modern devices have an efficiency of over 90% when the charge is between 5% and 10% of their nominal power. In a system with appliances of varied power (remote homestead, for example), the inverter often has to operate at partial charge when it is only powering lighting or other small appliances, and in this case, its efficiency at low power is very important. Manufacturers can supply devices incorporating two inverters to overcome this problem: these devices can operate between 0 and 100 W on a small circuit and automatically switch to a more powerful circuit when demand exceeds the requirements of the first circuit.

Consumption in standby mode

This is a very important characteristic for inverters that operate only occasionally but remain permanently under power. Often the energy used during standby is greater than that used by the appliances. An efficient inverter of 500 W/12 V, for example, consumes 0.4 A (around 5 W) on standby, which amounts to 9.6 Ah/day or 115 Wh/day. The best solution, if possible, is to stop the inverter between two utilisation cycles. To economise on energy, modern devices use consumer detection techniques to turn on the supply of all power once an appliance is switched on. The inverter, for example, will operate periodically for a very short period and its consumption is measured; if consumption exceeds the standby level, then the device remains switched on and delivers 230 V AC. When the consumer is disconnected, consumption falls and the device detects this and returns to the standby mode.

Different types of stand-alone inverters
Sine wave inverter–charger

Modern sine wave inverters for stand-alone installations use the same techniques as those developed for connection to the grid. The circuits are simpler to devise, since protection and synchronisation necessary for the grid are not needed. There are small inverters available on the market (100–500 W) using digital technology at relatively high frequency (30–100 kHz). On less expensive devices, there is no output filter and the high-frequency signal that is always present may therefore possibly disturb the appliance. More powerful devices using this technique generally have a filter eliminating high-frequency harmonics. For the supply of an off-grid dwelling, the system chosen is often a hybrid one incorporating an auxiliary generator that will supply power in the absence of sunlight and during seasonal variations. In this case, it is more advantageous to use an inverter/charger that will function as a battery charger when the generator is operating. The use of a traditional car-type charger to recharge the batteries with a generator is not recommended as this will often produce a default sine wave in order to complete the charge, and the charger often produces only half or a third of its nominal current. A reversible inverter of 1500 W generally supplies 40 A at 24 V, which is much more than a large traditional charger and enables the batteries to be recharged much more quickly. Also, a generator used to recharge a battery with a small charge is highly inefficient. If the inverter is not reversible, an electronic charger that accepts a wide range of input voltage may be added.

Square-wave inverter
This is the oldest and simplest technique to generate an AC wave. In this case, two transistors in the primary circuit of the transformer are controlled by a 50 Hz oscillator. If it is completely square, the signal generated will produce uneven harmonics that will often not be able to power inductive charges without a problem. Devices using this technique have no regulation of the output voltage: this will, therefore, vary with the input charge and voltage. Currently, the choice of using the square-wave inverter can only be justified if the appliance operates satisfactorily with this wave. Since the price of more efficient inverters has come down considerably, there is rarely justification for choosing this type of device.

Pseudo-sine wave inverter
Formerly these were the most efficient types of device, but they have recently been overtaken by modern sine wave models. The signal produced is a double square (positive and negative) with passages through zero: the passage through zero at each way reduces the harmonics compared to a pure square signal. The square wave of variable width depending on the input charge and voltage enables the precise adjustment of the output voltage. The variable width pulse also enables the inverter to be operated at lower output voltage in standby mode and reduce energy consumption: as soon as an appliance is switched on, the circuit detects the increase in consumption and starts the inverter operating at 230 V AC.

Inverters for stand-alone systems: summary of selection criteria
Before choosing an inverter, it should be ensured that

- a DC solution, often more economical in energy, does not exist;
- the possible consumption in standby mode does not outweigh the advantages of the solar installation;
- the inverter is able to start the appliance (can only really be tested by trial);
- its efficiency is adequate for the charging operation;
- the charge accepts the inverter's distortion (sine wave shape);
- the variations in output voltage are accepted by the charge;
- the inverter is protected against overcharging on both DC and AC sides and against overheating;
- the inverter cuts off the appliances in the case of low DC voltage (protection of the battery).

5.1.4 Other basic components

We describe in this section other elements indispensable to the good operation of a stand-alone or grid-connected PV system, such as lightning protection, switches, fuses and measuring components to monitor the installation.

5.1.4.1 Lightning protection

Lightning protection is indispensable to guarantee a reliable supply of electricity. The number of breakdowns recorded increases with altitude and above 1000 m, it is strongly recommended to install additional protection besides those generally incorporated in charge controllers.

Lightning damage affects first and foremost electronic equipment, regulators, inverters, lamp ballasts and monitoring equipment. The panels themselves are rarely affected, and if there is damage, the bypass or blocking diodes and connection box are the first affected. The cost of damage varies according to the size of the equipment affected but normally exceeds several thousand euros if the system is difficult to access. The advice that follows represents a minimum and further details are given in the literature.[13]

Three principles must be respected to achieve protection against lightning:

- conduct the lightning strike by the most direct route;
- minimise the surfaces of earth loops;
- limit the overvoltage wave by surge protectors.

This has the following implications:

- protection by external installations (possibly a lightning conductor) for the direct effects;
- the installation of surge protection to avoid any indirect effects.

These two protection systems should be linked to a single earth to be fully effective.

Protection against direct strikes

The structures that need to be protected are in general any large installations, public systems (mountain restaurants or refuges), professional installations (transmitters, beacons, etc.), and any sites that are exposed or at risk. We will not go into detail of the construction of different types of lightning conductors, which must be carried out by specialists who will provide a guarantee of their suitability.

Protection against indirect strikes

A strike on or near an installation can induce overvoltages that will destroy electrical equipment. For protection against them, several measures are indispensable:

- single earth point;
- equipotential network of earthing for all electrical equipment and conductors in the building;

---

[13] IEA PVPS Task 3, *Common Practices for Protection Against the Effects of Lightning on Stand-alone Photovoltaic Systems,* Available from http://www.iea-pvps.org

- installation of surge protection between conductors and earth of all equipment;
- cabling so as to avoid loops that may generate overvoltage during rapid variations of the magnetic field;
- shielding of data and telecommunication cables.

Earthing

The purpose of the earth network is to conduct current to the earth, and it must be equipotential in order to avoid local overvoltage when it is conducting a strike. The best system is a single and, if possible, meshed network of bare copper electrodes with a minimum cross-sectional area of 25 mm^2. If these conductors cross the ground between two connections, they could possibly serve as an earth. If not, the earth needs to be installed separately, for example, by a ring loop of the same bare copper electrode buried in the soil around the building. We will not give here details of the different earthing systems used by telecommunications installers as they depend on the type of soil, the type of installation and the ease of mounting.

Equipotential bonding

For the protection of persons, the equipotentiality of earthing must be insured so as not to create dangerous overvoltages when strikes occur. A meshed and not a star-shaped structure is advised with a link to the shortest possible equipotential bonding bar.

Surge protectors

A surge protector is a non-linear element limiting voltage below a given value. Several different components can carry out this function.

- A *spark gap* operates in a few hundred nanoseconds, and some models can allow up to 10 kA to pass. Its disadvantage is that once it is operated, a voltage of 24 V can suffice to continue the flow of current; often another varistor is installed or a fuse in series to interrupt this current.
- A *varistor* is a semiconductor that accepts current in up to 50 kA, but ages with time and at each strike, eventually turning it to a conductor. It is recommended that varistors with state indicators are installed, which will also protect the equipment if the varistor short-circuits through age.
- The *bidirectional Zener diode* protects equipment very rapidly but cannot dissipate much energy. It will generally be installed in equipment as a final protection.

The parameters of surge protection are determined according to the lightning risk as defined by European standard IEC 61024-1 'Protection of structures against lightning' at four levels of efficiency (Table 5.11).

Protection systems will be installed before the equipment to be protected, respecting cabling symmetry and the meshing of the earth cables. The distances between the surge protectors and the active conductors on the one hand and the earth of the equipment to be protected on the other hand should be less than 50 cm (Figure 5.18).

Table 5.11 Surge protection according to the level of protection

Parameters	Symbol	Unit	Level of protection			
			I	**II**	**III**	**IV**
Efficiency	E	%	98	95	90	80
Peak current	I	kA	200	150	100	100
Total charge	Q_{tot}	C	300	225	150	150
Pulsed charge	Q_{imp}	C	100	75	50	50
Specific energy	SE	kJ/Ω	10,000	5,600	2,500	2,500
Average rigidity	di/dt	kA/μs	200	150	100	100
Rolling sphere radius	R	m	20	30	45	60
Down conductor spacing	D	m	10	15	20	25

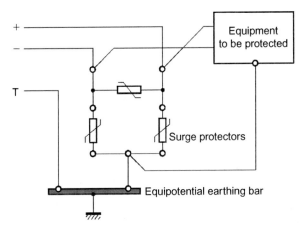

Figure 5.18 Installation of surge protectors

The maximum distance between a surge protector and the equipment to be protected should not be more than 10 m, otherwise surge protectors should be installed at each end of the cabling.

Routing of cables and screening
It is best to avoid the formation of loops or to limit their surface as far as possible by the meshing of earth cables. The shielding of connections is an excellent way of limiting overvoltages by induction, but is expensive. Another method is to install the cables in metal conduits or to use the panel support itself, bonded to the earth, as shielding.

Earthing of a DC conductor
For a better protection against lightning, it is preferable to bond one of the two DC electrodes to earth. To avoid the risk of electrolytic corrosion for certain

equipment, it is preferable to bond the positive electrode to earth, especially if the nominal voltage is high.

This measure is normally applied in 48 V telecommunications. But, in general, many appliances working on 12 or 24 V already have their earth bonded to the negative electrode, and if the ambient conditions (moist and saline air, for example) are not too unfavourable, to bond the negative pole to earth is more practical. This earthing should be made at a single point, if possible at the charge controller. The other protective devices, circuit breakers and fuses will then be installed on the other polarity. Figure 5.19 shows a diagram of a typical small all DC system; it is assumed that the environment is dry and distant from the sea, and the negative is therefore earthed.

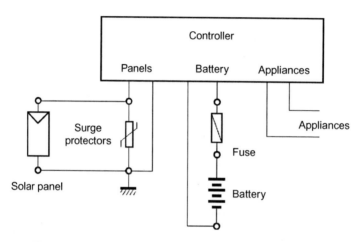

Figure 5.19   Small autonomous system with protectors

If an inverter was added to supply 230 V AC appliances to this small system, the earth of the small network would need to be bonded to the same point as the earth of the charge controller.

5.1.4.2   Recommendations for lightning protection – summary

- Direct protection for public and professional systems, and high value or exposed systems.
- If there is direct protection, panels should be installed in the area protected by the lightning conductor.
- Bonding of the earth by 25 mm² bare copper cable between panels and regulator.
- Cables shielded or laid in metallic conduits bonded to earth at each end.
- Earth cables of all devices inter-connected.
- Earth point of building at bottom of trench.
- Earth point close to charge controller.

- Positive electrode to be earthed if atmosphere is humid or saline or voltage is >48 V.
- Fuses and circuit breakers on electrode not bonded to earth.
- Surge protectors close to the equipment to be protected and connections <50 cm from conductors.

5.1.4.3   Fuses and circuit breakers

Electrical distribution from solar energy requires the same protections as a classic grid. However, special protections on the DC side are needed because direct current (which does not alternate and pass through zero) is more difficult to interrupt if an arc occurs. At 12 V DC, the danger of an arc is low, but at 24 V, with a series regulator, panels are sometimes in open voltage at more than 40 V and an arc can occur if two cables are badly insulated or if a connection has deteriorated with time. In systems with DC voltage of more than 100 V, a poor choice of protective equipment has been the cause of several fires.

- In Switzerland, in a 500 kW system at several hundred volts, an operator was measuring a short-circuit current of the panel array and when he tried to reconnect to normal mode by interrupting the measurement, the current created an arc that spread and destroyed several distribution boxes.
- In a small 3 kW system connected to the grid and operating at 100 V DC, a connection probably worked loose, an arc was created in the distribution box and the fire destroyed the installation and the whole roof of the building. In this case, the main problem was a poor choice of insulation materials in the distribution box.

5.1.4.4   Programmable switch

Programmable switches are often used in automatic systems: they are generally time operated but can also be operated by light levels or by detection of persons or movements.

These devices are based on the 230 V AC versions or are specially manufactured for solar and to consume a minimum amount of energy. Figure 5.20 shows an example of a series regulator incorporating a programmable clock for operating a refrigerator. The system here is designed to cool wine in a vineyard shop so as to be ready for the time when tastings are offered: the clock disconnects the output at night and reconnects it at 9 o'clock in the morning so the chilled wine is available from 11 o'clock. The regulator[14] includes two indicators for the charge current from 0 to 10 A and the battery voltage with an expanded scale between 10.5 and 14.5 V, the best range for measuring the battery voltage.

Programmable clock

These devices enable appliances to be switched on and off according to a programme, for example, for lighting at night (2 h in the morning, 2 h in the evening).

---

[14] http://www.dynatex.ch

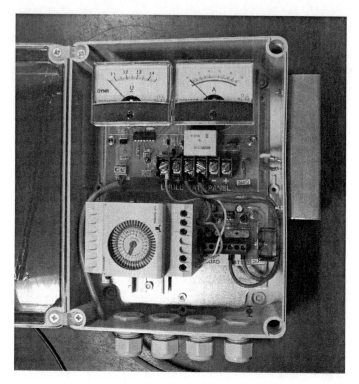

Figure 5.20   Charge controller with incorporated clock

There are economical models with a mechanical clock programmable by pins set at 15-min intervals, as well as digital models.

Time switch
This device enables the circuit to be switched for a limited time: the typical example is a light in a corridor or on a staircase. These circuits are usually activated by a pressure switch that enables several to be installed in parallel, which is useful if an appliance needs to be switched on from several different places.

Twilight switch
This device that is used everywhere for urban street lighting exists in 12 or 24 V DC versions, sometimes with all regulation built in. Some versions have been specially developed for the lighting of bus stops with switching triggered by the movement of persons, time-switched lighting and a clock limiting operation to hours of the bus service.

Movement detector
This device switches on a light when a person or other warm object (car, for example) enters the field of an infrared detector. It generally incorporates a twilight switch and a time switch limiting the duration of switching. Some models are available in 12 or 24 V DC. This type of device should never be used on a 230 V AC circuit produced

by an inverter: most of these small 230 V AC circuits are powered without transformers by lowering the AC voltage through a condenser. The current consumed is very reactive and can easily destabilise an inverter: we have measured a standby consumption of 65 W when two detectors and a time switch were connected to the output of a 250 W inverter.

Intelligent switch

For small networks of rural PV electrification, a Spanish manufacturer[15] has developed programmable switches with multiple functions: the operation of an appliance can be selected according to the state of the battery, the time and other parameters. Examples of potential use or uses are water pumping or ventilation when the battery is full, which enables excess solar energy to be used that would otherwise be lost, and the switching on of machines only if the battery is fully charged and in the middle of the day, when sunshine is at its maximum.

Monitoring

We give below some recommendations for the choice of equipment and methods to ensure the good operation of a stand-alone PV system. This list is centred around the most important component in need of monitoring: the battery. A battery needs to be monitored differently according to its state; during charging, the voltage gives an imprecise idea of its state because this depends on the current and state of charge. So either the battery must be measured at discharge or in the evening (without sunshine) when only a small current (1–2% of its capacity) is available, and its internal resistance does not upset the measurement. However, the measure of voltage will vary according to the age of the battery, and 0.2–0.4 V can be lost. With a small current (C100), a new tubular battery will have a capacity of 30% at 11.8 V, whereas a few years later this capacity will be measured at 11.4 V.

Indispensable equipment

It is assumed that these devices are not integrated into the charge controller, otherwise their use will be duplicated.

Log book – manuals

The first essential piece of equipment is a log book or any other means of recording all the information concerning the system from the beginning, and the date of any measures taken (change of battery, additional appliances, etc.). The log book should be kept with the manuals and equipment instructions.

Acidometer (for open batteries)

If the battery is open, an acidometer costing only a few euros will be adequate to monitor its capacity and state. Table 5.12 gives an example of a sheet for periodic monitoring of a stand-alone system. A density that hardly changes between the discharge state and the charged state is a clear indication of an aged battery that has lost its capacity.

---

[15] Trama Tecno Ambiental S.L., C/Ripollès, 46, 08026 Barcelona, Spain.

*Table 5.12   Table for noting state of battery measurements
(12 V, six elements)*

Date	Voltage	El1	El2	El3	El4	El5	El6	Remarks

Multimeter

A multimeter is also strongly recommended (indispensable if the battery is sealed) because it enables all voltages necessary to diagnose any problem with the system to be measured.

Recommended equipment

Measurement of current

An ammeter to measure the current of the panels enables the monitoring and possible fault detection of the panels.

Measurement of energy

An ampere-hour meter provides, in addition to measuring the current of the system, the measure of input–output capacity (in Ah), which enables the battery status to be forecast by comparing demand with production.

Professional equipment

For professional systems, which need a highly reliable supply, data loggers are very useful. The consumer can access data accumulated over several months or regularly contact the system by modem to remotely monitor the health of the installation. Typical measured data are battery voltage, panel current, appliance current, solar radiation, ambient temperature, panels and batteries.

5.2   Appliances for stand-alone systems

We describe in this section appliances used in PV systems. Many DC appliances have been developed for use in cars, caravans and boats, and we give here some of the most useful appliances for PV applications. The selection of appropriate appliances is important because they will affect the whole project. A solar system is designed for a particular use, which will determine all the other components. According to individual requirements, the solar panels will be sized, the storage and the charge controller will be chosen to match and the choice will be made whether to use DC or AC appliances.

5.2.1   The golden rule: economy of energy

The DC appliances (including inverters) used in stand-alone installations should have maximum efficiency and be able to operate reliably within the voltage range of the battery.

- For systems with lead batteries, 12 V nominal for example, the usual range is from 11 to 14 V at 25 °C.
- For systems with NiMH batteries, the appliance should be able to operate at up to approximately +25% of the nominal voltage, and below it down to 0 V. If the undervoltage is not acceptable, it will be necessary to add a load shedder to cut the output below an acceptable threshold.

Stand-alone PV systems have led to efficiency improvements in practically all DC appliances specially developed for this market.

Important

It is always best to choose appliances with high efficiency that often last longer because they are better designed, especially since the cost of the energy used is high. The additional investment required for high-efficiency appliances will have an immediate effect on a reduced solar power and storage requirement. This is often difficult for users who are already well equipped to understand, for example, the difference in the cost of panels and necessary storage to supply an excellent or a poor refrigerator will often compensate for the purchase of a new refrigerator.

5.2.1.1 DC or AC

In the design of a PV installation, it is generally better to look for appliances operating with DC, or to adapt them rather than to add an inverter and a transducer for 230 V AC. This is also valid for small machines such as milking machines, cereal mills and refrigerators. In general, DC motors of a few kW have a much higher efficiency than those running on AC, and an inverter always loses at least 10% of its energy in heat.

For the supply of chalets or isolated homesteads in temperate latitudes, it is often preferable to replace an inverter designed to feed relatively small appliances (small machines) by a small generator, which is often less expensive and also enables batteries to be charged in winter.

For systems with many lighting points, or when there is a considerable amount of cabling, it is better to work in 230 V AC, which enables the use of compact fluorescent lamps (less expensive) and reduce the amount of cabling. Small sine wave inverters that are efficient and economical can be found on the market for such uses.

Before selecting an AC appliance, it is essential to confirm that the inverter can supply it without problems; a poor choice can either prevent operation or cause breakdown. For example, it is essential to avoid small appliances without a transformer, and using a condenser and rectifier for the operation of their electronics. In a country inn with a dormitory, the owners had added a time switch and two infrared presence detectors/switches in order not to leave lights on; these three appliances with a purely reactive supply consumed 65 W (24 V DC of the inverter) without any light on, with the small 250 W inverter destabilised with its cooling fan on; these three appliances, intended to economise energy, consumed more than the three 13 W low-energy lamps connected.

5.2.2   Lighting

Lighting is the main use for stand-alone systems. The type most used is the fluorescent lamp that has the highest efficiency. Other sources of lighting such as incandescent bulbs and halogen lamps should be reserved for short duration use. Sometimes sodium vapour lamps are used for street lighting; their efficiency is higher but their mono-chromatic orange light and their price restrict their use. Lamps with white light elec-troluminescent diodes have been available since 2006 with efficacy comparable or superior to those of low-power fluorescent tubes (approximately 55 lm/W), which will certainly enable the future development of highly efficient lighting.

5.2.2.1   White LED lamps

These are manufactured from blue diodes covered with a fluorescent pigment intended to produce a wider light spectrum equivalent to white light. Their main advantage is their life expectancy, which should exceed 50,000 h, and their high efficacy, which is improving every year. Several laboratories are working to improve this parameter with the aim of exceeding 150 lm/W within a few years.

Their disadvantages are as follows:

- a source of light with a narrow angle (approximately 70°) but suitable for spotlights or pocket torches;
- price still high;
- a low unit power (of typically 1 W) requiring a group of several units to replace a traditional lighting source;
- the cooling of the light emitting diodes can sometimes be difficult;
- the need to have an accurate supply limiting the maximum current to guarantee long life.

However, their use is becoming increasingly frequent for occasional lighting applications such as brake indicators on cars, pocket torches and miners' lamps – and traffic signals where their high efficiency produces substantial energy econo-mies with consumption 8–10 times lower than for incandescent traffic lights. With energy economies like this, the cost of the lamps is amortised in a few years and their advantage of long life expectancy also reduces the cost of replacement. Today spotlights with white LEDs can be found with efficacies of the order of 20 lm/W – for example, a spot of 3.5 W, which is the same size as a 10 W halogen spot. LEDs with efficacies exceeding 100 lm/W are available in the form of components.

5.2.2.2   Fluorescent lamps

Fluorescent lamps offer a wide choice of lighting solutions. Table 5.13 shows luminous flux and efficacy values of the best fluorescent tubes of various dimensions as well as some values for incandescent bulbs, halogen, LED and sodium vapour lamps. We have used the values of Osram models, but these comparative data are

Table 5.13   Performance of low-energy lamps compared to traditional types

Type of lamp	Power (W)	Dimensions (cm)	Flux (lm)	Efficacy (lm/W)	Efficiency (rel.)
Incandescent	25	Dia 6 × 10.5	325	13	0.8
Incandescent	40	Dia 6 × 10.5	580	14.5	0.9
Incandescent	60	Dia 6 × 10.5	980	16.3	1.0
Halogen	10	Dia 0.9 × 3.1	120	12	0.74
Halogen	20	Dia 0.9 × 3.1	350	17.5	1.07
Halogen	50	Dia 1.2 × 4.4	1000	20	1.23
Halogen	100	Dia 1.2 × 4.4	2300	23	1.41
White LED	1	0.15 × 0.15	100	100	6.14
Straight fluorescent	8	Dia 1.6 × 29	430	54	3.31
Straight fluorescent	13	Dia 1.6 × 52	950	73	4.48
U-shaped fluorescent (PL)	9	2.7 × 14.5	600	67	4.11
U-shaped fluorescent (PL)	11	2.7 × 21.5	900	82	5.03
U-shaped fluorescent (double PL)	10	3.4 × 9.5	600	60	3.68
Straight fluorescent	18	Dia 2.6 × 59	1450	81	4.97
Straight fluorescent	36	Dia 2.6 × 120	3450	96	5.89
Sodium	18	Dia 5.3 × 22	1800	100	6.14
Sodium	36	Dia 5.3 × 31	4800	137	8.41

valid for other lamp manufacturers. All the models shown are low voltage (12 or 24 V); the fluorescent and sodium vapour types must be supplied by electronic ballasts or an inverter. This table shows that efficiency increases with power for all types of lamp. In comparison, we have taken as reference the 60 W incandescent bulb that is the most frequently used lighting level for a small lamp. The efficiency values of fluorescent lamps are given for operation on the mains of 50 Hz frequency with an inductive ballast; for DC operation, we have assumed that the losses from the ballast operating at high frequency (20–50 kHz) corresponded more or less to the gain from operating at these frequencies, which is around 10–15% compared to 50 Hz.

For a fluorescent tube to emit light, an electric current has to flow between the two tube electrodes: this current is carried by the plasma, which arises in the gas filling the tube. The plasma can only be established if the voltage reaches a higher level, which depends on the length of the tube and the gas used; for example, for a standard 120 cm cube, at least 700 V is needed for the current to flow. To lower this voltage, the two electrodes are heated so that electrons are liberated more easily, which also facilitates cold starting. To start a tube at 230 V AC and 50 Hz, an inductor is inserted on one pole and a starter as indicated on Figure 5.21: when the appliance is switched on, the starter that contains a gas of the same type as the tube, ignites and heats, its bimetal switch closes, which induces a high current to flow through the heating electrodes. When the starter switch closes, the current declines; the thermal switch cools and opens, which causes a high voltage at the tube terminals due to the series inductance; then either the tube lights or the cycle begins again. The switching can thus last a few seconds depending on the momentary phase and voltage when the sequence occurs, which explains the flickering when the tubes are switched on.

In low DC voltage, a high-frequency ballast (inverter) is used to replace the starter and the inductance, which, when it is not charged, supplies a voltage high enough to start the tube. There is a wide selection of 12 and 24 V DC fluorescent tubes on the market: they are generally in the form of strips in an aluminium or plastic housing designed for the caravan market. Electronic ballasts also exist on their own, which can be used to supply U-shaped (PL type) tubes that can be more easily incorporated in lamps of traditional shape.

Criteria for choosing fluorescent lamps
Low-voltage DC lamps

- Circuit with pre-heating of electrodes guaranteeing a high number of cycles (for example, minimum of 5000). Circuits without pre-heating blacken the tube at the ends with deposits from the deterioration of the electrodes.
- High-frequency oscillator producing a wave close to a sine wave (peak factor below 2 and wave symmetry better than 60/40%): a recent study has shown that the life of such lamps was considerably higher. Symmetry is also one of the important factors in the ageing of the tube: an asymmetrical wave has a continuous component, which is the second factor destroying the electrodes and blackening the tube at the ends. Also, lamps with a sine wave produce much less radio interference, which is important for products sold in Africa where AM radio is virtually the only type receivable in the countryside.
- High-efficiency circuit (>80%), resistant to voltage (10–15 V) and ambient temperature variations.
- Solidly built and the tube protected: in Africa, the main cause of breakdown is the breaking of a tube with a flyswatter (insects are attracted by the light and cluster on the tube).
- Circuit resistant to the ageing of the tube and able to operate without a tube or with the tube half or not lighted. When the tube ages or is at low temperature, it is more difficult to start and may sometimes only half light, and then the electronic oscillator is not correctly charged and its internal components (resistances and transistors) risk overheating.

Standard AC lamps (system with inverter)

- Lamps with electronic ballast are recommended because they consume less reactive current. If for reasons of cost lamps with reactive ballast are chosen, it is absolutely necessary to balance the reactive current and use inductive and capacitive lamps in parallel. In all cases, distortion of the inverter must be tested and its consumption measured to avoid surprises and breakdowns.
- With compact fluorescent lamps, the best supply is provided by a DC/DC BATNET inverter, which has the lowest peak current consumption (see Section 5.4.3).

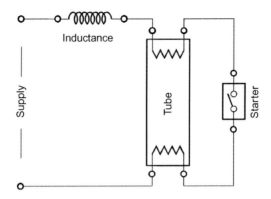

Figure 5.21   AC/low-frequency fluorescent lamp

5.2.2.3   Halogen and incandescent lamps

In comparing luminous efficiency (Table 5.13), it will be noticed that incandescent lamps are four to six times less efficient at equal power than fluorescent lamps, and it therefore makes sense only to install these lamps for very short duration use (a time switch in a corridor, for example). Halogen lamps are very slightly better, with an efficiency, at equal power, 20% higher than that of incandescent bulbs, but these also should be kept for short duration use.

5.2.2.4   Portable lamps, solar lanterns

Portable lamps are a particular type of DC fluorescent lamps: they also incorporate a waterproof battery, usually of lead, and a charge and discharge battery controller.

Additional choice criteria

- Avoid models incorporating the solar panel with the lamp: the battery in a panel-lamp case can reach 70 °C in the Sun (measured on a lamp of the major construction company in Africa) and age very quickly.
- If the battery is of the NiMH type, ensure that there is load shedding at low voltage, otherwise the poorly supplied ballast will cause the tube to age too quickly.
- Ensure that the charge regulator is the series type so that the recharging can be done from another system or a mains charger. For a lead battery, check that there is also a discharge regulator (low-voltage load shedder).
- Ensure that the possible continuous consumption of the battery (by the regulator) does not adversely affect the storage.

5.2.3   Refrigeration and ventilation

5.2.3.1   Refrigeration

Refrigerators for stand-alone installations use compressors operating with a DC motor and a compartment with reinforced insulation. Their price is much higher than standard 230 V AC models (around two to three times more expensive) but the energy consumption of the best models is lower by half. To improve the efficiency of refrigerators or freezers, following steps are necessary:

- place them in a cool situation, with as small as possible a difference between the inside and outside temperatures;
- make sure that the heat exchanger is well ventilated so that it can operate efficiently;
- add cold accumulators (ice cubes or cold blocks) in the ice compartment so that the compressor works longer and less often, which improves its efficiency;
- ensure that any food stored is as cold as possible to avoid the refrigerator needing to chill it;
- avoid using them in winter in temperate climates.

Also available are vaccine refrigerators for rural medical dispensaries; these have been tested by WHO[16] who can provide test measurement results.

It is not advisable to use a 230 V AC refrigerator powered by an inverter in small systems (see Section 5.1.4). Even if the power of the compressor is only 50–100 W, the inverter will need to be capable of starting a charge easily 10 times higher depending on the state of mechanical charge of the motor/compressor. Also, the DC consumption of the inverter could take up an important share of the energy, and 230 V AC models are often less well insulated than models specially developed for solar applications, and this is true even for 230 V AC class A models.

5.2.3.2   Ventilation

For ventilation in small spaces there is a large range of DC fans of the type used in electronics. Some are even installed with solar cells in a ventilation duct for a caravan or boat.

For larger requirements, the drying of hay for example, many industrial fans can be obtained with a DC motor. In such cases, the system for starting the motor is more difficult to resolve than the DC modification. If an AC model is chosen, it must be ensured that the inverter can properly start it. For hot countries, 24 V DC ceiling fans are available, which are around 40% more efficient than their equivalents in 230 V AC.

5.2.4   Pumping and water treatment

The pumping of water is one of the priorities of solar energy in Africa. For users in Europe, the usual requirements are providing water pressure to a dwelling, a caravan or a boat.

5.2.4.1   Direct solar pumps

A solar pumping installation in a hot country must be carefully designed (see Section 5.4.4): even if the technology is perfectly suited, a pump always requires maintenance and so calls for adequate training for people supervising it. Also, the

---

[16] The Cold Chain Product Information Sheets, SUPDIR 55 AMT 5, Expanded Programme on Immunization, World Health Organization, 1211 Geneva 27, Switzerland.

pump must be adapted to local pumping conditions: it must be considered how the water table or water course may change when considerable amounts of water is pumped, and what would be the variations of level and water quality.

We will not go into the advantages and disadvantages of systems with or without batteries; before installing a DC pump directly to a solar generator, it must be ensured that it will accept considerable variations in the generator current and voltage during the day.

A wide variety of pumps is available for solar applications, and we describe these below.

Piston pumps

These are driven by an efficient DC motor that can reach 60% for pumps with powers of some hundreds of watts. They are used for pumping from a nearby lake or river or reservoir. There are models for direct coupling to the solar generator or supplied from batteries. They are fairly tolerant of variations in water quality and of running dry, and are mainly used for irrigation, filling of reservoirs or pressurising small supplies. They are used directly on power from the solar panels, but a current booster has to be added for starting and the panel power needs to be at least 20% higher than the nominal power of the pump. If constant working at maximum efficiency is preferred, a battery model should be used that will also enable small levels of solar energy insufficient to start the pump to be recovered at the beginning and end of the day.

Centrifuge pumps

These are intended for relatively deep wells (with solar systems typically up to 120 m). The pumps are driven either by a watertight submerged DC motor or on the surface through a shaft or by a submerged AC motor powered by an inverter. The pumping height is proportionate to the power of the pump, which limits this type to relatively large power (>kW, tens of l/min). Average efficiency is the worst of all systems available, typically between 30% and 50%. These pumps behave erratically if the water level varies considerably, as they cannot work dry; they are also very sensitive to water quality. If there is a mechanical problem, the whole system is at the bottom of the well, making maintenance and checking difficult.

Submerged membrane pumps

These pumps are attractive up to a depth of around 60 m for small requirements (a few litres per minute). Operating with a watertight DC motor, their efficiency is up to 50%. The pumps are fairly tolerant to variations in water quality and can run dry. One problem encountered is that some models cannot tolerate too high entry pressure (variation in the height of the water table). The main disadvantage is a more limited life and higher levels of maintenance.

5.2.4.2   Water distribution under pressure

The main need for pumping in temperate latitudes is to provide mains water pressure for a dwelling. The system uses either a pump controlled by a pressure switch, starting when a tap is opened (Figure 5.22), or a tank in the attic filled from a well by the pump. Both systems have advantages and disadvantages.

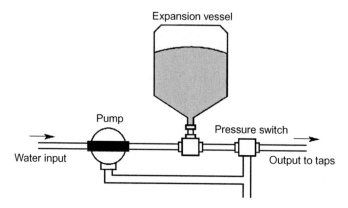

Figure 5.22   Small water main under pressure

- The system under pressure provides a higher pressure, which is better for the operation of a shower or for obtaining water rapidly. Another advantage is that there is not a large tank to be cleaned in winter. The disadvantages are that the pump starts as soon as more than a few litres of water (depending on the size of the expansion vessel) are run, and that the taps are continually under pressure.
- The system with a tank is simpler to install: the pressure on the taps is not critical, the pump can be switched on either by an automatic system with a float in the tank or manually. The disadvantages are low pressure and the need to clean the tank in winter.

Small water mains under pressure are usually used for boats and caravans: in some models, the expansion vessel is reduced to a pump output tube in flexible plastic material that can expand and absorb pressure variations. The bigger the size of the expansion vessel, the less often pump has to work, which reduces sound nuisance. Some pumps include the pressure switch that simplifies cabling.

Heating circulator pumps

The third use of solar pumps is the circulation of water from solar thermal collectors. The pump can either operate directly from sunlight, in the hope that its output is adequate to the thermal performance of collectors, or be switched on by a regulator measuring the temperature difference between the collector and hot water tank. Generally centrifuge pumps are used for this purpose: models between 8 and 40 W are normally sufficient for family requirements. The same type of circulating pump can also be used for wood-fired central heating in a chalet.

Water purifiers

This type of equipment is absolutely necessary in hot countries if water is to be drunk without first boiling it. There are techniques that enable water from a well to be treated without chemical additives to produce adequate drinking water for the consumer. A recent study has shown that the solar energy required for treating water is 20,000 times less than that needed to boil it.

These systems work by forcing the water through a particle filter, one or two charcoal filters and a UV lamp to kill bacteria. Recently developed techniques of depositing silver on the charcoal filter have led to higher efficiency, because ozone is generated, which enables the treated water to be stored longer.

5.2.5 Hi-fi, TV, computers and peripherals

5.2.5.1 Hi-fi and TV

There is a wide range of 1 V DC radio and audio equipment designed for use in cars, which is perfectly suitable for solar systems. For a classic 230 V DC music centre on a system with an inverter, a sine wave model is recommended to avoid harmonic distortions that would otherwise be difficult to filter out.

Modern LCD television sets do not require much energy if they are not too big. For 12 or 24 V DC equipment, the choice is limited.

A DC/DC converter with 300 V output can also be used to more easily supply appliances with a switching power supply.

5.2.5.2 Computers

Most laptop computers need to be recharged at voltages of 15–20 V, which calls for a special DC/DC converter. The other possibility is to use the mains supply of the computer connected either to the DC/DC converter producing 300 V, or to go via an inverter. To evaluate true energy consumption, it is advisable to measure the appliance, taking account of the charge/discharge cycles of the built-in battery, which is usually lithium.

For printers, the most economical types are inkjet models (12–150 W); laser printers use 300–1500 W.

5.2.6 Connecting and cabling of appliances

Few suppliers provide special connectors for the solar appliances. One,[17] however, does offer a range of plug sockets and suitable switches: the contacts are of a good size (will take more than 50 A) and cannot be confused. It is even possible to fit a miniature fuse within the plug (I_{max} 16 A). At the same time, the supplier sells regulators using these connectors, which makes maintenance and checking easier, as the inputs and outputs can be disconnected without accessing the cabling.

In choosing cabling accessories, the most important criteria are as follows:

- easy access for verification, cabling not too dense;
- high-quality non-corroding terminals (brass, nickel, gold);
- sufficient insulation for double the nominal DC voltage, fire-retardant materials.

5.3 Stand-alone applications in developed countries

The uses of PV energy and the questions that it raises are different in developed countries from developing countries. In the developed countries, it is mainly used today by individuals or collectives to generate electricity to feed into the grid using

---

[17] Alternativas CMR, Spain.

solar panels on the roofs of private houses or public buildings. By extension, large power stations are also being built, which produce sufficient to power whole communities. Stand-alone solar systems, which are less widespread, are normally used to power remote residences or for professional use. In hot countries, the context is different insofar as grid-connected PV hardly exists for financial reasons; but stand-alone PV on the other hand is essential to supply remote communities, which will probably never be connected to a mains grid. Another characteristic, which differentiates the less developed countries, is the higher ambient temperature that calls for particular precautions and seriously limits the life of lead batteries.

5.3.1   Stand-alone PV habitat

Some locations are difficult to access and therefore expensive to connect to the mains electricity grid – particularly in the mountains or on islands. But there are other remote rural areas where PV can be justified, simply because the distance involved makes connection to the grid more expensive than a solar installation.

The more rugged the terrain and the further from the grid the location, the more attractive PV becomes. Energy requirements must also be entered into the equation: the higher they are, the more justified is a link to the grid. The two solutions must be evaluated and compared. It is clear, for example, that a solar installation would not be the same for a house lived in all year round as for a chalet only used in summer and for some weekends in winter. The weekend house in Switzerland, which we describe in detail in Section 5.6.2, demonstrates this clearly.

5.3.1.1   Composition of a PV system

Installations of this type have installed power ranging from 200 Wp for the most modest to 3 kWp for the largest, with between 2 and 30 m^2 of crystalline silicon PV modules. These panels will be ideally installed facing south and sloping at 60° (the ideal pitch for France and Switzerland, for example). When the dwelling is remote, there is often space around it, and therefore frames to support the array can be set up on the ground or on a terrace not too far from the rest of the installation to avoid cabling losses (Figure 5.23). Otherwise, according to the way the buildings are configured, the panels can also be installed on the roof or on the facade (see Section 5.5.7 for installation details).

The batteries used are generally the fixed lead-acid type, with liquid electrolyte, with a large reserve and recyclable stoppers. These represent the least expensive and most durable solution, provided maintenance guidelines are followed (see Section 5.1.1). Some maintenance will be possible since the building is inhabited. Sealed batteries, which need no maintenance, are obviously more practical although more expensive. And since they do not undergo stratification, they should last at least as long with a good system of regulation. The installed capacity ranges typically from 100 to 2000 Ah for normal domestic applications.

The charge regulation is done as normal, with an equalisation charge to complete the charging of the battery (see Section 5.1.2). Possibly a double user output could be attractive: one drawing directly on the batteries for vital appliances like

Figure 5.23   Remote rural homestead powered by a PV system [photo Apex BP Solar]

the refrigerator, the other with load shedding for less crucial appliances. In this way, the consumer can manage his priorities by giving privilege to certain appliances if the weather is bad and/or consumption is higher than usual (see Section 5.1.2, Discharge control). Manual configuration of the appliances on the regulator is usually better than automatic re-switching at the end of load shedding, especially if the buildings are only intermittently occupied. What is the purpose of turning on lighting in the absence of users?

5.3.1.2   Appliances – DC or AC?

A PV generator always works on a DC voltage that is a multiple of the battery voltages (12, 24 or 48 V), while most domestic appliances use 230 V AC (at 50 Hz in Europe) since they are designed to operate on the mains. But does this mean that all the power produced by the panels should be systematically converted into AC?

This can be done if, for example, the whole internal electrical installation is already in place and entirely cabled in AC, and if the appliances are very economical in energy, this can avoid work, but it is unusual (it can happen that at the last moment the consumer cannot benefit from the connection that he was counting on). A householder connected to the mains is less concerned with the energy consumption of appliances because his electricity is less expensive. And the first parameter to be considered in the design of a domestic PV installation is energy economy. The choice of voltage obviously depends on the type of appliances that are to be supplied (lighting, domestic appliances, music centre, etc.). DC appliances, when they exist, are almost always preferable because their use avoids the energy losses of the inverter, and they are generally designed for low consumption. This is true for lighting, television and certain refrigerators (see Section 5.2.3). But some equipment like computers do not exist in 12 or 24 V DC, they must inevitably be supplied through a DC/AC inverter or a DC/DC transformer adapted to their operating voltage since they use a switching power supply.

Therefore, what is needed is either DC cabling for small appliances of less than 200 Wp, with just lighting, television and perhaps a small refrigerator; or mixed cabling for more substantial dwellings, made of two sub-networks: one with a direct feed from the regulator for the DC section, and the other equipped with a DC/AC inverter at the input. The capacity of this inverter must take into account the total power of the AC appliances likely to operate at the same time in the house. It must be able to supply this power permanently, as well as the start-up power, which can be substantially higher.

5.3.1.3   Consumer behaviour and backup energy source

Consumers of entirely stand-alone solar installations must be ready to adapt their behaviour from that normal with traditional mains electricity. This is because these stand-alone installations are designed for a daily maximum consumption that should not be exceeded. Unneeded appliances should be switched off and no appliances should be used longer than necessary. In winter particularly, when the sunshine is at its minimum and the nights at their longest, the need for lighting is greater. In these circumstances, it is important to monitor consumption closely so as not to suffer any shortage. This is not so necessary in summer, when the available energy is much more abundant.

Another way of reducing the risk of shortage is to have available a small backup generator allowing the batteries to be recharged if necessary (see Section 1.2.1, Hybrid stand-alone systems). Its VA capacity will be calculated with reference to the time required for weekly recharging, for example, of half the battery capacity.

Example

300 Ah–24 V battery. To supply a half its capacity in 3 h, it requires a current of 150 Ah/3 h = 50 A, so the charger required is 50 A × 24 V × 1.25 = 1500 VA, assuming a charger efficiency of 80%.

An important point to consider is the quality of the voltage supplied by the generator: most small generators supplying a few kilowatts produce a wave that is too square to be usable with a conventional charger (50 Hz transformer and rectifier). In this case an electronic charger with a switching power source should be used, which accepts a wider range of input voltage to complete the charge of a battery.

5.3.2   Stand-alone professional applications

Well before the upsurge in grid-connected PV systems, professionals were already looking to solar energy to provide a stand-alone energy supply. In this section we will give some examples of sectors where PV has been used and the reasons for its success.

The equation is quite simple: PV system + remote appliance = peace of mind for the operator. For a local community, a maintenance company or a motorway concessionaire, PV electricity can provide a reliable energy source without maintenance, almost anywhere. Obviously, there is a limit to the amount of energy that can be supplied, but when one thinks of the costs involved, for example, in the maintenance of a large number of time-stamped ticket machines or a telecommunications relay situated on a rocky peak, it is clear that any solution that results in fewer visits, at least for several years, is very attractive.

When the quantity of energy required is below the threshold of economic connection to the grid (largely dependent on the distance involved), a stand-alone source of energy must be sought: expendable or rechargeable batteries taken to the site or a renewable source such as solar, hydraulic or wind. To make a valid comparison, the operator must include all the costs involved in any one solution. In the case of supply from batteries, the management of the stock of batteries, site visits to replace them, recycling costs, etc. should be considered. In the case of the PV solution, the operator will have higher investment costs but very low operating costs. The case study in Section 5.6.1 includes this type of comparative analysis.

5.3.2.1   Telecommunications applications

PV was first used on the ground in the area of telecommunications relays in the early 1970s (following its use on satellites), and it has continued to be widely used in this sector until today. These relay stations transmit data such as telephone, radio and television signals (Figure 5.24). The need for them has further increased with the development of mobile telephones and the related GSM (Global System for Mobile Communications) transmitters.

To maximise the range of the signals and the extent of cover, these relay stations are often installed on high ridges or mountain summits. It follows that they are often far away from populated areas, in locations that are difficult to access or simply not served by the electricity grid. But they need a reliable source of electricity because their operation must be permanent and guaranteed. At first, diesel generators or non-rechargeable batteries were used and these were associated with high maintenance costs on account of repeated visits, sometimes by helicopter.

In comparison, a PV system offers high reliability and virtually no maintenance. The only problem is battery life duration, which depending on local weather conditions will require maintenance visits every 2–5 years, or even less frequently. The optimisation of this parameter can be done by comparing the cost of replacement and of maintenance visits with the extra cost of a longer-life battery. These installations often have a fairly high installed power, from 1 to 4 kWp (10–40 m^2 of PV modules), since energy requirements are often several kWh/day.

*Figure 5.24   Telecommunications relay station supplied by a PV system
[photo Total Energy]*

In very cold regions (far north of Canada, high mountain locations), the behaviour of the batteries in the cold will be a major factor and they should be fitted with thermal insulation, since their capacity falls in winter during periods of low sunlight. Sometimes the addition of a top-up generator is needed. Then the system becomes a hybrid one (see Section 1.2.1).

PV, besides being used for relays, also has more modest, although still indispensable, telecoms applications such as emergency telephones on the sides of roads and motorways, and more generally in applications linked to transport.

5.3.2.2   Transport

Except for some prototype ultralight racing cars, designed for prestige competitions, PV has not yet been used for the propulsion of vehicles, as this requires too much energy. At most, one could envisage its use to power the small motor of an electrical bicycle or as an extension of power for motorised wheelchairs for handicapped people. Electrically propelled boats have recently started to use solar energy to extend their range. But the transport sector includes much isolated fixed equipment needing an electrical supply: road and rail signals, air and maritime beacons and buoys, emergency telephones, weather monitoring, etc. (Figure 5.25). Today solar panels can be seen on the side of every road and motorway, powering indicator signs, small weather stations, and emergency telephones, but less known are maritime and air applications. Companies responsible for marine beacons have used solar panels for more than 25 years, particularly on buoys, with more or less difficulty. These applications have, in fact, served rather as a laboratory for developing solutions adapted to maritime usage. Also, many red warning beacons for air traffic at 20–40 m above the ground are now supplied by PV modules to ensure operation in all circumstances.

Figure 5.25   Maritime signalling buoy (Apex BP Solar)

We should also stress that PV is not just used as a substitute for existing solutions, it also participates actively in innovation. One new application currently being developed illustrates this phenomenon well: the tracking of vehicle fleets. This is of concern to many operators, especially for tracking railway wagons and lorry trailers. Unlike wagons used for transporting passengers, goods wagons have no energy supply. It often happens that they get lost in railway sheds or shunting stations when information is not followed up. By mounting a stand-alone box with a GPS (Global Positioning System) receiver and a GSM transmitter on each wagon, they can easily be tracked and localised. The operator can also be kept informed on the progress of his goods when the wagon is part of a train, and can make appropriate logistical arrangements and improve his service to clients. These tracking devices are easily made independent by a small PV system of less than 10 Wp, and the whole system can be combined into one unit, facilitating its installation and virtually doing away with maintenance.

This economy of maintenance almost always justifies the use of solar PV in the transport sector, which is characterised by very extended equipment fleets. Sometimes an independent energy supply is essential: the emergency telephone networks on roads and motorways must be available for use at all times even in case of accidental interruption of the mains electricity (for example, when a tree falls on a power line). It is sometimes a question of life and death. Operators, therefore, must rely either on traditional batteries or increasingly on solar for these exacting applications.

5.3.2.3   Remote measuring and monitoring: networks, safety, meteorology, etc.

Similar considerations apply to gas and electricity distribution networks where the safety of persons and property is often at risk. Electrical energy is not always available (or, at least, not in the desired form), and PV is now widely used, particularly for remote verification and monitoring of high-pressure gas pipelines, and errors on electrical medium and high-tension lines.

The devices used are of modest power, less than 100 Wp, and often much smaller (1–5 Wp).

In the safety domain, they are also used for ionising lightning conductors, alarms, video surveillance of sensitive sites, and fire detection (Figure 5.26).

Applications are also developing in meteorology, and increasingly weather stations are being installed in local communities and industrial parks. Further applications overlap with transport requirements, in, for instance, the detection of black ice and fog. In the field of environment, there is increasing interest in the quality of air and water, and measuring and analysing devices of all kinds are available to measure different parameters. When mains electricity is not available and the measuring equipment does not require too much energy, PV is an ideal solution (although often some kind of heating system is included for use in winter to keep the sampling devices frost free, which means that their consumption is too high for solar energy). The power of these devices varies but is between 1 and 100 Wp, since it needs to be small enough to fit in a box or on a mast.

5.3.2.4   Cathodic protection

PV is also used on major public works including submerged or buried metal structures (bridges, viaducts, etc.). To prevent their degradation by corrosion, these metal structures must be given negative potential by placing them in contact with an electric current from an anode buried nearby. The parts to be protected thus constitute the cathode of a sort of battery (hence, the name of cathodic protection) of which the electrolyte is the moisture in the soil itself.

As the structures they protect are large, PV systems of this type are often of a considerable size (several tens of kWp). This application is not limited to highly

Figure 5.26   PV supply for road sign (Total Energy)

industrialised regions. Many are installed in desert areas, for example, to protect pipelines in oil-producing countries.

Water management

This is another sector where solar can be very useful. Water management also requires remote monitoring and data transmission: flow rates in water conduits, levels in water towers, monitoring of river levels to prevent floods, etc.

But in this sector PV is above all used for pumping (on the surface or at depth) and water treatment. Although PV pumping is particularly widespread in developing countries (see Section 5.4.4), it is also used in temperate countries, for example, for livestock farming in dry regions where water is only available underground. Solar generators in these cases are of 2–5 kWp, without batteries, but with an inverter or an electronic booster to directly supply the pump motor (as an example, an installation with 4 kWp of PV modules, with a 4 kVA inverter and a three-phase pump of 1600 W, cost €52,000).

Water treatment systems such as chlorination or UV sterilisation often use PV energy in rural areas when the amount treated is relatively small. In these cases, installations from 100 to 500 Wp are adequate. But many sewage farms with bacterial beds, biological discs and activated mud require larger installations of 5–10 kWp. In these cases it is necessary to compare the costs of solar investment (approximately €100,000) with those of connection to the grid.

5.3.2.5   Agriculture

We mentioned in the previous section the role that can be played by PV pumping in agriculture. It is also used for meteorological observations, in particular in the horticultural and market garden sectors that use greenhouses. For efficient greenhouse cultivation, farmers increasingly use automation for the management of watering and temperature, with the help of a small monitoring station.

One of the oldest solar applications is the electric fence: when an isolated battery in a field is kept on a charge by a solar panel, usually no larger than 10 Wp, this avoids much recharging. These devices are not permanent fixtures and are generally brought indoors in winter when the livestock is no longer at pasture.

There are also programmable electric valves, powered by small photocells, controlling automatic systems for watering cattle, as well as PV bird scarers and low-frequency vibrators planted in the ground to scare moles and field mice.

5.3.2.6   Urban applications

Even in the urban environment, it is not unusual to find places where connecting to the mains is not convenient. In general, electricity is only available on the public highway at night, in the form of street lighting. But the connections are not always simple because of administrative or technical problems, for example, when the town is not the owner of the equipment to be electrified and is not expected to meet the energy bill. This is also true of street improvements, where it can prove more economic to set up a totally stand-alone system (a parking ticket machine, for example) than to dig up the pavement over 30 m to run a cable.

What is more, there are many places where energy is not readily available, such as rural bus shelters, pedestrian streets and isolated parking lots.

PV may well be competitive for safety and convenience measures, which a community may wish to install, such as lighting or road markers (with low-power sources). Applications include streetlamps, the lighting of monuments, signs, etc. This lighting is automated so as to only operate at night, being switched by twilight detectors (see notes on Twilight detectors, presence detectors), clocks or a combination of the two. Despite that, they consume considerable amounts of energy because of their operation over a long period, and often require 100 Wp or more of PV power per light point, even when the energy chain is optimised with low-energy lamps and high efficiency of energy conversion. In the case of bus shelters or little used roads, this consumption can be considerably reduced by adding a presence detector (see below), and economies can reach a factor of 10 (for example, if the lighting is on for 1 h instead of 10 h/day). The disadvantage of these systems is that it is rather difficult to forecast the exact consumption, and there is a risk that anticipated energy needs will be exceeded.

Twilight detectors, presence detectors

These are essential devices, and very useful for reducing consumption to what is strictly necessary. Why light a bus shelter if no one is in it, or if it is broad daylight? A presence detector is in fact a *movement detector*, using an infrared cell sensitive to thermal fluctuations caused by the presence of a person (wavelengths of a few μm). The *twilight detector* generally uses an independent cell with the sensitivity of the human eye (amorphous silicon or selenium). But with solar lighting this function can be integrated into the lighting regulators, with the solar panel itself acting as detector: measuring its open circuit voltage gives a rough approximation of the ambient solar radiation. These detectors are mounted in relays and act as a switch for the lighting supplied (see Section 5.1.4).

Among other urban appliances, we may mention parking ticket machines, street signs and other signage (see Section 5.3.2.2). As we have seen, energy independence for this equipment is a definite advantage for the operator. However, it remains true that these urban PV applications often face the technical difficulty of controlling the luminous environment. Parking ticket machines, for example, are often situated in narrow streets, under trees or on a terrace shaded by the blind of a cafe. This usually means that the solar panel must be oversized or even specially developed for the application (for example, with more than 36 cells, to guarantee a charge at low light levels).

Also, fairly strict installation guidance must be given to avoid the most unfavourable situations; in fact, it is often best to analyse the installation sites on a case-by-case basis.

5.3.3   *Portable electronics and leisure applications*

Solar power is an autonomous source of energy: the energy is produced and consumed in the same place. When the required supply is very small, solar power can be used for portable devices in the same way as a battery. Unlike a battery, a photocell needs light, and it provides low intensity energy, but almost unlimited in time. It is therefore suitable for use in portable devices such as

- calculators,
- watches and clocks,
- small measuring devices (medical, sport, weather),
- LCD displays, electronic labels,
- some toys and gadgets (models, flashing badges, etc.).

These applications are microsystems with powers of the order of a few microwatts that have been made possible by the development of very low power electronics. Amorphous silicon cells are the most suitable for them because they can be adapted to the sizes and voltages of these small electronic circuits, and are the only ones that will operate under artificial light. Storage is provided either by supercapacitors or by NiMH batteries. Sometimes the supply is direct, without storage or just with a buffer capacity to start the circuits. This is the case with the pocket calculator: as light is necessary to read the screen, the PV cell is always under power at the time of use.

There are many other potential applications for solar energy to power consumer goods in developed countries, and one wonders, for example, why there are not solar panels on mobile telephones, organisers and laptop computers. The explanation is simple. These products are generally used inside, in relatively low light levels, and use much more energy than the simple calculators described above (at least a few watts, and sometimes for several hours per day). The energy supplied by a solar panel in these conditions would not be significant compared to the energy consumed. However, with the standby power of mobile phones falling, many models could have their autonomy considerably extended by an integrated solar cell. The reason that telephone manufacturers do not include this is no doubt because the quality to price ratio is not yet better than that of the Li-ion rechargeable batteries, which have made huge progress, particularly for use in this market. A photocell would also need to be better adapted to the product, for example, being thinner and more flexible. Future developments in thin-film technology may make this possible, notably for the watch market, as they are the only cells that can really be curved. But it is clear that solar energy has many other potential applications in portable electronics.

Outside, where light energy is much more abundant than indoors, PV modules are widely used in the leisure sector: sailing, caravanning, hiking and gardening. It is not a major sector in terms of market share, but one where PV can provide a unique service. Here, as elsewhere, solar energy can often provide a solution where no other form of energy is available.

We could mention the example, somewhat extreme but genuine, of an expedition to the Far North. The explorers had taken a radio transmitter to stay in contact with their base, and various sources of energy to supply it. It was found that after a month, with the constraints of weight, cold and safety, only the PV system, composed of a small module recharging a waterproof battery adapted to extreme cold, still enabled them to communicate.

More usual applications for PV modules in the leisure sector are as follows:

• pleasure and competition boats;
• caravans, camper vans, mobile homes;
• garden lamps and fountains;
• motorised gates and swimming pool covers (Figure 5.27).

However, in the area of garden lighting and marker lights of all kinds, gimmicky products that do not operate well or have a limited life expectancy should be avoided. Contrary to what one might think, solar lighting does not operate according to the principle of one recharge during the day followed by a discharge at night: primarily because the weather is not always favourable, and also because if the battery is discharged every day, it will be subjected to too many cycles and will

*Figure 5.27   Motorised swimming pool cover powered by a PV system
[photo DEL]*

not last long. A well-designed and durable solar lamp is not going to be cheap (see Section 5.4.2).

The PV market for sailing and caravanning is active because PV modules are well adapted to requirements. The electrical appliances are already DC because they are powered by a battery most of the time. They include small lighting points, and some audio, navigation electronics, GPS, radio transmitters. Another form of energy such as gas is recommended for refrigeration as it requires too much electricity. Consumers do not require total autonomy, as batteries can be recharged in ports and on camping sites, but the energy from the solar panels enables these recharges to be spaced out, and bring greater comfort and security.

The PV modules used are generally from 50 to 100 Wp, sometimes more for long-distance sailing, and large surfaces on boats can be used for solar panels to increase installed power. For these modules to be suitable for use on sailing boats, they must be able to tolerate a saline atmosphere and possibly be able to adapt to the slightly curved surface of the deck or a catamaran float. Some manufacturers have therefore designed modules that can be slightly curved, with added protection against a saline atmosphere, especially the metal parts and electrical connections. Crystalline silicon is best for this type of application, because weight is an important parameter so efficiency has to be high. Consequently, the curvature of these panels can never be very great, because if a crystalline cell is bent, it snaps.

Whether on the roof of a vehicle or the deck of a boat, the modules must be mounted securely and if possible with good ventilation of the cells so that they do

Figure 5.28   Photovoltaic pump in Africa (Total Energy)

not overheat, which reduces their performance, see Section 3.1.2. Adaptation kits for mounting and cabling can be found today for most configurations.

The regulation of this type of system must be carefully carried out to ensure long battery life and avoid the inconvenience of a deeply discharged battery.

5.4   Stand-alone applications in hot countries

5.4.1   Essential needs

More than 2 billion human beings are not connected to an electric grid, and there is little chance of this situation changing any time soon; thus, the need for decentralised energy is huge. For the countries of the South, PV is particularly suitable because the differences in sunshine between summer and winter are not as marked as in temperate countries (Figure 5.28). For these hot countries, the main priority needs for electricity are for

- the provision of lighting to enable activity in good conditions in the early evening,
- access to information by television and radio,
- minimal refrigeration requirements for the preservation of medicines (Figure 5.29),
- ventilation,
- pumping and treatment of water,
- the operation of farm machinery (grain mills, milking machines, etc.) to make daily life easier.

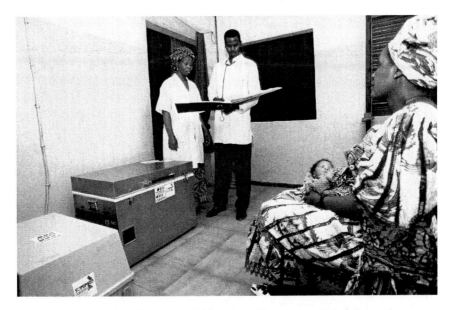

Figure 5.29 Vaccine cold box in a dispensary (Total Energy)

We need to abandon preconceived ideas if we are to bring our knowledge and technology to developing countries; our priorities are often not the same as those of people living in remote communities: for example, television and ventilation are often perceived as more important needs than lighting. One also needs to name who will take the decision in the choice of particular equipment: in West Africa, women will be more likely to choose appliances such as a cereal mill to make their domestic tasks easier, whereas men would choose a television set. Major rural electrification programmes are financed by NGOs (non-governmental organisations) and the World Bank with mixed results.

5.4.1.1 Importance of training and financing
If the user has not financed the system or contributed to the financing of it himself, it is likely to deteriorate, and soon after the last Western technician leaves the region, it will break down, be dismantled and parts sold off. The users must have received technical training: in Mexico, it was explained to the users of a large system that the battery level had to be checked and water added if necessary; the users, who were farmers, understood that, like plants, the batteries needed water and gave them a good watering every evening!

5.4.2 Small individual and collective systems
5.4.2.1 Solar home systems
Many programmes of rural electrification install large numbers of small individual domestic systems called solar home system, usually comprising a 50 W panel, a battery from 50 to 70 Ah, a small regulator and some lighting (Figure 5.30). These regulators sometimes provide outputs at 9 or 6 V to supply small appliances.

Figure 5.30   Domestic solar system in La Réunion

The main defects of these systems are that the battery only lasts 1 or 2 years in a hot country, the regulators are sometimes poorly sized and soon break down and, finally, the user does not have sufficient resources to maintain and renew components as they wear out.

An alternative to individual systems may be to develop small networks operating at 24 V DC that offer many advantages over SHS; we will describe these in more detail in Section 5.4.3.

5.4.2.2   Solar lantern

The solar lantern represents the most elementary SHS: the 'lamp' appliance contains all the elements of the system apart from the solar panel. The panel must always be separated from the battery-regulation block to avoid transmitting its thermal losses to the battery. The regulator must be of the series type to accept other types of charges: for example, AC mains or the cigarette lighter socket of a car. The latter source can maintain a 12 V battery but its voltage will not be adequate to complete the charge. The most frequently used battery is a sealed 12 V/7 Ah lead battery. The regulator must have a discharge control that cuts off the lamp below 12.5–12.8 V. If solar lanterns are used from a 12 V system, DC/DC converters must be added to provide 15 V for their recharge. There are also charge distributors enabling several lanterns to be connected to one large solar panel. This design was developed in the 1980s at the time when small solar panels were all crystalline and were much more expensive by the watt. Today, a small 4–12 W amorphous panel is generally used, which is competitive at this power range. Figure 5.31 shows one of these solar lanterns.

Figure 5.31 Solar lantern and 4 W amorphous panel (Dynatex)

5.4.2.3 Solar charging unit

A system often used with more or less success is a solar recharging unit for lead batteries. In principle, it should be able to recharge the batteries within a day, but this is often difficult given the unreliable nature of solar energy. More than 1000 systems have been installed in Thailand for example, with mixed results. Basically, the design is sound: each house has a lamp with a low-voltage disconnect, and the 4 kW recharging stations for 24 batteries have regulators for each battery with a charge indicator and well-sized cables and battery terminals. The system is designed to supply capacity of 50 Ah to each battery every 10 days on the assumption that the user consumes around 5 Ah/day. The 12 V/110 Ah battery chosen is a model used in lorries, the load-shedding regulator being adjusted to cut-off at around 50% of the remaining capacity.

As with all community systems, the key to success depends on the organisation and maintenance of the system as well as the commitment of the users to involve themselves in the project. In certain regions where the participation of users or the organisation is not adequate, the systems go downhill rapidly due to the theft of panels or their lack of maintenance, with vegetation soon covering part or all of the PV array, regulator parts stolen or the equipment being badly used. In a number of systems where the regulators were stolen, villagers connected the batteries directly to the panels, which, when there were not enough batteries, had the effect of over-charging them and in doing so producing a lot of gas corroding the connections. In some cases, old batteries were used as supports to compensate for the shortening of the cables. Sometimes the users connect smaller batteries that cannot tolerate a daily charge. All the systems that functioned correctly without the assistance of an adequate organisation were due to the initiative of a user who understood how to deal with small breakdowns and manage the system with other villagers entrusting him to charge their battery.

5.4.3   *Rural electrification in small 24 V networks*

The main attraction of installing systems used by a whole community is that the energy base can be used for craft or agricultural activities and so become indispensable to the life of the village. When a PV system becomes essential to everyday activity, people learn to use it and maintain it in a good state and with experience, to better manage the energy at their disposal. This type of system, therefore, has the best chance of surviving and of bringing a real improvement in living conditions to communities too remote to be connected to the grid and with little hope of ever being connected.

5.4.3.1   Instructions and training

To guarantee the success of rural solar PV in hot countries, users must change their mindset from simply plugging in an appliance to the nearest socket. The approach must be to try to respond to the local expressed need by stressing:

- a high efficiency for a reasonable and sustainable cost;
- highly ergonomic and adapted to the level of users;
- exceptional reliability enabling day-to-day operation without outside technical support.

Following these recommendations, it is usual to suggest systems with multiple sources using several DC or AC voltages to supply all appliances as efficiently as possible. The risk of connecting an appliance to the wrong source is considerable, and cabling and labelling must be carefully carried out.

Finally, to ensure a long life for the system, the training of users must be carried out from the start of the project, with local labour being used so that the knowledge of the system installation remains with future users. A report from the International Energy Agency contains a list of recommendations for the design of PV electrification projects (*Managing the Quality of Stand-Alone Photovoltaic Systems – Recommended Practices*) available on the IEA – PVPS website.[18]

5.4.3.2   Advantages of 24 V DC

Plentiful and powerful appliances
At this voltage, accessories of around 1 kW power can be found, which enables the operation with improved efficiency of appliances not available in 12 V. Also, many appliances exist in this voltage, widely used in lorries and small electrical transporters.

'Low-voltage' battery
A 24 V battery is made up of twelve 2 V elements in series, which is a quantity still small enough to avoid drifting between the elements and the necessity of using a complex regulator, which controls individually each 2 V cell.

---

[18] http://www.iea-pvps.org/products/rep03_15.htm

Compatibility with 230 V AC

In 24 V DC, the usual switches and commutators developed for 230 V AC operate without a problem, which is no longer the case from 36 V onwards.

Small networks

The distribution of current between households near to each other is possible to supply small appliances such as lamps, fans, TV and radio. However, the great attraction of 24 V is to be able to operate machines that have a much better efficiency with DC motors than with AC or when driven by petrol or diesel engines.

5.4.3.3 24 V agricultural or craft equipment

The examples and comments below largely drawn from the experience of Yvan Cyphelly of the CMR company who has worked for many years in the African Sahel region. We acknowledge his outstanding insight into the needs of rural African society.

High-efficiency motor

Any equipment installed in the Sahel needs to be virtually impermeable to dust and to be able to operate at an ambient temperature of 45 °C. These two conditions place huge limitations on the use of traditional motors or machines, which would never survive there. Therefore, high-efficiency motors must be chosen that can do without ventilation and can transmit the heat they generate to a solid construction by thermal contact and through the transmission axle if it is a good conductor. To reach the 80–90% efficiency level necessary, a DC motor with permanent magnet and mechanical switching can be used where axial dispersion reaches 85% and provides excellent heat transmission. The other advantage is an excellent use of copper, reducing ohmic losses and limiting the weight to 3.5 kg for 1.5 kW of power. The future probably lies with a motor with electronic switching that today can reach an efficiency of 87% for a ferrous magnetic circuit, which could improve to over 90% if the use of iron is abandoned.

Power of machinery

Machines designed for milling, hulling, refrigeration, circular saws, chain saws, power drills and other small machines can very often operate with motors from 1 to 1.5 kW. As current will not reach 100 A at 24 V, their cabling is unproblematic. Motors with electronic switching will be preferred for all machinery operating at different speeds, for their losses are practically proportionate to the power used. For requirements at a nominal power, mechanical switching can suffice because losses are independent of speed, and therefore proportionately minimal at full power.

For small power requirements (<800 W), a wide range of motors/reduction gears with electronic switching designed for portable tools working on batteries are available today. This type of motor can be adapted to drive sewing machines, milling lathes, small drills, etc.; 12 V and sometimes 24 V models can be found, which makes their connection to the local micro-network easier.

Cereal mills

A very useful accessory is the hammer mill that can mill millet, sorghum or maize grains, considerably lightening the workload of African women. Aid organisations have tried to introduce mills operating on diesel engines, but these are unreliable in a hot and dusty environment. The great advantage of DC motorisation is that it enables a motor to be used, which turns at the nominal speed of the mill, 4000–5000 rev/min, the correct speed for bread flour. Diesel motors and AC motors (maximum speed 2800 rev/min) can carry out this work without the addition of transmissions with belts and pulleys, which are dangerous and unreliable in a dusty atmosphere. Also, a diesel engine often weighs more than 100 kg, which makes it difficult to transport, and maintenance must be done on the spot at high cost in remote locations. An equivalent DC motor weighs 19 kg and can be carried in a backpack. The other advantages of solar DC motorisation are lower maintenance and incidental costs as well as simplified operation.

Maintenance and incidental costs reduced
Reduced maintenance costs, no oil, filter or fuel needed to be changed or filled up. Motor brushes to be cleaned every 6 months and batteries changed every 3–4 years. Much less noise, no pollution, vibration or exhaust gases.

Simplified operation
The mill is started simply with a gravitation rheostat or an electronic starter, activated without any mechanical effort, without the use of handles or a belt. The mill can be run for each individual user, no need to keep it running under the observation of a miller.

Figure 5.32 shows a mill manufactured in Senegal: only the drive and regulator parts had to be imported, the remaining parts are made on the spot, which enables it to be serviced locally and supplied with spare parts.

The same type of drive can be found for machines to hull grain husks.

Milking machine
DC motorisation can also be used to drive the pump of a milking machine. A company specialised in this type of product[19] has developed a very efficient machine that uses around three times less energy than a machine with a traditional three-phase motor. It offers two models of 750 and 1120 W operating on 24 V DC, which enable one to three animals to be milked for the first and four or five for the second at 2200 m altitude. The machine uses a sophisticated vacuum regulation to save energy: an electronic sensor measures the depression in the vacuum container and the circuit controls the operation of a motor to stabilise its value. It only needs 12 Wh of energy to produce 1 l of milk. Other advantages of this machine are a lower noise level and reduced maintenance, as the motor turns less quickly than in a three-phase machine.

Cold machines
Industrial refrigeration enables agricultural production to be maximised and, in particular, the income of producers to be diversified. A DC motor is used to drive a refrigeration unit for the conservation of milk: in order for it to be usable in the

---

[19] Brückmann Elektronik, Bahnhofstrasse 17, 7260 Davos, Switzerland, http://www.brueckmann-el.ch

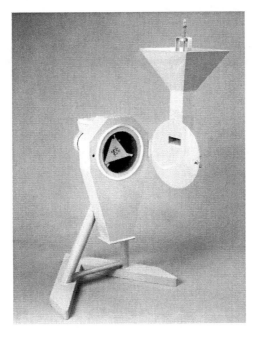

Figure 5.32   Cereal mill manufactured in Senegal

food industry, the milk needs to be rapidly cooled after milking and kept at a few degrees until the collecting lorry arrives. The system uses a DC motor with a rheostatic starter for the compressor and the cold box installed in a container. The solar panels are mounted on the roof of the container, which enables it to be moved if necessary. Figure 5.33 shows the interior of the container. A major food processing company uses these systems in Senegal in regions not connected to the electric grid.

5.4.3.4   Applications without motorisation

24 V DC solar PV can also be used to power applications such as electric welding units, electronic equipment or appliances specially developed to have a very high efficiency. Among the latter, we have already mentioned the BATNET in Section 5.1.3, and we give below its main advantages.

BATNET

The BATNET is a DC/DC converter that enables a 12 V or 24 V DC supply to be converted to 300 V DC, the voltage corresponding to the peak voltage of 230 V AC. It is mainly used to supply 230 V AC low-energy lamps or small appliances equipped with a switching power supply (chargers for mobile telephones and small appliances), but it is not suitable for appliances with a transformer operating at 50 Hz.

Figure 5.33   Autonomous refrigeration container

230 V AC low-energy lamp

The low-energy lamp uses a switching power supply transforming at start-up the mains AC voltage into DC voltage before activating a high-frequency oscillator (25–50 kHz), which supplies the fluorescent tube. The advantages of using these high frequencies are as follows:

- improved light efficiency, the fluorescent coating emitting around 10–15% more light at 25 kHz than at 50 Hz (mains frequency);
- the avoidance of scintillation phenomena occurring at 50 Hz, which are tiring for the eyes;
- miniaturisation of the electronic ballast enabling compact lamps to be produced;
- the possibility of instant starting compared to the fluorescent tube on the mains with an induction ballast and starter.

The main disadvantage of switching power supplies is a distortion of the supply current at 230 V/50 Hz, which only takes current during part of the wave and thus causes the distortion of the voltage and a big increase in peak current (Figure 5.34). This high peak current also seriously limits the use of low-energy lamps on a DC/AC inverter: a good 1 kW inverter will only be able to supply 300 W of low-energy lamps.

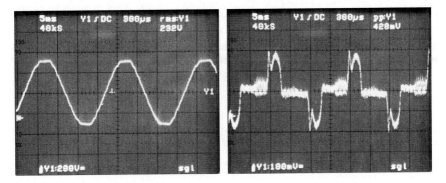

Figure 5.34   Voltage wave of 230 V AC mains and current wave of 20 W low-energy lamp

Advantages of 300 V DC

When a switching power supply is connected to a DC source of adequate voltage, the advantages are as follows:

- the charge of the first stage of supply occurs practically at continuous current;
- there is no peak current;
- ohmic losses between the supply and the appliance are considerably reduced.

Supply of distant appliances

One particularly interesting application of 300 V DC is lighting some distance away in a small isolated settlement. In Africa, the cheapest available cable is often telephone cable with a 0.14 mm^2 section, which has a resistance of around 82 Ω for 300 m. At this distance, with a 40 W lamp, the current at 300 V DC is 132 mA and the fall of voltage 11 V, which corresponds to 1.5 W of losses. If the same lamp is supplied with 230 V/50 Hz, the reduction in voltage rises to 50 V and the loss to 3.2 W. With this high loss of voltage, the lamp cannot start and if it does start, it will produce considerably less light.

Safety

As the BATNET comprises a high impedance connection with the battery, it is not dangerous to touch one of the 300 V terminals. Only a very small DC can flow to the earth, which is hardly perceptible. If both terminals are touched, there will be a shock, but far less dangerous than with 230 V/50 Hz, which is particularly dangerous to the heart. In DC, 2.5 times more current is needed to produce heart damage compared to AC 50 Hz.

5.4.4   PV pumping

Pumping is definitely one of the PV applications that makes it particularly valuable in Africa and indeed everywhere where there are difficulties in supplying drinking water. It was in fact one of the first applications of the source of energy, which is widely available in various countries (Figure 5.28). Today, worldwide, one person in five (nearly 1 billion people) lack access to safe drinking water. Water consumption varies widely: a North American consumes on average 700 l of water a day, a European 200 l, an urban African 30 l and a Haitian 20 l. In the Sahel, the amount of water bought from the public fountain is often 10–15 l/person/day.

5.4.4.1   Principles and composition of a pumping system

As the need for water is greater in hot countries and during dry periods, generally when there is plenty of sunshine, the production of solar energy coincides with the need for water. The solar pump was born from this convergence and it has been in use for more than 30 years. But solar energy should not be restricted to the simple function of pumping. The installation of the drinking water supply is a complex operation. When the pump is driven by solar energy, it is tempting to describe the whole drinking water supply operation as 'solar pumping'. The danger in this is that the quality of the other infrastructure may get overlooked, since the tendency is to concentrate on the pumping part of the operation. Yet each element in the system has its own importance, and any weak link can compromise the end objective, damaging the reputation of solar energy, which, since it is in the spotlight, is all too often blamed for any failure in the system. These thoughts, and the section that follows, are taken from a summary by Hubert Bonneviot, published by Energies pour le Monde.[20]

A drinking water conveyance system comprises the following:

- a water source (well or more often a borehole);
- a mechanical pump;
- a PV array.

It also requires the following infrastructure:

- a water tower (tank situated higher than all consumers);
- a piping network;
- public drinking fountains, and sometimes individual house connections.

There is generally no need for a battery, since the tank fulfils the role of storage. On the other hand, a starting booster may be necessary to prime the pump at the start of the day as soon as there is sufficient sunshine (Figure 5.35).

5.4.4.2   Sizing

The water source must be studied with care: the depth of the water table, its capacity, its development over time must all be known, and the borehole sunk according to the nature of the soil and drilling rules.

The tank has two functions:

- gravity feed to the water points (like any other water tower), whether the pump is in operation or not;

---

[20] *Adduction d'eau potable avec pompe photovoltaïque – Pratiques et recommandations de conception et d'installation* by Hubert Bonneviot, Fondation Énergies pour le Monde.

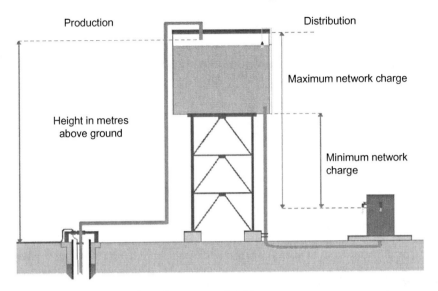

*Figure 5.35   Diagram of a solar energy drinking water conveyance system
(PV panels not shown)*

- the storage of water, to act as a buffer between periods of pumping and periods of drawing water.

Its volume can generally be calculated on the basis of the daily output of the pump, taking, for example, 120% of the production of the most favourable day. Its height is the result of a compromise between the pump's production capacity and the distribution pressure requirement.

On the production side, the solar panels need to be sized to operate the pump according to the following four parameters:

- the daily demand for water;
- the height to be pumped (between the level of the pump in the borehole and the entry to the tank);
- levels of sunshine (and temperature);
- the efficiency of the type of pump chosen.

The calculation of the peak power of the generator can be made by a simple formula: *the energy expended on pumping in one day corresponds to the work needed to raise the daily production of water V to a height H (equal to the total height from the level of the water in the borehole to the high point of the tank).*

Volume V (m³) has a mass $M = 1000 \times V \times d$ (kg) representing a weight of $1000 \times V \times d \times 9.81$ N, in other words $9810 \times V$ N, since the density d of water is 1.

The work, or hydraulic energy (J), is thus:

$$E_h = 9810\, V \times H \tag{5.7}$$

Translated into Wh (1 h contains 3600 s and 1 J = 1 Ws, therefore 1 Wh = 3600 J):

$$E_h = \frac{9810V \times H}{3600}, \text{ or, more simply, } E_h = V \times H \times 2.725 \qquad (5.8)$$

This hydraulic energy is supplied by the pump that receives electrical energy delivered from the panels (E_{prod}). If R is the combined efficiency of the pump and the PV generator (including the temperature effect), the result is $E_h = E_{prod} \times R$.

We can then apply the formula that links the peak power of the panels P_p (W), the electrical power produced E_{prod} (Wh/day) and the daily total solar irradiance E_{sol} (kWh/m²/day, or an equivalent hours, see Section 5.5.4): $E_{prod} = E_{sol} \times P_p$, and deduct the theoretical power of the pump in Wp

$$P_p = \frac{V \times H \times 2.725}{E_{sol} \times R} \qquad (5.9)$$

where

V = daily volume in m³,
H = pumping height in m,
E_{sol} = daily total solar irradiance in kWh/m²,
R = combined efficiency (generator, electronics and electric pump).

It will be noted that if, with a given power, a volume V can be pumped to a height H, it will also be possible to pump half the volume to double the height, or conversely double the volume to half the height.

To estimate the efficiency of the pumping, see Table 5.14, which gives the efficiency as a function of the generator power.

Table 5.14   Indicative total efficiency of a pumping system according to generator power

Power of PV generator	<800 Wp	800–1200 Wp	2–3 kWp	3–5 kWp	5–10 kWp
Total efficiency (%)	40	20 or 40	30	35	40

Note

The power range 800/1200 Wp straddles the use of volumetric and centrifugal pumps, with efficiency increasing from single to double.

Example of calculation

If we take a requirement of 80 m^3 of water per day, with a pumping height of 15 m and solar irradiation of 5 kWh/m^2/day, and we assume an efficiency of 30%, the peak power necessary is $P_p = (80 \times 15 \times 2725)/(5 \times 0.3) = 2.2$ kWp (or 10 m^2). The efficiency of 30% was therefore a valid assumption (power bracket 2–3 kWp).

Obviously, these calculations are estimates that must be refined and confirmed by a more thorough study.

5.4.5   Hybrid systems

When a small electrical network is installed in the village or an isolated community, a hybrid system using several sources of energy becomes virtually indispensable. PV systems may be linked with hydro or wind power, but in most cases, the second source will be a diesel generator. Having a generator producing current on demand enables the batteries to be fully charged regularly, and periods of unfavourable sunshine to be compensated by a greater use of the generator. The disadvantage of such a system is that the generator creates noise and pollution, is expensive to maintain in good running order and has to be supplied with diesel, which is sometimes difficult in isolated locations. Figure 5.36 shows a PV and thermal system supplying a very isolated inn in the Spanish Pyrenees.

5.4.5.1   Sizing

The main difficulty for designing a hybrid system is the sizing of the components: this must be done in a pragmatic way, which means that the first elements should be resized according to the results obtained from calculation. The following elements (energy requirements, supply of the generator, solar generator, batteries) should be known or carefully estimated before the design.

Energy requirements

This is the most difficult element to estimate: the uses of the electricity produced must be known with as much precision as possible in order to prepare a division of resources between the different generators of the system. Variations of consumption in the foreseeable future also need to be estimated, because consumers normally use little power at first when the system has just been installed, but later buy more appliances and the total consumption increases.

Supplying the diesel generator

Considerations to be taken into account are access to the site, type of vehicle used (car, truck, 4 × 4), the local price of diesel, the availability of a competent mechanic to service and maintain the generator. From these, initial elements can be

Figure 5.36   PV system for a remote inn in the Pyrenees

calculated or the operating costs of the generator and of the energy produced can be estimated, taking into account its maintenance and replacement after so many thousand hours of operation.

Solar generator

If the solar system is oversized, the advantage will be high reliability and additional availability of energy, but the final cost of the energy will be high; if the system is undersized, the diesel generator will be often used with all its disadvantages. It also depends on the financial situation of the client, his interest in producing 'clean' energy, and all the climatic and solar radiation factors of the location.

Batteries

The sizing of the batteries is critical because the capacity of a group of batteries cannot be easily modified (see Section 5.1.1). One very important parameter is the ambient temperature that affects the choice of size: in hot countries, the main cause of degradation is internal corrosion, and it then becomes more attractive to choose a smaller sized battery and to carry out major daily cycling to rapidly consume the charge. A large battery, on the other hand, would never see its cycling potential fully used because corrosion would set in beforehand. We have already seen in Section 5.1.1 that there are other approaches to maximise battery life, and it is useful to consult the scientific literature before designing a large system, as technology in this domain is progressing rapidly.

5.4.5.2   Operation of the system

A study of 44 systems of rural micro-electrification[21] has highlighted the key elements in the success of such schemes. The most important aspect is not technical but socio-economic: networks functioning well and most appreciated by their consumers were those which had a good organisation controlling the system from the start. These organisations (called SOTEC (*Socio-technical and Economical*) in the Spanish-speaking countries) take care of the technical running of the networks, the training of consumers in energy economy and choice of suitable appliances, monitoring individual consumption of energy and collecting payment for energy consumed and the level of services provided (generally a tax based on total consumption and peak power). The whole system remains in the ownership of the SOTEC that often receives aid at the start of the network but subsequently has to balance its budget to maintain and renew worn-out components of the system.

In Spain, many isolated villages and hamlets have been equipped with hybrid systems for electrification. They all have local SOTECs that receive initial subsidies to promote the installation of these micro-grids. In Aragon, SEBA (*Serveis Energetics Basics Autonoms*) has brought electricity to a number of individual and collective sites with some special features in the way that the networks are managed.[22]

Energy limited
The consumer undertakes by contract not to consume more than a certain quantity of daily energy for a given peak power: a special meter keeps him informed on the state of his reserve before cutting him off if he exceeds his quota.

Energy sharing
If the consumer visits a neighbour at a festival, for example, he can unplug the small memory circuit of his meter (disconnecting his house from the network) and bring it to make a contribution to the energy consumed at his neighbour's house.

Energy management
When the batteries are properly charged, the users receive an extra free energy quota.

Some appliances can only operate at certain hours or when the batteries have reached a certain charge level: for example, the operation of a washing machine is restricted to the middle of the day when sunshine is at its highest.

---

[21] X. Vallvé, *et al.*, 'Key parameters for quality analysis of multi-user solar hybrid grids (MSGs)', 17th European Solar Energy Conference, Munich, Oct 2001.

[22] For information, email: tta@tramatecnoambiental.es

5.4.5.3   Example of system

SEBA installed a hybrid system at Artosilla in the commune of Sabiñánigo, Huesca Region, Aragon. Table 5.15 summarises its specifications, and Figure 5.37 shows a block diagram of the system components.

Table 5.15   Hybrid system, Artosilla, Spain

Cost of the system	€120,000
Financing	Municipality 35%, Aragon 46%, EU 19%
Consumer costs according	33 kWh/month and 1.1 kW peak, €14.22/month
to contract	67 kWh/month and 2.2 kW peak, €19.40/month
	100 kWh/month and 4.4 kW peak, €25.33/month
Maintenance costs	€142/month(estimate)
Grid voltage	230 V, 50 Hz
Solar power	7.264 kW
Generator	Vanguard F12 propane, 10 kVA
Batteries	48 V–2080 Ah/100 h, BP Powerbloc S-2100
Regulator and inverter	MPPT and 7.5 kW inverter
Monitoring	Data logger
Energy monitoring	Individual meters/limiters (sharing possible)
Guarantee and maintenance	15 years service by SEBA
Social aspects	Community of 8 consumers. Young artists moved in permanently since electrification
Other information	Trama Tecno Ambiental*

*Trama Tecno Ambiental S.L., C/Ripollès 46, 08026 Barcelona, Spain http://www.tramatecnoambiental.es

The normal maintenance operations are carried out by technicians who visit every 6 months: they download data from the data logger onto a portable computer for later examination in case a problem is suspected. One of the consumers has the task of cleaning the panels when necessary, carrying out basic monitoring and sending certain data to SEBA every 3 months to verify the performance of the system.

5.5   Design of a stand-alone PV system

5.5.1   *Procedure*

We described in Chapter 1, the stages for calculating the cost and sizing of a PV system. Exact sizing is a relatively complex process because there are many parameters to take into consideration, a good dose of the unforeseeable (the weather), and above all multiple interactions between the different choices. For example, the energy consumption of the charge controller must be added to that of the appliances to define the total consumption of the system. But the choice of charge controller depends on the size of the PV array, which in turn is itself determined by the total consumption. Thus, the design of a PV system is the result of an optimisation of repeated estimates. The diagram in Figure 5.38 summarises the steps to follow in the case of a simple DC PV system (without energy conversion or auxiliary source of power).

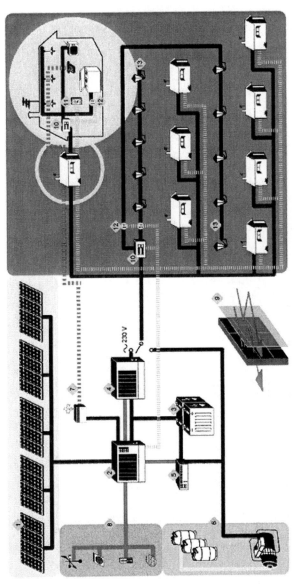

Figure 5.37 Block diagram of the Artosilla system

1. Solar array;. 2. Charge regulator and general control; 3. Battery bank; 4. DC/AC inverter; 5. AC/DC battery charger; 6. Diesel generator; 7. Data exchange, telephone, system information; 8. Weather data acquisition; 9. General lightning protection; 10. Electricity counter; 11. Load controller (low priority); 12. Load controller (high priority); 13. Street lighting

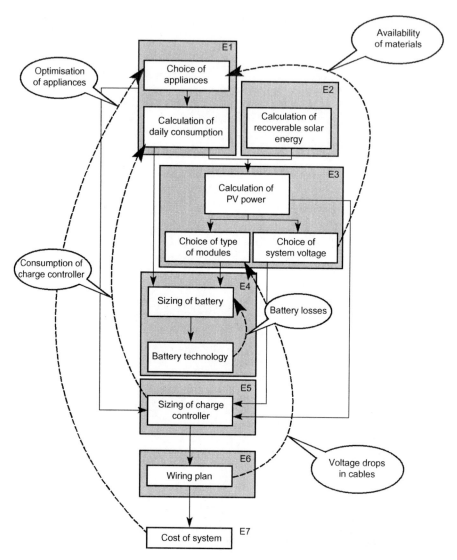

Figure 5.38   Simplified diagram and sizing of a stand-alone DC PV system

Solar energy professionals use optimisation software (sometimes online) to define solutions to meet their clients' requirements. Some programs are available for purchase or use on the Internet, notably the Swiss software Meteonorm 2000 (version 4.0)[23] and PVsyst,[24] already described in Chapter 4, and the Canadian RETscreen.[25]

[23] http://www.meteotest.ch
[24] http://www.pvsyst.com
[25] http://www.retscreen.net

These software programs include various possibilities (such as how to calculate losses caused by shading), but they all follow a similar procedure, which can be summarised in seven stages.

- Stage 1: determining the user's needs – voltage, power appliances and duration of use.
- Stage 2: calculating the recoverable solar energy according to location and geographical situation.
- Stage 3: defining the PV modules: operating voltage, technology, total power to be installed.
- Stage 4: defining battery capacity and choice of technology.
- Stage 5: choosing a charge controller.
- Stage 6: wiring plan: defining wiring accessories, cable section, etc.
- Stage 7: cost of the system.

5.5.2 Evaluation of requirements (stage 1)

It should be recalled here that the designer of PV system must constantly keep in mind that solar electricity is expensive; therefore, *energy economy* must be prioritised for all appliances. Even if they cost more to buy, the long-term expense will be much less because fewer PV modules and batteries will be needed to supply them (see Section 5.2.1). Also, appliances must be put on standby as soon as possible so that they only consume energy when it is absolutely necessary.

In fact, there often need to be several versions of the client's project specifications, each consuming less than the previous one, before deciding on the definitive PV system (see in Figure 5.38 the 'return' from stage 7 to stage 1).

5.5.2.1 Appliance voltage

Since PV energy is in direct current, the ideal is that the appliances should also be in 12 or 24 V DC. The higher the power of the PV array, the higher the voltage chosen should be (to avoid too high amperage, see Table 5.18). So it is not always easy to make an a priori choice. Once the power and therefore the desirable voltage of the PV system have been determined, the choice of appliances can sometimes be reconsidered (see Figure 5.38).

Sometimes, the choice is not possible when the appliances only exist at a certain voltage. This is even more true when the appliance only exists in 230 V AC, which is the case with most domestic appliances (see Section 5.2). Obviously, the one with the lowest consumption will be preferred, but an inverter will then be indispensable. That will increase consumption (in standby and/or in operation), which must be accounted for in the electricity balance.

5.5.2.2 The appliance's energy requirement

An accurate calculation of energy requirements is necessary for the design of a good system, bearing in mind that each additional requirement will be translated into an increase of power to be installed: more panels, more batteries.

For calculating the energy required for each appliance, there must be a clear understanding of the difference between *power* and *energy*, so we will repeat it here. The type of energy we are dealing with is electrical energy, but the following definitions are valid for all forms of energy.

Power is an instantaneous value (like an output).

Energy is the value over a period of time (like a volume).

These two values are thus related by time. Energy is the product of power by time (see Appendix 1 for units of measurement): $E = P \times t$.

Examples

A solar panel produces 88 W at a given moment, a modem consumes 120 W when active.

Four panels generated 250 Wh yesterday, my portal consumes 0.5 Wh for each opening/closing cycle, my EDF meter shows that I have consumed 550 kWh in 4 months.

This relationship enables the daily energy consumption of an appliance to be calculated, which is the product of the power consumed by the duration of use per day. As a PV system generates energy during daylight hours, it is natural to take the period of 24 h as a unit of time. The *electrical energy consumed* (E_{cons}) in 24 h should be noted for each appliance, in Wh/day. In everyday language, this is the *daily consumption*.

Note

'Per day' always means here 'per period of 24 h'.

To calculate the total consumption of all appliances, the electrical energy consumed in 24 h by each appliance or each electrical function is added up:

$$E_{cons} = P_1 \times t_1 + P_2 \times t_2 + P_3 \times t_3 + \ldots \tag{5.10}$$

In practice, Table 5.16 can be used. It must be filled in with scrupulous care.

Note

The consumption of power tools in 220 V AC via the DC/AC inverter is calculated as follows (the inverter having an efficiency of 85% at the power of 500 W:

$$500\,W \times 0.5h/0.85 = 294\,Wh$$

This assumes that the inverter is only started during the operation of the tool it is supplying. If it had to stay on permanently, it would be necessary to add its standby consumption, which is its standby power multiplied by 24 h (strongly advised against).

Table 5.16 Table for calculating energy consumption (Wh) (example)

Appliance	Number	Voltage	Power (W)	Conversion efficiency DC/AC (%)	Duration of use/ day	Daily consumption (Wh/day)
Lamps	5	24 V DC	10	–	3 h	150
Radio transmitter (on standby)	1	24 V DC	2	–	24 h	48
Radio transmitter (when transmitting)	1	24 V DC	160	–	2 h	320
Power tools	1	220 V AC	500	85	30 min	294
TOTAL	–	–	672	–	–	812

When all the appliances are working on the same voltage, the daily consumption can also be measured in Ah, a unit that is more practical for all systems connected to a battery. To obtain the figure in energy (Wh), this figure is simply multiplied by the system voltage. Table 5.17 shows an example of an emergency telephone, the electronics of which operate between 10 and 15 V. It needs to run on a 12 V nominal lead battery, without any cabling losses, since the panel, battery and regulator are all contained in one assembly. Once the problem of the voltage is resolved, everything can be calculated in current: current consumed, charge currents, operating currents, etc.

Table 5.17 Consumption calculating table in mAh (example of an emergency telephone)

Function	Number	Current (mA)	Duration of use/day	Daily consumption (mAh/day)
On standby	–	0.5	24 h	12
In operation	3 conversations	150	5 min/conversation	3.75
Test call	2	100	15 s/call	0.833
TOTAL				**16.6**

Section 5.5.4 contains details relating to the various losses of voltage and current.

We may note in passing that, in this example, the standby consumption is the largest in mAh (see final column of table), even if its current is the smallest. Everything that is permanently connected always creates a large daily consumption.

If the consumption varies over time, this must be taken into account. In the case of weekend consumption, this 2-day consumption is spread over the 7 days of the week by multiplying the result obtained E_{cons} by 2/7. And it is this average value that will be used to size the panels since energy is captured even in the absence of the owners (see case study of the house in Switzerland in Section 5.6.2). Another thing to be borne in mind: seasonal variations. In agriculture particularly, some equipment is stopped and stored during winter (electric fences, drinking troughs, etc.). Also, water consumption may be different and so pumping requirements will differ. So there is a 'winter' and a 'summer' consumption. This is the case notably with our livestock farm in Morocco (see Section 5.6.3).

5.5.3   Recoverable solar energy (stage 2)
5.5.3.1   Orientation and pitch of the modules

The position of the PV modules in relation to the Sun has a direct influence on their energy production. It is very important to situate them in the best position in order to make maximum use of their possibilities. *Orientation* is the cardinal point that the panel faces (south, north, southwest, etc.). *Pitch* is the angle that the panel makes with the horizontal, measured in degrees (see Figure 5.39).

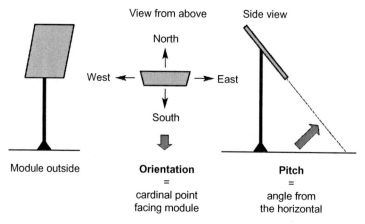

Figure 5.39   Defining the orientation and pitch of a panel

How should the panels be oriented and pitched?

When the choices are available, the ideal orientation of a PV module follows a very simple rule – *towards the equator*, in other words:

- orientation to the south in the northern hemisphere;
- orientation to the north in the southern hemisphere.

The determination of *pitch* is a bit more complicated. But first take the case of a stand-alone system that requires energy at a more or less constant level throughout the year. Since winter is the least sunny time of the year, production must be optimised during this period. The panels must be able to recover the energy of the Sun when it is low on the horizon (see Section 2.2.3). It follows that in Europe, for year-round utilisation, the ideal pitch is approximately equal to the latitude of the location +10° (for a south orientation). For France and Switzerland, panels therefore need to be oriented to the south at a pitch of 60° to the horizontal.

When the system only operates during the summer period, a pitch of 20–30° is preferable, particularly for a southerly orientation.

Orientation to the south is not always possible for a house because of its location. While north, northeast and northwest are generally excluded as being too unfavourable, it is possible to place modules facing east and west. In fact, compared to an ideal orientation of 30° S, not more than 15% will be lost in annual production for orientations to the east, west, southwest and southeast if the pitch does not exceed 30° in relation to the horizontal.[26]

5.5.3.2   Meteorological data

Let us first recall the physical characteristics of the incident solar energy we are going to exploit. There exists between the *cumulative solar radiation* and the *instantaneous solar radiation* the same dimension of time as between electrical power and energy defined in Section 5.5.2: the surface exposed to the Sun receives, at a given moment, a *solar irradiance* in W/m^2, which is a flux, a power per unit of the surface (see details in Chapter 2). This flux varies with the passing of a cloud and according to the time of day. By the end of the day, this flux has produced a daily energy or *cumulative (or integral) solar radiation* in $Wh/m^2/day$ ($Wh/m^2/day = Wh/m^2 \times h/day$), the multiple of radiation by time. As instant radiation is variable, the daily energy is obtained by calculating the whole curve of radiation as a function of time (area of the radiation curve over the day).

Meteorological stations can now supply quantities of solar radiation statistical data in the form of $kWh/m^2/day$ (the references are given in Section 2.2). It is these total daily data that are mainly used for the sizing of a PV system.

Knowing in detail the production of a PV panel hour by hour is only useful when one needs to estimate the losses due to shading. The maximum current output of a panel or PV array should, however, be known in order to size the charge controller, but that depends little on the geographical location (see Section 5.5.5).

If the situation of the array does not suffer from shading, fairly accurate sizing can be carried out with 12 solar radiation values: the average daily solar energy values for each month of the year for the plane of the PV modules (same orientation and pitch). The data will be taken from the nearest weather station to the installation site. Appendix 2 contains statistics from selected weather stations in Europe and worldwide.

---

[26] *Perseus* Guide (see Bibliography).

For a quicker sizing, the *lowest value* of the operating period of the system will be used. For a year-round utilisation, this is often the lowest value in December, at least in Western Europe. Map A2.1 in Appendix 2 enables a direct reading to be taken along isometric curves: if the PV system is to be based in Rennes, for example, the value of 1.4 kWh/m^2/day will be used (orientation south, pitch 60°). For summer use, for example from May to September, the sizing should be done with the value of May, if that is the lowest during this period.

5.5.3.3   Shading

Sometimes modules are placed where buildings or other obstacles such as mountains or trees can partially shade them from the Sun. Unfortunately, the effect of shading on the radiation received is very difficult to estimate intuitively, and we cannot suggest a simple method for estimating it, even crudely.

However, it is very important to pay particular attention to partial or even occasional shading: if one single cell is shaded, the current of the whole chain of cells in a string is reduced, and this can have serious consequences if the panels are not equipped with shunt diodes to prevent hotspots (see Section 3.1.4). Care also has to be taken when panels are mounted in rows; to limit the effect of one row shading another, all the panels in the lower row should be connected in the same string so that any shading only affects a single string of cells.

The effect of nearby shading on direct solar radiation can also be estimated. This requires an exact knowledge of the surrounding obstacles in three dimensions: they can then be entered on the curves showing the course of the Sun at different times of the year (which is known precisely and only depends on latitude, longitude and altitude). Generally losses are concentrated on months of the year when the Sun is lowest on the horizon. Calculation is quite difficult (we will not describe it here) and only takes into account direct radiation losses. However, near or distant obstacles can also occlude part of the diffuse radiation, throughout the whole year. This effect is complex and often underestimated, especially in regions with a high proportion of diffuse radiation (middle latitudes).

It can never be too often repeated that the site must enjoy a good exposure to sunlight if unpleasant surprises are to be avoided. But when faced with a serious shading problem, it is much more sensible to make use of specialised software such as PVsyst.

5.5.4   Definition of PV modules (stage 3)

5.5.4.1   Calculation of system peak power

If the Sun is the only source of energy in a stand-alone system, with no backup generator, PV modules have to supply all the energy consumed, including losses at all levels. We must not forget that the battery is a buffer storage capacity that enables solar energy to be stored for use outside periods of PV production. But it will never be a source of energy as such. Some operators believe that by increasing the capacity of their batteries, they will be able to supply more appliances. This is true in the short term, but if they consume more than they produce, the batteries will be fatally discharged sooner or later.

Daily electrical production of a module

A PV module is usually described by its peak power P_p (W), in STC conditions (1000 W/m^2 at 25 °C with a solar spectrum of AM 1.5, Section 3.1.4).

Thus, when exposed to these STC conditions, it will produce at a given moment an electrical power equal to this peak power, and if that lasts N hours, it will produce during this time an electrical energy E_{prod} equal to the product of the peak power and the time elapsed:

$$E_{prod} = N \times P_p \tag{5.11}$$

Electrical energy produced (Wh) = number of hours' exposure in STC conditions (h) × peak power (W)

But radiation is not constant over the course of the day; therefore, this law cannot be strictly applied. Another widespread erroneous calculation: the panel produces 50 Wp; therefore, during a 10 h day, it will produce 500 Wh! This overlooks the fact that radiation over the course of the day is far from being permanently equal to 1000 W/m^2. It must be remembered that this standardised value of 1000 W/m^2 corresponds to an intense solar radiation, full cloudless sunshine, which only occurs in Europe at noon on the finest spring days (in summer the haziness of the sky reduces the amount of radiation received).

In order to calculate what a PV module produces during a day of sunshine of a given type and total solar energy expressed in Wh/m^2/day, this solar energy will be estimated as the product of the instant radiation of 1000 W/m^2 by a certain number of hours known as *number of equivalent hours*. Figure 5.40 explains this equivalence: the areas under the curves are the same – that of real radiation, of rounded shape, and the square equivalent curve. Since the reference radiation has a value of

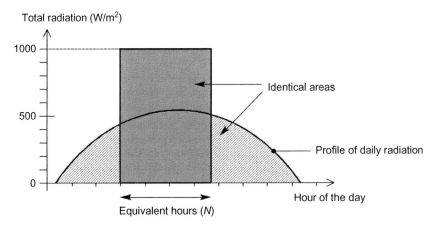

Figure 5.40   Representation of the number of equivalent hours of a day of solar radiation

1000, the number of equivalent hours is numerically equal to the total solar energy as expressed in kWh/m^2/day.

$$E_{sol} = N_e \times 1000 \qquad (5.12)$$

Daily solar energy by unit of surface (Wh/m^2/day) = number of equivalent hours (h/day) × 1000 (W/m^2).

Example

During one day at the Trappes weather station in December, facing south at 60° pitch, the Sun provides 1.12 kWh/m^2/day. This day is the equivalent of 1.12 h of a radiation of 1000 W/m^2: 1120 Wh/m^2 = 1.12 h/day × 1000 W/m^2.

We can then assume that the power of the panel is directly proportionate to instant radiation, which is more or less true if the panel has sufficient voltage (we will return to this approximation in 'estimating losses'). The peak power of the solar panel can then be multiplied by the number of equivalent hours to obtain the production of the PV module during the day:

$$E_{prod} = N_e \times P_p \qquad (5.13)$$

Electrical energy produced during the day (Wh/day) = number of equivalent hours (h/day) × peak power (W)

Since $N_e = E_{sol}/1000$, the energy produced can be calculated, provided the values used are accurate:

$$E_{prod} = E_{sol} \times P_p \qquad (5.14)$$

Electrical energy produced in the day (Wh/day) = daily solar energy (kWh/m^2/day) × peak power (W)

It should be recalled that peak power is the product of the voltage V_m and the current I_m at the point of maximum power of the module. Therefore, by dividing this equation by the voltage V_m, the production can be expressed in Ah/day by using the current at maximum power I_m. Theoretically, Q_{prod} is a capacity or a 'quantity of electricity' but more simply it is often called 'electrical energy' like the Wh:

$$Q_{prod} = E_{sol} \times I_m \qquad (5.15)$$

Electrical energy produced in the day (Ah/day) = daily solar energy (kWh/m^2/day) × current at STC peak power of module (A)

But these calculations are only valid for an isolated panel in ideal conditions. They do not take into account the inevitable losses of the total system in real conditions. These losses have various causes and affect certain parameters of the system.

Electrical losses

We will identify these one by one to make it easier to estimate.

We will take into account all the sources of loss in the system, including those due to the battery, wiring, etc. (except those caused by shading, which we have already discussed in the earlier section). These must be included in the sizing of modules because they provide all the energy consumed, including energy that is lost.

Note

Any electrical losses of the converters (DC or AC) are not taken into account at this stage, but they are included in the calculation of consumption of appliances (E_{cons}).

Types of losses

Starting with the input of solar radiation, we find the following:

a. losses due to *dirt on the panel* or snow, or sand, or even by glass placed in front of it, which modify its charge current, the voltage not being affected.

    Then, there are falls in voltage between the panel output and the battery input (line losses):

b. at the terminals of the *blocking diodes* (Figures 3.17, 5.11 and 5.12),

c. at the terminals of the *series regulator* if one is included, because it includes in-line electronic switches (Figure 5.12),

d. at the terminals of *the cables* depending on their length, their section and the current carried.

    One further loss directly affects the voltage of the panel:

e. the reduction in voltage when the temperature rises (Figure 3.9), as the peak power is calculated at a temperature of 25 °C.

    The battery also plays a role, since it does not restore energy at 100%, so the following need to be considered:

f. the energy efficiency of the battery: ratio between the energy restored and the energy supply.

    When the regulator is not of the MPPT type, a loss arises through voltage mismatch:

g. in a system with a classic regulator, the voltage is imposed by the battery (plus the line losses) and so the PV module does not work at its maximum power point.

    Furthermore, there can be a difference between the reality of the calculation shown in (5.15) because it assumes that the power of the panel is proportionate to the radiation whereas, in fact, it is the current that is proportionate (see Section 3.1.2, Influence of illumination), so sometimes the following must be considered:

h. the losses at the beginning and the end of the day when radiation is low and the voltage inadequate to charge the battery.

Finally, there is a loss linked to the real power of the panel, which may be lower than that given in the manufacturer's documentation. We will not include this 'loss' in our calculations because it does not always occur, but one must be aware that it can happen. Verification of these values is not easy and must be done by a specialised laboratory. To avoid this type of uncertainty, some manufacturers show the power measured panel by a panel on their label on the back.

Quantifying losses

Some losses can be reduced by taking certain measures: the fall of voltage in cables (d) can be reduced to the minimum by good wiring (see Section 5.5.6). Dirt on the panels (a) can be avoided by regular cleaning, which will avoid 5–15% of losses. Snow generally slides to the bottom of the panel as soon as the Sun warms their surface. Sand problems can be dealt with by placing the panels slightly off the ground to avoid accumulation of wind-blown sand at the foot of the frames.

High temperature loss (e) will only affect systems in hot countries, where heat effects can be reduced through a good ventilation system for modules. In temperate countries, heat is less critical for stand-alone systems because it occurs in summer at a time when the energy balance is in excess on account of higher solar radiation.

Losses arising through voltage mismatch (type g) will generally be completely avoided by the use of an MPPT regulator, the function of which is precisely to achieve a balance between the battery and the panel by aligning itself on the maximum power point of the panel (see Section 5.1.2, MPPT regulator).

The technology of the modules is also important. Amorphous silicon panels, for example, react better to low light levels than crystalline silicon modules and losses of type h will not occur. Their voltage also varies less with temperature (loss e), Section 3.2.2.

A further important point: line losses of types b, c and d, those due to temperature (e), low light levels (h) and losses linked to panel-battery voltage mismatch (g), only concern voltage. Once the wiring has been optimised, if the PV module cannot cope with the remaining falls in voltage, in other words, if its voltage is too low or if its voltage falls too rapidly with solar radiation, the system will not operate correctly or possibly not at all (the battery will not charge). It is therefore essential that the modules can cope with these voltage losses. Extra modules will not compensate for this shortcoming.

Concretely, it is best to take the following steps (unless a good MPPT regulator is available):

- Take necessary precautions to limit falls in voltage: adequate wiring, series regulator for 24 or 48 V DC systems, good ventilation.
- Calculate the fall in voltage between the panels and the battery: for example, 0.8 V in the series diodes + 0.5 V in the wiring + 1.5 V loss through heating above the average temperature of the site (see Section 3.1.2).
- Choose modules whose voltage V_m at peak power is above or equal to the maximum voltage of the battery + this loss of voltage: $14\ V + 2.8\ V = 16.8\ V$, for example (see definition of V_m in Section 2.4.1).

- Finally, calculate the PV array according to current at this maximum power (A), and in battery capacity (Ah), not considering the voltages but only the losses affecting the current.

To simplify, let us repeat that as a general rule, *PV modules supplying a 12 V nominal system should have a voltage at maximum power point at least equal to 17–18 V for operation in hot countries, and 15–16 V for operation in temperate countries* (for a 24 V system, double these values).

The current losses, which will inevitably remain (types a and f), are introduced into the energy calculation in A or Ah in the form of a coefficient C_1 that we will call 'current loss coefficient'.

Evaluation of current losses
For losses due to dirt on the panels, the following values can be generally used:

- 0.9–0.95 as a general rule,
- 0.95–1 for panels cleaned regularly,
- 0.8–0.9 for horizontal panels that are not cleaned,
- 0.92 for panels placed directly behind glass[27] (the loss in this case is 4% of reflection for each glass surface, therefore 8%).

It should be recalled that the efficiency of the battery (loss f) is the ratio between the capacity restored and the capacity charged. For lead batteries used in PV one can generally assume an efficiency in Ah of between 0.8 and 0.9 according to the battery model and reliability required.

To sum up, with modules that have a sufficient voltage reserve to cope with the losses in voltage described earlier in this section, if the effects of a and f are combined, the coefficient C_1 will vary between 0.65 (= 0.8 × 0.8) and 0.9 (= 1 × 0.9) depending on the type (without glazing on the modules).

Practical calculation of PV power
These losses will be directly introduced into the calculation of the electrical production of the modules. To do this, we take the formula (5.15) and add the coefficient C_1:

$$Q_{prod} = C_1 \times E_{sol} \times I_m \tag{5.16}$$

Energy produced during the day (Ah/day) = current loss coefficient × daily solar energy (KWh/m²/day) × maximum STC power of module (A).

To calculate the power necessary for the system, we use the above formula in reverse, replacing energy produced by energy consumed (total daily consumption defined in stage 1). What we are trying to do is to answer the question: what peak power is needed to provide electrical energy corresponding to the needs of the system?

To do this, we use the weather data determined for the system according to stage 2 of the procedure.

---

[27] Provided ventilation is adequate to avoid heating by the greenhouse effect.

Important

To be certain of having enough power in all seasons, this calculation will be made in the most unfavourable sunshine conditions (in winter for Europe, usually in December). The exception is inter-seasonal storing; see the case study described in Section 5.6.1.

Therefore, from (5.15) we get:

$$I_m = \frac{Q_{cons}}{E_{sol} \times C_1} \tag{5.17}$$

Current at maximum STC power of the module (A) = electrical energy consumed per day by the system (Ah/day)/(most unfavourable daily solar energy (kWh/m^2/day) $\times$ current loss coefficient).

If the daily energy is expressed in mAh, the module current will be calculated in mA.

Example of calculation

- Daily consumption from Table 5.16: 812 Wh/day or 34 Ah/day divided by 24 V (desired nominal voltage)
- Current loss coefficient: 0.75
- Daily solar energy in Paris in December (facing south at 60°): 1.12 kWh/m^2/day
- Current I_m needed: $I_m = 34/1.12 \times 0.75 = 40.5$ A
- With modules with a maximum voltage V_m of 34 V, the PV power of the system will be at least $P_p = 40.5$ A $\times$ 34 V = 1377 Wp ($\doteq$ 1.38 kWp)

5.5.4.2 Module technology

The most appropriate technology for modules depends above all on the power to be supplied, and also on the type of climate, on cost and sometimes on aesthetic considerations.

Amorphous silicon has a particular appearance, and it has particularly good performance in low light levels and under diffuse light conditions. On the other hand, its solar efficiency is only 7–9% as against 13–20% for crystalline silicon (see Section 3.2.2), which may cause a space problem because more panels are required.

It will therefore be reserved to particular cases:

- low power (less than 10 Wp) in a temperate climate (see Section 5.6.1);
- low-cost applications (electric fence recharging, some lamps in Africa, sports applications);

- portable or flexible products;
- certain architectural applications on account of its uniform aesthetic aspect.

Amorphous panels of more than 40–90 Wp can be found at competitive prices. However, with their low efficiency, more extensive surfaces are required and support structures will cost more. Therefore, this technology is rarely used for stand-alone systems of any size.

Most PV applications of more than 50 Wp are therefore equipped with crystalline silicon modules of a power generally between 50 and 150 Wp.

In all cases, it must be ensured that the type of panel is suitable for the appliances by reviewing all the electrical parameters, with particular importance being paid to the following points:

- adequate voltage (see calculation of losses);
- type of guarantee offered on peak power (sometimes the guarantee only covers 80% of the nominal power and this must be borne in mind in the calculations);
- type of climate;
- ease of installation, etc.

5.5.4.3 Operating voltage and structure of PV array

Nominal voltage of PV system
The voltage of the PV system (12, 24, 48 V or more) depends on

- the type of appliances,
- the PV power of the system,
- the availability of materials (modules and appliances) and
- the geographic location of the system.

In stand-alone systems, for a given power, a low voltage implies high current, which produces ohmic losses in the cables. A 100 W–12 V appliance already represents a current of 8 A. Obviously, wiring of sufficient diameter will be chosen to limit these losses (see Section 5.5.6), but it would be foolish, for example, to wire a 3 kWp PV array in 12 V, which would give an output current of 250 A! There would have to be many thick cables, and the regulator would need to be sized to tolerate such a current. Changing to 24 V DC directly reduces these three constraints.

For a modest installation, with, for example, some lighting points and a television set, the voltage could stay at 12 V. But as soon as there are more powerful appliances (refrigerators, pumps, etc.) and/or if the PV power exceeds 500–1000 Wp, the voltage must increase to 24 V or even 48 V DC. We have seen earlier that a voltage of 24 V DC is often used in hot countries for medium-sized systems (see Section 5.4.3). At this voltage, it is possible to use mains switches of the 230 V AC type, which is not possible even at 36 V. A voltage of 48 V DC is only exceeded in particular cases: connection to the grid, PV array of more than 10 kWp, supply of large agricultural machinery, etc. Table 5.18 shows the voltage most suitable for most purposes.

Table 5.18   Voltage recommended for PV systems according to their power

Power of PV system	0–500 Wp	500 Wp–2 kWp	2–10 kWp	>10 kWp
Recommended voltage	12 V DC	24 V DC	48 V DC	>48 V DC

Once the voltage is decided, it must be verified that the appliances are available in this voltage. If not, it is always possible to install DC/DC converters (see Section 5.1.3). Their efficiency must then be taken into account, the daily consumption of the system recalculated and the PV power increased accordingly.

Designing the PV array

Once the required PV power has been decided, an array of modules is designed in series/parallel or only in parallel depending on the voltage of the modules and the array to be constructed. The number of modules should of course be rounded up to their higher whole value and sometimes to the higher even number when they have to be wired in pairs. To return to our calculation example: we needed 1377 Wp at 24 V nominal. Assume that the modules chosen are 47 Wp–12 V. We will need to install 30 modules to have 1410 Wp, wire them in pairs in series to make 15 strings of 24 V, and install these 15 strings in parallel.

Another example: if 150 W–24 V panels were available, we would need 9 (1350 Wp) or 10 (1500 Wp) modules. If the sizing is 'generous' and we have already built in a safety margin for certain appliances, we could perhaps make do with 9 modules, otherwise we should plan for 10. In both cases, the panels will all be installed in parallel.

These series-parallel installations were described earlier in Chapter 3 (Figure 3.16).

5.5.5   Sizing storage and the regulator (stages 4 and 5)

The sizing of a battery is selecting an amount of 'buffer' storage to overcome temporary climatic variations by the hour, the day (it will allow nocturnal consumption), and days of bad weather.

Figure 5.38 shows the amount of backtracking that must be done during a sizing operation. It is the same for the batteries: the choice of nominal capacity mainly depends on the technology (sealed or open batteries, AGM, gel, etc., see Section 5.1.1) because of variations in capacity with temperature, the number of cycles, life expectancy, etc.

Battery choice also depends on budgeting strategy. Here again the policy to adopt is different in hot countries from temperate countries. The battery is often the least durable component of a PV system, and will therefore need to be replaced before the panels. If higher reliability and longer life is the objective, for example, because the location is of difficult access, a large battery will be preferred for greater autonomy without solar input and a longer life. Tubular batteries, for example, can last 15 years and provide 1200 cycles at 80% discharge. Even if the battery is expensive, it may well be more economical than expensive site visits. However, for systems less intensively used or situated in very hot regions, cheaper batteries may well be a better solution (solar plate batteries, or even car batteries, available everywhere), because it is probable that corrosion on the terminals will set in quite quickly whatever the technology. So the price should be low, with replacement planned every 2 years or so (see Table 5.1 and the section on battery life).

5.5.5.1   Autonomy without solar power

The number of *days of autonomy*, N_{da}, is the period that you need the system to operate when no power is produced by the PV panels. This is the basis used to size the battery.

There is one exception however, when *inter-seasonal storage* is required, the battery is oversized to back up the panel in winter and its capacity will need to be larger than the simple requirement of autonomy without solar power (a specific case of this type using sealed batteries is described in Section 5.6.1).

The duration of autonomy required is linked to the probability of having a series of bad weather days in succession, with very little solar radiation. This depends directly on the meteorology of the location. For normal use in temperate climates, it is usual to assume an autonomy of 5–8 days, reducing the risk of failure to less than 1%. This risk can be even further reduced for highly sensitive systems by assuming 10 or even 15 days without solar input. In tropical countries where the weather is more regular (with significant solar contributions even on rainy days), this autonomy can be reduced to 2–4 days. Meteorological databases (see Section 2.2.3.2) include statistics on the number of days without sunshine.

5.5.5.2   Calculation of battery capacity

The nominal capacity of the battery is generally given for discharge in 20 h (C_{20}) at a temperature of 25 °C.

The capacity necessary for an operation of N_{da} days and the daily electrical consumption of Q_{cons} is

$$C_u = N_{da} \times Q_{cons} \tag{5.18}$$

Useful capacity of the battery (Ah) = number of days of autonomy without solar power × daily consumption of the system (Ah).

This useful capacity C_u is not the nominal capacity C_{20}, but the capacity really available to the system at any time. To calculate the nominal capacity as a function of this desired capacity, we must take into account the temperature and/or the depth of discharge authorised. Let us examine how to do this.

Depth of discharge

A battery must not be discharged beyond a certain point or it will risk being damaged.

Reminder

E_{CH}, a number between zero and one, expresses the state of charge of a battery, and thus the depth of discharge P_D, proportion of the capacity of discharge, is expressed as

$$P_D = 1 - E_{CH} \qquad\qquad (5.19)$$

A battery charged to 70% ($E_{CH} = 0.7$) is at a depth of discharge of 30% ($P_D = 0.3$).

Care must be taken with this parameter when the battery is intensively used, that is, when autonomy without solar input is short, 2 days, for example. The battery will then be submitted to fairly frequent cycling, at night to 20–25% and during periods of cloudy weather. An absence of sunny weather for 2 days is more probable than for 8 days. Attention must be given in this case to the number of cycles that the battery can tolerate during its life and raise the threshold of the depth of discharge to increase the number of cycles. We have seen in Section 5.1.1 on batteries that the number of cycles will be in roughly inverse proportion to the depth of discharge: for example, a battery able to supply 300 cycles at 100% discharge should be able to supply 600 cycles at 50% discharge with a good charge controller.

In practice, in the absence of problems with low temperatures and for normal usage (4 days of autonomy), we can apply the coefficient $P_D = 0.7$–0.8 according to the battery model: 0.7 for batteries that can tolerate small number of cycles (car battery, plate battery), and 0.8 with a high number of cycles (tubular and gel batteries). If the battery needs to cycle more (2 days of autonomy or less), P_D could be reduced to obtain a longer life. On the other hand, if the battery is very unlikely to be discharged (more than 8 days of autonomy), P_D could be taken at 0.9 or even 1 (100% discharge authorised, in other words, 1.85 V/element, nominal end of discharge without damage to the battery).

Temperature effect

If the system needs to function at low temperatures (isolated professional applications in temperate countries particularly), this will be the main cause of reduction in capacity.

The chemical reactions of charging and discharging the battery are slowed by the cold, which has the effect of lowering the capacity of the battery (Figure 5.5).

To determine the resulting reduction in capacity, we need to consult the discharge curves at different temperatures provided by the battery manufacturer, such as the ones in Figure 5.41. Note that these curves are not universal, they vary considerably according to the battery model and, in particular, according to the composition of its electrodes.

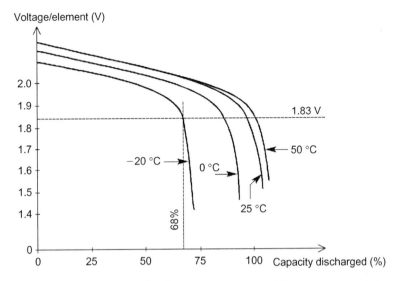

Figure 5.41 Determination of reduction in capacity with temperature (selected example: sealed lead battery type AGM-Hawker)

According to the minimum temperature the battery will encounter on site and the minimum voltage that the system can accept (some electronics disconnect at 11 V, for example, or 1.83 V/element for six elements), the temperature reduction capacity coefficient R_T can be determined from these curves. In the case represented in Figure 5.41, the battery will have a capacity of 68% available at $-20\,°C$ if it is discharged down to 1.83 V/element, so R_T will be taken as equal to 0.68.

Calculation of capacity with reduction coefficients
In order to take into account the phenomena of temperature and depth of maximum discharge, the nominal capacity is calculated as follows:

$$C_{20} = \frac{C_u}{P_D \times R_T} = \frac{N_{da} \times Q_{cons}}{P_D \times R_T} \qquad (5.20)$$

Nominal capacity C_{20} (Ah) = number of days of autonomy without solar power (days) × daily consumption (Ah/day)/maximum authorised depth of discharge/temperature reduction coefficient.

5.5.5.3 Choice of type of battery

We have just seen the influence of the parameters of cycling and life expectancy on the choice of type of battery.

The other parameters that must be considered are as follows:

- maintenance,
- the rate of replacement,

- cost,
- availability,
- recycling.

If the user is present on the site of the PV system or close to it or if maintenance visits are possible, an open battery can be used, which calls for monitoring of levels and densities of the electrolyte. A sealed battery requires no maintenance (except for its terminals, see Section 5.5.7, Installation and maintenance of batteries). The cost effectiveness will depend completely on the budget that will be drawn up according to conditions of use. The practical case studies described in Section 5.6 describe problems arising and solutions recommended.

5.5.5.4   Charge controller sizing

Before sizing a charge controller, decisions must be taken on the type to be installed and the associated options. Section 5.1.2 gives all the details on the different characteristics of a charge controller (see, in particular, Table 5.9).

Choice of technology
The first question to be answered is whether load shedding is required or not.

A simple *charge controller* ensures that the battery is well charged and protected against overcharging, but it does not deal with any possible discharge problems. It is not equipped with a disconnect device to cut-off all or part of the appliances in the case of a discharged battery. This type of regulator is generally adequate in all cases where the risk of accidental discharge is very low, such as those with generous sizing of modules, programmed consumption with a low probability of exceeding it, or a system already provided with a 'low battery' monitor. This is the case with many stand-alone professional appliances, which are unable to be supplied by battery through a solar system: their electronic chip includes a monitoring function that measures the voltage of the supply from any source, and when the threshold of low voltage is reached, it activates corrective action (cut-off or alarm).

A *charge/discharge regulator*, on the other hand, is usually required for domestic appliances, because they can easily exceed anticipated consumption levels. It is then best to 'disconnect' the installation, in other words, to cut-off at least part of the appliances (the least important) to allow the battery to recharge.

Parallel installation of regulators
Most manufacturers offer regulators at 20 or 30 A charge current, but if this level is exceeded, it is best to divide the array into several equal parts and connect a regulator to each section of the array before the connection with the battery. It must be verified that the technology of the regulators permits this type of connection and that there is no interference between sub-systems: for example, with series regulators generating a PWM wave at the end of the charge, when the first regulator switches to PWM mode, it is important that this pulsed voltage does not disturb the other sub-systems that could then fail to function correctly leading them to overcharge the batteries.

The choice of regulation technology – shunt, series or MPPT – should be guided first by the power of the PV system and the type of battery to be charged.

A shunt regulator, which must dissipate the power of the panels if the battery is overcharged, is better suited to small systems, and the series regulators to larger systems. Also, the latter cause a greater fall in series voltage between the panels and the battery. See Table 5.9 for a comparison between the technologies.

It is important that the regulator should be compatible with the model of battery used: the equalisation charge is only relevant for open batteries, and cut-off thresholds also depend on the technology used (Table 5.5).

Sizing

Once the best technology has been identified, the charge controller should be sized according to the following essential parameters: voltage, input current and output current.

- *Nominal voltage* (12, 24 or 48 V DC): it must be the same as that of the array.
- *Input current*: this is the maximum charge current that the modules are capable of producing at any time and can be accepted without problem by the charge controller. To estimate this current, the safest is to take 1.5 times the total short-circuit current of the modules for a shunt regulator and 1.5 times the total current I_m at maximum power point.

And for regulators that also provide discharge protection (disconnect function):

- *Output current*: this is the total maximum load that the appliances can draw simultaneously. This value depends on how the appliances are used: which appliances will operate at the same time? Are there transitory points of peak consumption? Some appliances (incandescent lamps and motors particularly) consume considerably higher power at start-up than their permanent consumption. Good regulators accept high transitory current (see their technical specifications), but it is always best to carry out a test.

Example

Returning to the example of Section 5.5.2: we have 1410 Wp at 24 V DC with thirty 47 Wp 12 V DC modules (in 15 strings of two modules), to supply 812 Wh/day in Paris (Table 5.16). The regulator chosen is a series model, and we must therefore calculate the input amperage by taking 15 times the maximum power current of a module and multiplying it by 1.5. The result is $1.5 \times 15 \times 47$ W/17 V $= 62$ A, assuming that these modules have a voltage V_m of 17 V. Note that if the installation had had the same power, but at 12 V, the total charge current would have doubled.

For the output current, if we assume that all the appliances operate at the same time, they would consume 32 A at 24 V permanently, taking into account the efficiency of the inverter (172 W/24 V $+$ 500 W/(24 V $\times$ 0.85) $=$ 7.1 A $+$24.5 A), which is considerably lower than the 62 A input current. And as for transitory peaks, we assume that the power tools use three times their nominal power – 73.5 A at 24 V – the total peak current would be 80.6 A.

The series regulator should therefore be a 24 V–60 A model. Obviously, it must be checked that it has the necessary LED indicators and protections (input and output fuses, protection against overvoltage and reverse polarity). It should also be able to cope with a transitory current of 80 A.

Then one would also look to certain options that are not indispensable but are sometimes recommended:

- an independent temperature sensor if the battery and the regulator are not at the same ambient temperature,
- an independent measure of voltage if the regulator and the battery are not close to each other (measuring by the battery supply cable would be unreliable because of the fall in voltage),
- meters to monitor the battery voltage and the PV array amperage.

5.5.6   *Wiring plan (stage 6)*

Once the system has been decided on, it must be properly installed, and this is what we will deal with in the second part of the Section 5.5.7.

But the first thing to be tackled is the wiring plan, to ensure the coherence of the whole system. We have already seen in the estimation of losses (see Section 5.5.4) how punishing the voltage drop in wiring could be.

It is also important to verify that the diameters of wiring chosen are compatible with the terminals of the chosen components: modules and regulator particularly. It can happen that they will not accept wiring of the section selected, in which case an intermediate junction box or additional terminal strip will be needed.

Before calculating the wiring sections, an overall electrical plan of the installation should be made. The location of the components should also be planned as accurately as possible so as to be able to work out the distances from modules to regulator, regulator to battery and regulator to appliances.

It should be remembered that for the regulator to provide an accurate measure of the battery voltage, it should be placed as close to the battery as possible.

On the overall electrical plan, the length of each cable should be noted and the amperage that it will have to carry (see, for example, the overall plan of the system of the chalet in Section 5.6.2).

The information that follows concerns wiring; it is general and applies to the installation of all the electrical components of the system.

5.5.6.1   Choice of wiring sections

The fall of voltage in a conductor is given by Ohm's law:

$$dV = R \times I \quad \text{where } R = \rho \times \frac{l}{s} \tag{5.21}$$

where
R = resistance (Ω),
l = length (m),
s = section (mm^2) of conductor,
ρ = specific resistance roughly equivalent to 20 mΩ mm^2/m for copper.

5.5.6.2   Direct current

The first rule to follow is to estimate what line losses are acceptable (Table 5.19); the fall in voltage must be measured on both terminals for current going and returning to the appliance. The current should also be limited to 7 A/mm² to avoid overheating of the conductors. We give below some figures for a nominal voltage of 12 V; obviously the values given will be proportionate for higher nominal voltages: for the connection between panels and regulator, around 0.5 V can be accepted at the normal current of the panels whereas for the regulator-batteries connection it should remain below 0.05 V if accurate regulation is to be maintained.

Table 5.19   Ohmic losses in wiring (copper conductors)

Wiring section	mm²	1.5	2.5	4	6	10	15	25
Resistance	mΩ/m	13.3	8.0	5.0	3.3	2.0	1.3	0.8

Fall in voltage per metre of double wiring								
Current 1 A	mV/m	26.6	16	10	6.6	4	2.6	1.6
Current 3 A	mV/m	79.8	48	30	19.8	12	7.8	4.8
Current 5 A	mV/m	133.4	80	50	33.4	20	13.3	8
Current 10 A	mV/m	266	160	100	66	40	26	16

Length of wiring corresponding to 5% loss at 12 V nominal								
Current 1 A	m	22.5	37.5	60	90	150	225	375
Current 3 A	m	7.5	12.5	20	30	50	75	125
Current 5 A	m	4.5	7.5	12	18	30	45	75
Current 10 A	m	2.3	3.8	6	9	15	23	38

For the wiring of appliances, a maximum loss of 0.5 V is acceptable (approximately 4%). In star wiring, it is usual to install several cable sections in parallel from the battery and then finish with a single cable for the last appliance. A good rule is to use sections of 2.5 mm² as basic wiring and to install as many cables in parallel as is necessary for minimum losses.

For connections between the solar panels and all exterior wiring, it is best to use flexible multi-core wiring with insulation resistant to UV radiations (rubber, for example).

Figure 5.42, based on the same calculations of ohmic losses, provides a convenient chart to directly determine wiring sections (for a fall in voltage of 3–4%):

- first find the point on the right-hand scale of the current carried by the cable (ensure correct voltage);
- on the scale in metres on the left, mark the distance that the cable has to cover;
- draw a line between the two points: at the intersection on the central scale, read off the section of wiring to use.

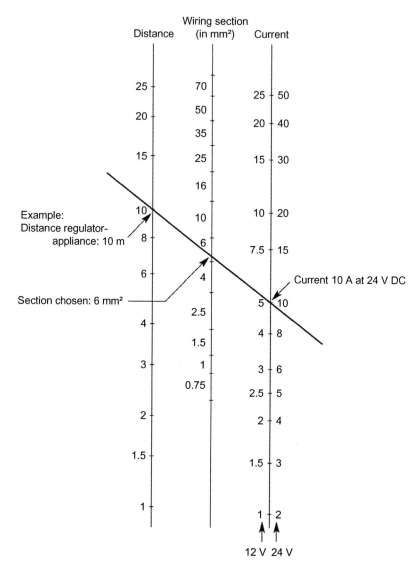

Figure 5.42   Chart to determine wiring sections in direct current [source G. Moine]

Then the exact loss can be calculated using Table 5.19.

5.5.6.3   Alternating current

For distribution in AC in a system with an inverter, all the 230 V AC wiring must conform to the standards of the country concerned. Details may be obtained from the electricity companies.

5.5.7 *Installation and maintenance of a stand-alone system*

The installation of a PV system does not differ greatly from that of a traditional electrical installation, but the peculiarities of direct current and the low voltage linked to large currents impose a number of special precautions. Also, since solar panels need to be installed outside, a whole series of problems linked to the environment may appear: corrosion or ageing, depending on the degree of salinity, materials and installation methods used.

The commissioning of a PV installation is part of the minimal training of the user: all the components must be reviewed and checked and basic measurements must be carried out on the batteries. All measurements carried out must be entered into the log book along with any other relevant observation. The difference from a classic maintenance operation is that the check must include verification that the technical specifications have been respected; during subsequent controls, values measured at the commissioning will be used as reference to determine the state of the system.

Maintenance of these systems is much reduced, with personnel in charge of maintenance mainly being responsible for monitoring the batteries, since open batteries require regular checks, especially in hot countries.

We give below the rules for the installation and maintenance of all system components.

5.5.7.1 Installation and maintenance of modules

The installation of PV modules is carried out in two stages: mechanical installation and electrical connections – in this order, naturally. Even so, a number of precautions must be taken to ensure that the installation is not electrically dangerous for installers or users, that it provides a reliable service without breakdown, and that it is durable with as little maintenance as possible.

It is also worth thinking of insurance to cover possible damage by a third party (vandalism or theft).

Mechanical installation

We saw in the preceding chapter (see Section 4.2.2) three types of mechanical mounting for PV panels: mounting on a roof or facade, integration into a building and mounting on frames (Figure 5.43). These methods are valid for both grid-connected and stand-alone installations.

If several frames are placed one behind the other (in rows, see Section 4.4.1), especially in temperate climates where the Sun is low, care must be taken that panels do not cast a shadow on the row of panels behind them. One way around this is to raise the panels in the row behind (Figure 5.44).

For small stand-alone systems, the panels may also be mounted on masts or boxes.

Mounting on a mast

This can be very convenient in locations where there is little space on the ground. But it is mainly used either to prevent theft or to avoid obstacles that could cast

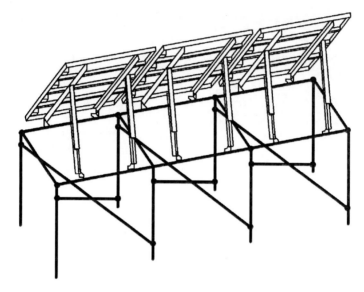

Figure 5.43   Example of frame support for PV modules (Solarex)

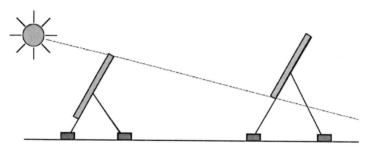

Figure 5.44   Placing of rows of modules to avoid shading

shade. It is normally reserved for small arrays (<5 m^2). The fixing devices must be robust to resist the wind, suitable for the areas and weights of the modules, and with clamps suitable for the mast diameter. Suitable masts can be found at mechanical or building suppliers: the best type to choose is hollow cylindrical poles, with an inspection aperture so that cables can be run inside. Figure 5.45 shows examples of how modules can be fixed to masts.

Mounting on a cabinet
This is a way of integrating small PV modules, especially amorphous silicon modules, with an electronic appliance. It may be useful to mount them flat on the top of a small cabinet or a box containing the electronic measuring device or the emergency telephone, which it is to supply: they are invisible, and therefore

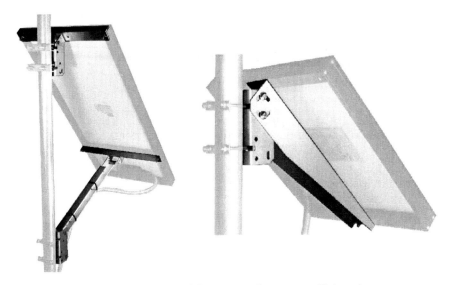

Figure 5.45 Modules mounted on masts (Solarex)

unlikely to be stolen and require no particular support – fixing with waterproof adhesive or fixing screws attached to the back of the module. The fixing must be flexible and non-corrosive. This type of installation is used in the telemetry system described in Section 5.6.1.

Sometimes the module can be even mounted inside a cabinet with a transparent lid. If the unit has a good index of climatic protection (IP 65), non-encapsulated modules could even be used, since weather protection will be provided by the cabinet.

Electrical installation of modules

Safety

A PV module generates voltage in the presence of light. Before being connected, it is in open circuit and therefore produces a voltage at least equal to 1.5 times its nominal voltage: 22 V at no-load is a usual voltage for a 12 V nominal module. Additionally, for systems comprising several panels in series, it is very easy to generate an electrical arc. It only needs a poor connection or bad installation to start an electrical arc that will only disappear when the terminals are destroyed or it gets dark. Direct current can also be dangerous to the human body: at high intensity it can cause serious burns. Installations operating at less than 50 V do not present a major hazard. But from 120 V, special protective measures must be taken.

It must always be remembered that a PV array generates a voltage close to the maximum as soon as the Sun rises, even in cloudy weather. Extreme care must be taken during wiring operations, especially on arrays operating at several hundred volts.

Important

To prevent risk, cover the modules with an *opaque* cloth during all wiring operations so that no electricity is produced.

Junction boxes

Most PV modules are equipped with one or two junction boxes at the back, as we have explained in Section 3.1.4. Figure 5.46 shows one of these output junction boxes with bypass diodes and its wiring.

The rules for wiring modules are as follows:

- the output cable glands should be situated at the bottom as far as possible, or else on the side, but never on top, to avoid penetration of water into the box;
- the cable sheath must penetrate the junction box and cable gland be tightened onto it (Figure 5.46);
- any unused cable glands must be blocked (either with a stopper provided by the manufacturer, or with resin);
- the output cable from the module must follow the 'water drop' rule in forming a U so that any moisture will run to the bottom (Figure 5.46);
- in case of doubt, the polarities should be checked with a voltmeter inside the junction box (even when covered, the module will have a polarity);
- once the cables are connected, the connections, the cable gland outlets, and the cover of the box before closing should all be sealed with a 'cold melt' protective resin.

The wiring section should follow the sizing rules defined in Section 5.5.6.

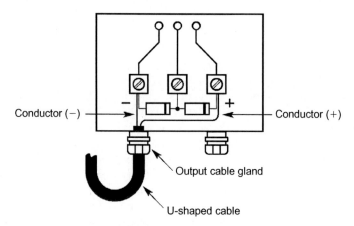

Figure 5.46   Wiring of the junction box of a module

If the cable needs to pass through an external wall, prepare it so that it also forms a loop (hanging drop) and passes through the wall at an upward angle

(Figure 5.47), and inject silicon around the hole in the wall to ensure that it is watertight. Any water running down the cable will tend to collect at the loop and not penetrate the wall.

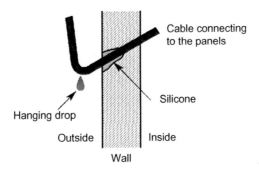

Figure 5.47   Cable passing through an exterior wall

Series installation

As we have seen in Figure 5.46, a module junction box almost always includes two cable glands, and this is useful for series installation. Figure 5.48 shows series wiring: (a) with cable glands at the bottom, better for avoiding water penetration, and (b) with the cable glands installed laterally (more practical).

> **Note**
> In both cases, the sheath covering the conductors between the two modules is an insert.

Parallel installation

This type of installation has already been discussed in Section 3.1.6: it is important to mount the diodes on each series string before connecting in parallel. To do this, the junction box already equipped with diodes can be used, as shown in Figure 3.17: the small cable glands receive the module cables and the large one is used for the common output cable to the regulator. This box is often placed under the modules that it is linking, also with output cables at the bottom. It is best to allow one box per frame. Figure 5.49, taken under the module array of a pumping installation, shows the common connection of the modules to one box (at left). To the right is the pump starting box. Note the looped cabling and the module output boxes with opposite cable glands (Figure 5.48).

If there is a large number of modules, more than can be wired in a single junction box, an extra junction box should be provided to connect the cables from the other boxes.

Otherwise, if the regulator includes several 'panel' inputs, it will be possible to connect in parallel groups of modules in parallel or different series strings, which

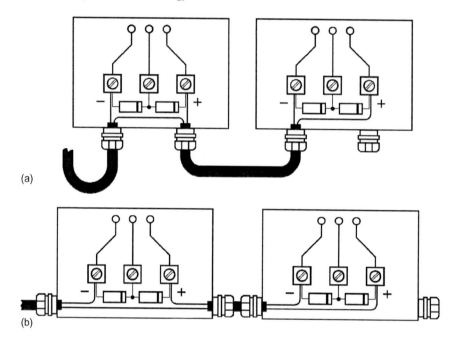

Figure 5.48   Series wiring of modules

Figure 5.49   Wiring unit on pumping system

would avoid this extra junction box. The amperage that each of these inputs can accept should be checked. For example, a 60 A charge/discharge regulator may include four 15 A inputs, which will enable the joined cables from four groups of five 3 A modules to be connected to it, each group being mounted on the junction box with diodes.

Reminder

When the modules are connected in this way by boxes including series diodes, it is unnecessary to also install an anti-return diode in the regulator.

Module maintenance

Maintenance of the modules consists in ensuring that nothing is blocking solar radiation and that the electricity produced is correctly transmitted to the charge controller. The modules should be cleaned with clear water without detergent. It should be verified that the modules are intact: no water penetration, no brown cells (see Section 3.1.4, Hotspots and bypass diodes). Surrounding vegetation should be cut back if necessary to avoid any shading of the panels.

The electrical connections and mechanical fixtures should be checked, as well as the watertight seal of the junction boxes (see checklist in Appendix 3).

5.5.7.2 Installation and maintenance of batteries

Good ventilation should be always provided for batteries to avoid the accumulation of explosive gas. If the battery is not in a special place, it is best to put it in a plastic container of the type used for storage for protection in case of acid leakage, which would attack any organic material. Sealed batteries are often indispensable for portable applications.

For systems comprising a large number of elements, they should always be arranged symmetrically to facilitate wiring. They should be mounted on a rack supplied by the manufacturer or on simple wooden racks. The batteries should be aligned with their connections, and stoppers easily accessible. Figure 5.50 shows an example of a large battery in a system of rural electrification in the Spanish Pyrenees. The system comprises two rows of twelve 2 V/750 Ah elements installed in series to provide a nominal voltage of 48 V (total capacity 36 kWh). The wiring here is very simple, from 2 V element to 2 V element with a series connection from low to high at the far right, the + and − outputs are at the far left. Besides the power cables, additional connections for the temperature sensor and a separate voltage meter can be seen. This example shows easy and optimal access to the batteries, which enables the state of the terminals, electrolyte levels and, if necessary, the specific gravity and the voltage of each element to be easily checked.

For banks of professional sealed batteries, manufacturers can sometimes provide racks to store them flat and facilitate wiring. Horizontal storage improves the

Figure 5.50   Bank of tubular batteries

maintenance of the electrolyte, which does not undergo variations in pressure due to gravity.

Figure 5.51 shows an example of a 12 V battery made up of six 2 V/500 Ah elements supplied by a major Japanese manufacturer. The rack needs to be solid to support a weight of over 100 kg.

In large systems, the high number of batteries represents a safety risk. A single battery contains enough energy to cause a fire in the case of a short circuit and produce a considerable amount of hydrogen if the regulation is faulty. The training of personnel maintaining the battery bank is very important.

Parallel connection
It is only possible to connect batteries in parallel if they are identical: in particular, two batteries of different ages should never be connected in parallel because the older will make the newer one age rapidly.

When connecting in parallel, it is recommended to put a fuse in series with each battery or string of batteries: for checking, the fuse is removed, the battery can easily be isolated and measured, and the fuse acts as a protection for the string if a battery short circuit occurs.

Serial connection
The precautions for connecting in series are even more drastic: it is only possible for absolutely identical elements. If two batteries of different capacities are connected

Figure 5.51   Bank of sealed batteries

in series, as they are fed by the same current, the smallest will be overcharged at the end of charge, and at discharge, it would be in a situation of deep discharge and consequently failure.

On the other hand, to obtain a certain capacity, it is better to connect large elements in series than small elements in parallel. For example, to obtain 12 V/500 Ah, it is better to connect two batteries of 6 V/500 Ah in series than two batteries of 12 V/250 Ah in parallel.

High voltage

For banks of batteries comprising a large number of elements in series, the voltages of each element must be closely monitored, especially if there is a high number of charge/discharge cycles: for a battery bank used for an emergency supply comprising more than 100 elements in series, discharges are rare and partial, recharging is slow and the batteries are kept 'floating', which allows a slow equalisation of the elements. In a solar system, there are usually many more cycles and there is a risk of overcharging the weaker elements at the end of the charge and completely emptying them to the point of reversing their polarity at discharge. In a large system installed in the Mediterranean region in the 1980s, the overcharging of an element of a 200 V battery bank caused a fire. To avoid this inconvenience, a charge controller can be used that measures all the elements and transfers energy between them to balance them (see Section 5.1.1).

Safety and information

Systems in hot countries

Lead batteries are widely used in rural locations in hot countries. Many systems have been subsidised by NGOs who have also studied the behaviour of users and the reliability of the installations. Much practical information can be found in these studies on ways of educating users and on the measures to be taken to guarantee a long system life.

A study in Mexico[28] has shown, for example, that advice to "check the level of batteries" was understood as "batteries need water like plants", which led several users to regularly add water to the batteries even when it wasn't needed. "Distilled water" was understood as "clean water" and the users added water from various sources after boiling it to disinfect it.

The installation of PV systems in rural areas must always be accompanied by the training of maintenance personnel to monitor the installations. In the operational budget, there must be an entry from the beginning for follow-up and maintenance and one for replacing the batteries after a few years. Recycling of the batteries must also be budgeted for to avoid local dumping of worn-out batteries.

Direct current safety precautions

Batteries contain a considerable amount of energy, and in large systems, the high DC voltage poses a real risk: in DC, if an arc appears between two conductors, it is much more difficult to extinguish than in AC because the voltage does not pass through zero at each alternation. This danger is even higher on the panels where the open circuit voltage is at least 1.5 times the nominal voltage.

If two wires of opposite polarity touch, three things may happen:

- if they are close to the batteries, they usually melt and vaporise;
- if they are long, they may heat up and dissipate energy, eventually completely discharging the battery. For example, a system with a series regulator was installed in a mountain inn only open at weekends; blocking diodes were installed in the panel junction boxes and the battery connection was only made by a relay that was closed when the batteries were charging; during the week, a panel frame was damaged in a violent storm and caused a short circuit in the panel-regulator cable, and as the cable was sufficiently resistant to dissipate the energy, the batteries were entirely discharged;
- by becoming red hot, they can cause a fire.

It is important to install fuses on the battery terminals, and these fuses should blow even if a short circuit occurs at the most remote extremity of the wiring.

Precautions during handling of batteries

This advice is essential for installers and anyone maintaining batteries:

- use insulated tools to avoid any accidental short circuit;
- do not have metallic objects nearby, for the same reason;

[28] J. Agredano, *et al.*, 'Hybrid systems: the Mexican experience', *Proceedings of PV Hybrid Power Systems Conference*, Aix-en-Provence, Sep 7–9, 2000.

- take care when moving open batteries so as to not spill the electrolyte (which is acid), if possible leaving them in their container;
- wash hands after handling batteries;
- if the electrolyte touches the skin, rinse in abundant clean water;
- if it enters the eyes, seek medical help.

On the subject of short circuit, while connecting two 12 V–40 Ah batteries in series, an installer touched the opposite terminal from the one he was tightening with his wrench: the terminal completely vaporised and the battery was useless.

NiMH batteries

The installation of low capacity NiMH elements (button or stick type) follows different rules – basically those of electronics and disposable batteries. On the other hand, the recommendations given earlier in this section concerning serial and parallel installation, and maintenance of terminals are perfectly valid. To make an NiMH battery of usable voltage, 5–10 elements (6 or 12 V) can be connected in series: it is essential to prepare these elements so that the battery will last as long as possible, otherwise, when discharging and the voltage falls, there is a risk of reversing the voltage of the weakest elements, which will make them age very rapidly and cause the system to break down. Similarly, overcharging causes rapid ageing of nickel batteries and is the main cause of the 'memory effect' observed with NiCd batteries, which reduces the battery capacity after repeated overcharging. NiMH batteries are less sensitive to this problem but overcharging should be still avoided to prolong their life.

Battery maintenance

A checklist of procedures to be carried out by maintenance personnel is given in Appendix 3. For individual users, we give a simplified procedure stressing the most important points.

For all battery types

Observation of the battery's behaviour at the end of charge: with an 'on–off' regulator, the cycles should be slowed with the battery remaining fully charged for some hours without use. If the end-of-charge cycling remains rapid, there is a high probability that the battery has lost its capacity. In particular, with a mobile system (solar lantern, rechargeable radio, etc.), if the 'full-charge' indicator comes on rapidly after the device is connected to its charger, that indicates that the battery voltage is rising rapidly and that it will no longer accept the charge, because its capacity has considerably reduced.

With a PWM regulator at constant voltage, it is difficult to observe this behaviour; in this case, if the discharge cycles after a full charge get shorter, this indicates a reduction in capacity.

Open batteries

- Visual control of the electrolyte levels, if necessary, top up with distilled water (normally MIN and MAX levels are marked).
- If the charge controller is not fitted with an automatic boost charge device, this function can be replaced by a switch shunting the controller and allowing occasional overcharge of the battery to occur for 1–5 h (according to the level

of current), once a month (electrolyte should be topped up afterwards). This equalisation charge should be carried out after every major fall in voltage leading to appliances being cut-off. This will only be possible for users with a basic level of technical knowledge.

• Maintenance of connections and cleaning of terminals.

Sealed lead batteries

• Never carry out an equalisation charge that could dry out the battery, but use the charge controller with end-of-charge cut-off adjusted to the overcharge threshold specified by the manufacturer.
• Check the voltage and connections, and in the case of large systems, measure voltages of the individual elements to verify the balance of the battery.

NiMH batteries

Check the connections between the elements and the state of the battery housing; clean up any oxidisation.

5.5.7.3   Installation and maintenance of the charge controller and other components

We give below a number of recommendations that are applicable to the majority of charge controllers; however, it is important to follow the instructions given by the manufacturers, which may be different in some cases.

Mechanical installation of charge controllers

A charge controller should be installed near the batteries and in a location with easy access so that its indicators and measuring instruments can be monitored. If the installed power is high (>100 W), good circulation of air must be ensured to cool the blocking diodes and the transistor switches. The ideal situation for a small system is against a wall at eye level above the batteries. Cabling connected to the regulator will be fixed and clearly identified for future maintenance. The output cables will be always on the bottom of the regulator or possibly on the side (with additional precautions) to avoid any penetration of water running down a cable. If the cabinet is situated outside, the 'water drop' rule must be respected and the wiring should enter the cable gland via a U loop (Figures 5.46 and 5.47).

Electrical installation of charge controllers
Special connections

The remarks below are relevant for all types of charge controller and are not dependent on the technology used.

Thermal compensation: if the regulator is equipped with thermal compensation and an external sensor, the sensor will either be fixed to one of the batteries or attached to the battery cable to measure the temperature correctly at the battery level.

Separate measurement of voltage: if the regulator has cable entries to measure the voltage independently of the current entries, this measure will always be done directly on the battery terminals to avoid the fall in voltage caused by the series fuse. The wiring section is not important here, as the measuring consumes very little current: simply choose the most suitable wiring for the available connectors.

Shunt regulator

Shunt regulators should not normally be connected to solar panels without a battery already being connected (without a battery, they will oscillate and sometimes dissipate too much energy).

To connect the regulator-batteries and panels, proceed as follows:

1. prepare the two cables without connecting anything;
2. connect the battery cable starting at the regulator terminals;
3. complete the connection by connecting the battery;
4. connect the cable between the panels and regulator, starting at the regulator terminals.

Notes

- Note the state of the LEDs or measuring instrumentation that should indicate the state of the regulator during each operation.
- A 'full-charge' LED blinking rapidly often indicates an open battery connection (see manufacturer's documentation). If it occurs, check the wiring.

Series regulator

Series regulators are at lower risk of being damaged if they are left without a battery connection; however, in this case, any short circuit must be avoided if the battery cable remains unconnected, which could overcharge the regulation transistor.

To connect the regulator to the battery and panels, proceed as follows:

1. prepare the two cables without making a connection;
2. connect the battery cable starting at the regulator terminals;
3. complete the connection by connecting the battery;
4. connect the panels-regulator cable starting at the terminals of the panel (watch out for the open voltage), and make sure not to create an electrical arc that would cause rapid corrosion of terminals.

Notes

- Note the state of the LEDs or measuring instrumentation that should indicate the state of the regulator during each operation.
- A 'full-charge' LED permanently lit often indicates an open battery connection (unless the battery really is fully charged).

MPPT regulator

The technology used by MPPT regulators is usually closer to that of series regulators than shunt regulators; however, each manufacturer has their own technology, and general recommendation cannot be given: for these regulators the manufacturer's manual must be rigorously followed.

Connections between regulator and appliances
DC appliances

For wiring these, there is no difference between the various regulator technologies. The first thing to know is the load-shedding components used in the 'discharge protection' part of the regulator.

- If the output is cut by a relay, there is usually no priority potential for the earth.
- If the output is cut by a transistor (generally MOSFET), it must be established which pole is common to avoid inadvertently shunting the transistor and disabling the protection. Small regulators sometimes use a MOS transistor channel 'n' disconnecting the negative terminal: in this case, the negative terminal of the battery must never be 'bridged' with the negative terminal of the appliances, which could earth the positive terminal (see Section 5.1.4).
- In the most sophisticated regulators, the output is often controlled by one or several MOS channel 'n' transistors but connected in series from the positive terminal, which calls for a more complex command circuit.

For the wiring of appliances, local regulations must be respected to ensure the conformity of the installation:

1. Remove the fuse protecting the regulator output.
2. Prepare all the wiring of the appliances, switches and accessories.
3. Test with an ohm-meter between the $+$ and $-$ that the wiring of the appliances has no short circuit (appliance switches open!).
4. Replace the output fuse.
5. Test the appliances.

For large systems, it is useful to divide the appliances into stages or locations and to have a centralised electrical switchboard with circuit breakers for each division. Be sure also to respect local electrical regulations.

AC appliances

For this type of system, only the DC section, input to the inverter, is specific to PV and the remaining aspects of AC installation are traditional.

The inverter should be placed as close as possible to the batteries for highest efficiency. It will usually be connected directly to the batteries since it is equipped with its own low-voltage disconnect. The section of the wiring must absolutely conform to the level of current that is often very high: for example, for a 1 m connection with a 1000 W inverter at 24 V, the maximum current will be close to 50 A (taking efficiency into account); with a cable of 2×6 mm^2, the maximum losses will be around 1.4% and the heating of the cable within the recommended value of 7 A/mm^2.

A fuse or circuit breaker, if one is not already incorporated in the inverter, should be fitted in this connection so as to be able to switch off and protect the inverter.

Maintenance of regulators

In principle, regulators need very little maintenance. The elements to monitor are the tightness of connections on the terminals and the state of the wiring. The operating parameters should not change over time except after an overcharge caused, for example, by a lightning strike close to the system; in this case, it should be verified that the charge completes correctly and that the charge stops.

Maintenance of wiring

The state of the wiring should be checked especially at any junctions: in a temperate and dry climate, an annual check is adequate, but if the atmosphere is humid or saline, checks should be carried out more often. When testing the condition of the wiring and connections, any loosening of the terminals should be detected and remedied, to avoid any possible corrosion.

Maintenance of fluorescent lamps

See Appendix 3 – maintenance operations checklist.

5.6   Practical case studies

Of the four stand-alone examples described here, the first three (telemetering, the chalet and the farm) were all put together by the authors and should only be treated as examples. They may perhaps include some improbabilities, apart from the PV technology details, but they were inspired by real cases. We do not enter into details of the applications supplied, limiting ourselves to descriptions of relevant, typical and realistic characteristics.

The fourth case, on the other hand, is completely authentic. That is a case study of a small wastewater treatment plant in the Vaucluse, southern France.

All the PV generating solutions described here are functional and perfectly applicable.

5.6.1   Telemetering in Normandy

This case describes a very low power professional application in a region with low solar radiation. The description is in three stages, following the sequence of the project: first, we describe the *technical specifications*, then give a *critical analysis and calculations* to select the equipment, and finally show the *technical solutions* arrived at and their installation.

5.6.1.1   Technical specifications

This part describes the requirements from clients' point of view: their requirements, choices and technical constraints.

Context

The B&B company specialises in the treatment of effluent containing specific waste from various industries. This wastewater flows from its production sites to the treatment plant through a network of canals, mainly through isolated tracts of the Normandy countryside. In order to monitor in more or less real time the rates of flow in these canals, it was decided to install flowmeters with the data transmitted to a control point to centralise data on the volume of effluent to be treated. To do this, telemetering posts are set up, each provided with a submerged flowmeter and a GSM transmitter. They are capable of measuring and transmitting on a permanent basis. As most of the sites have no source of energy because of their remoteness, a stand-alone energy solution had to be found. However, the possibility of connecting to the mains will also be considered.

Energy requirements
Description of equipment chosen
The technicians chose equipment with the following properties:

*   DC supply,
*   low electrical consumption,
*   programmable.

This was so as to be able to use disposable batteries, solar energy or rechargeable batteries.

The flowmeter has only one operational function, and it consumes 250 mA at 12 V DC (or 3 W). Its electronic measuring device is equipped with a 'low battery' detector that sends an alarm signal over the GSM when the supply voltage drops to 10.5 V. This function allows remote monitoring of the good operation of whatever supply is being used.

The GSM transmitter consumes 30 mA on standby. Trials to establish the range and any obstacles between the measuring points and the data centre showed that 150 mA during transmission was adequate in all cases.

Since it cannot receive but only transmit, the GSM will not need to remain on standby to receive messages. Only the transmission mode at 150 mA will be used.

These two components can be commanded by an external card module not to remain permanently on standby, but only operate on demand. This control card is therefore added, consuming 50 mA in operation, and also providing data storage. It will be on standby most of the time, being 'woken up' by a clock card, consuming only 250 µA.

Optimisation of operation over time
At first, the client maintained that the flows must be measured permanently. At this stage of progress on the project, it is decided to try to reduce consumption, and decisions must be taken on what is really necessary and what is not.

Note

This stage of the project is essential: we will see later that it is this optimisation of the operation of equipment in time that has made the project possible. The total amount of energy consumed in a day was divided by more than 100! The role of the solar energy professional is essential at this stage to help the client to cast a critical eye on his real needs.

From the experience acquired, an analysis of the *real* need for information is made. The original idea of 'real-time measurement' had to be revisited, and questions had to be asked like: what quantity of data can be processed in a day, or at what frequency is it useful to measure the flow in each canal?

A study of already identified variations in flow and the operating conditions of the treatment plant lead to the conclusion that the measurement every 4 h would be sufficient for the adequate operation of the process.

The monitoring equipment is therefore put on standby for most of the time and is only turned on every 4 h to carry out the measurement and transmission of data. This 'optimised' design enables total consumption to be much reduced.

Optimised electrical specifications
Summary of selected conditions of operation:

- Nominal voltage: 12 V DC
- Operating range: 11 V–15 V
- Power consumed:
   - Clock card (24 h/24): 250 μA
   - Flowmeter + control card (six times in 24 h): 250 + 50 = 300 mA for 2 min
   - GSM transmission (six times in 24 h): 150 mA for 1 min
   - Total consumption:

$$0.25 \, \text{mA} \times 24 \, \text{h} + \frac{(300 \, \text{mA} \times 6 \times 2 \, \text{min}) + (150 \, \text{mA} \times 6 \times 1 \, \text{min})}{60}$$

$$= 81 \, \text{mAh/day} \qquad\qquad (5.22)^{[29]}$$

Peak consumption: 300 mA[30]
This electrical energy consumed over 24 h is described as Q_{cons} (Ah). Figure 5.52 shows consumption over time.

---

[29] The expression 'per day', when referring to consumption or solar radiation, always means 'per period of 24 h'.

[30] As the flowmeter and the GSM will not operate at the same time.

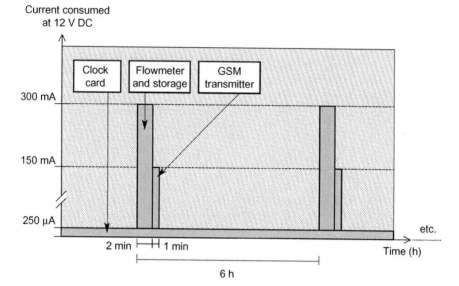

Figure 5.52   Consumption profile of flowmeter + transmission

Let us also recall what the system would have consumed if the equipment had been left on permanent operation (tinted area): the system would have needed $(300 + 150)$ mA $\times$ 24 h $= 10.8$ Ah/day, or 130 times more energy.

Location sites and constraints
Locations and constraints
There are 12 monitoring sites to be equipped. To understand how much energy will be used at each of them, we must examine their situation one by one.

- Sites number 1–9 are situated in the open countryside, on hills not easily accessible or on the edge of a track, and all have a good solar exposure.
- Site number 10 is near the treatment plant, 50 m from the caretakers' building.
- Sites number 11 and 12 are in woodland.

The design engineers suggested a cabinet to house the electronics and transmitter (the flowmeters are buried). It measures $50 \times 40 \times 30$ cm, and includes a locking door. Ideally, everything should fit inside it, including the energy source.

Theft and vandalism must also be considered, and to reduce temptation the equipment should be made invisible or impregnable.

Life expectancy and maintenance
The life of the flowmeters and the electronics used is estimated at 6 years. For safety and economic reasons, it is advisable to only plan for

- the first visit 2 months after commissioning,
- the second visit at the halfway point, after 3 years.

And no other maintenance visit. Ideally, the sites will thus remain completely isolated for 3 years so that no operational expenses will occur.

Climatic data

Besides the solar radiation data that will be provided by the company supplying the PV equipment, the client will obtain information on the climatic characteristics of the region concerned:

- extreme ambient temperatures: −15 °C, +35 °C;
- normal temperatures: −5 °C, +25 °C;
- wind (excluding exceptional storms): 110 km/h maximum;
- snow: 50 cm maximum (3–4 days/year);
- no saline environment, no particular chemical constraints.[31]

5.6.1.2   Analysis and technical solutions

The question to be asked now is the following:

What technology solution are the most suitable for these 12 sites to meet the technical specifications at the least cost?

We will first size and cost the solutions to help make these choices.

Electrical energy from the grid

It would be out of the question to supply these very low monitoring points from a medium voltage EDF grid by connecting from a line at the edge of the road: a special transformer would be needed, with independent metering as for an individual dwelling, without counting the cost of the trench necessary to reach the location of the flowmeter.

On the other hand, if a 230 V AC feed was available near at hand, it would be possible to install a buried cable. The economics of this solution depend on the distance involved. If the terrain is relatively easy, the cost of such connection is estimated at €50/m of trench. The maximum power consumed is 300 mA × 12 V = 3.6 W. A small 5 W transformer 230 V AC/12 V DC would be necessary, but its cost is low (€10–€20). The investment would therefore be €2500 for 50 m and €5000 for 100 m (depending on the distance between the monitoring point and the 230 V mains supply).

Battery solution without using solar energy

Let us see now whether it is possible to supply the DC system using disposable or rechargeable batteries continuously for at least 3 years (the minimum duration required).

Theoretically, the total consumption in capacity over 3 years is

$$\frac{81 \times 3 \times 365}{1000} = 88.7 \text{ Ah} \qquad (5.23)$$

The daily consumption is divided by 1000 to express the result in Ah.

---

[31] Solar panels should not be exposed to acid smoke, for example.

We should recall now that the useful capacity (C_u) expresses the capacity needed in real conditions, whereas the nominal capacity (C_n) relates to 25 °C and particular discharge conditions (generally discharge current C/20). Therefore, this capacity should be increased by a coefficient to take account of cold and other possible losses (depending on the technology chosen). We will use a coefficient of 0.7, bearing in mind that the minimum temperature of the location is −15 °C.

$$C_u = C_n \times 0.7 \tag{5.24}$$

$$C_n = \frac{C_u}{0.7}$$

$$C_n = \frac{88.7}{0.7} = 127 \quad \text{Ah}$$

The option of using just a lead-acid *rechargeable battery* is unfortunately not available: the best batteries have a self-discharge rate of 50% a year. Without the supply to recharge them, they cannot remain in a charged state for more than 2 years. It is therefore impossible to leave them 3 years without recharging.

If a *disposable battery* is used, it can be completely discharged, and therefore the useful capacity required for 3 years of operation is 127 Ah. The battery chosen should meet the following requirements:

- nominal voltage: 12 V;
- available capacity: 127 Ah between 11 and 14 V;
- behaviour in cold: 70% of minimum capacity at −15 °C;
- life: 3 years between −5 and +25 °C (−15 and +35 °C extreme temperatures).

The only disposable batteries of these capacities at affordable prices are zinc-air batteries and air-depolarised batteries.

The most suitable product in terms of capacity is a zinc-air 8.4 V, 130 Ah battery measuring 162 × 122 × 182 mm, weight 3.9 kg, unit cost €20.50. Two will be needed in series, plus a small 'step-down' DC/DC converter (see Section 5.1.3) to bring the voltage down to 11–14 V. Cost of converter is €15. Over a period of 6 years, the set of two batteries will need to be replaced once, with a technician visit (visit cost €40). The total budget of the solution over 6 years will thus be €137. In this case, it will be seen that the storage used is clearly superior to that of solar generation since the duration of operation required of it is much longer (3 years, as against around 10 days for a PV supply).

PV solution

The equipment can only be chosen with an accurate energy budget bearing in mind solar radiation data for the location and products available on the market.

Exploitable solar radiation

Two solutions are considered for the installation of the panel, the first to optimise the solar energy received, and the second, to make the panel invisible and so protected from theft and vandalism:

- 'optimal' installation: facing south and at 60° pitch to the horizontal;
- 'invisible' installation: horizontal on the top of the cabinet.

The solar radiation data should be those provided by the local regional weather centre. Figure 5.53 shows an extract of the map of France (from the *European Solar Radiation Atlas*[32]) with the site of the telemetry installations indicated.

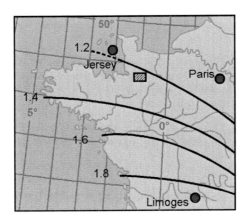

Figure 5.53 Geographical location of the flowmeters (hatched) (complete map in Appendix 2)

So the figures should not be overestimated, we will take the weather station of Jersey, on the pessimistic side of this sector (lowest solar radiation of the sector) (Table 5.20). We will therefore use the data provided by the *European Solar Radiation Atlas* in the two configurations planned for.

Table 5.20 Daily global radiation data at Jersey (average values in $Wh/m^2/day$)

	Jan.	Feb.	Mar.	Apr.	May	Jun.	Jul.	Aug.	Sept.	Oct.	Nov.	Dec.
Horizontal	806	1598	2882	4296	5335	5991	4549	5606	3376	2133	990	646
60° tilt south	1350	2430	3590	4330	4600	4810	4660	4290	3880	3090	1640	1230

---

[32] These radiation values are given in Appendix 2 and the references of the *Solar Radiation Atlas* in the Bibliography.

Pre-sizing

Let us first make an approximate calculation of the peak power of the panel and the battery capacity as described in Section 1.2.1, with estimated reduction coefficients.

The number of days of autonomy without solar contribution (N_{da}), an essential parameter for the reliability of the system, will be taken as equal to 10 days, bearing in mind the climate in Normandy (succession of days with very unfavourable solar radiation).

Reminder

The daily electrical consumption (Q_{cons}) is 81 mAh at the operating voltage of 12 V. In Wh, this gives $81 \times 12 = 972$ mWh.

The power required for the solar panel 12 V is calculated for the two orientations selected, with a reduction coefficient initially equal to 0.7.

We must remember that the daily solar radiation E_{sol} used for this calculation is the most unfavourable of the year (in December in the two cases, which concern us):

$E_{sol} = 0.646$ kWh/m^2/day for horizontal installation, and
$E_{sol} = 1.23$ kWh/m^2/day for 60° tilt south installation.
Therefore,

$$P_c = \frac{0.081 \times 12}{0.646 \times 0.7} = 2.1 \text{ Wp for horizontal installation} \tag{5.25}$$

$$P_c = \frac{0.081 \times 12}{1.23 \times 0.7} = 1.1 \text{ Wp for 60° tilt south installation} \tag{5.26}$$

Note

Since all the components of the system will be 12 V and close to each other, one can also calculate directly the STC load at the panel with the help of the following formula, with an approximate current loss coefficient C_l of 0.8:

$$I_m = \frac{Q_{cons}}{E_{sol} \times C_l} = \frac{81}{0.646 \times 0.8} = 157 \text{ mA for the horizontal installation}$$

and

$$I_m = \frac{Q_{cons}}{E_{sol} \times C_l} = \frac{81}{1.23 \times 0.8} = 82.3 \text{ mA for the 60° tilt south installation}$$

$$\tag{5.27}$$

As far as the battery is concerned, it can be roughly calculated as follows:

Capacity required for an autonomy of 10 days with an initial reduction coefficient of 0.6 is

$$C_n = \frac{10 \times 0.081}{0.6} = 1.35 \ \text{Ah} \tag{5.28}$$

We can already see that the solar installation will be small: a panel of around 1–2 Wp with a battery of 1–2 Ah.

Choice of technology and final sizing

We must now accurately identify the equipment and confirm the energy budget. First, we will consider if it would not be interesting in this case to work with interseasonal storage.

Inter-seasonal storage: An explanation of this principle of operation. Instead of selecting a panel based on solar radiation in December, the lowest in the year, we choose one suitable for more favourable solar radiation in October or November. And we compensate this reduction in panel size by a slightly larger battery. The battery is thus made to participate in the operation in winter: it partially discharges in November–December–January and regains its full charge in spring.

This solution is often more economical, because solar panels are expensive. Another advantage is that the battery works by discharging once a year to a level of around 50%, which cannot do it any harm.

Important

This principle can only be applied to sealed batteries that do not suffer from remaining in an intermediate charge state, their electrolyte being maintained in the separator, whereas an open battery must regularly receive a full charge to avoid stratification problems.

We will call 'winter deficit' D_w the total capacity that is 'missing' in the system energy budget if the power alone has to provide all the energy. The panel is adequate to balance, for example, the consumption in October but can no longer meet the demand in November, December and January. During this period, the battery reserves must be drawn on in the form of the missing 'winter deficit'. Subsequently, when solar radiation is more favourable, the battery will be completely recharged (usually by the end of March).

Solar panel: In this case study, this would permit us to use a very small size panel to integrate it into the cabinet and make it less obvious. There is plenty of space within the cabinet for the battery, so its capacity is not a problem. The client would welcome horizontal installation of the panel on the top of the cabinet because it is invisible this way and so does not encourage theft, and is easy to install especially since there is no constraint of southerly orientation.

Note

Any horizontal installation requires good drainage of rainwater, because any standing water can infiltrate and cause damage to the solar panels.

Amorphous silicon panels are efficient at recovering diffuse light and low level radiation that will be common in the region concerned (highly temperate climate with an often cloudy sky), and we will first need to refine the sizing using these panels.

In the power range requested, there is a 12 V panel typically supplying 1.7 Wp, with a guarantee of 100 mA/15 V STC (Figure 3.29). Its power is slightly below the 2.1 Wp originally estimated for a horizontal installation (or 157 mA of STC charge current), so the battery capacity needs to be increased to ensure sufficient reserve for the inter-seasonal winter storage.

Battery technology: Bearing in mind the capacity required, life expectancy, operating voltage, low temperatures in winter and the space available in cabinet, the best solution would seem to be a sealed lead battery. Gel technology would be suitable, providing effective protection against overcharging, and a model is available that performs well in cold weather conditions. It has a good average current charge efficiency, estimated at 90%. Its life is 10 years at 25 °C and 5 years at 35 °C. The actual situation will be between the two, as the ambient temperature only very rarely reaches 35 °C in this region, and is often below 25 °C, but the heating of the battery in the cabinet has to be taken into account (between +5 and +10 °C in relation to the ambient temperature if it is well ventilated). The required life of 6 years is therefore realistic in these conditions.

Energy budget and calculation of storage capacity: To calculate the battery capacity, a more accurate calculation of the energy budget in winter is necessary to establish the winter deficit.

To do this, we first need to evaluate the loss coefficients to be applied to the panel and the battery (see Section 5.5.4.1, Calculation of system peak power):

The drop of voltage in the wiring and the reduction because of the temperature are not critical in our case (the elements of the system are closely connected, and the rises in temperature will occur during the summer, when there is an excess of energy anyway). The panel voltage of 15 V is therefore adequate to cope with the remaining voltage drop, of 0.3 V in the diode of the regulator (which will be of the Schottky type), and the reductions in instantaneous radiation.[33] We will therefore assume a current loss coefficient of 0.8 (10% of losses through dirt on the panel and 90% through battery charge efficiency). The useful capacity of the battery here is determined by the cold: the temperature reduction coefficient R_T is calculated

---

[33] The operating voltage of the amorphous silicon panel chosen is little affected by low light levels, and it is therefore unnecessary to take into account a fall in production occurring at the beginning and end of the day.

according to the method given in Section 5.5.5 for a minimal voltage of 11 V (1.83 V/element) and an extreme temperature of $-15\,°C$ (Figure 5.41).

The calculating parameters are therefore

- solar radiation data: Jersey horizontal;
- daily consumption: 81 mAh;
- current produced by the solar panel: 100 mA;
- current loss coefficient: 0.81;
- battery reduction coefficient: 0.75.

Table 5.21 shows the monthly energy budget during the critical period (winter). The production and consumption of the system has been calculated for each month. To obtain the monthly electrical output of the panel, the daily production is multiplied by the number of days in the month.

Table 5.21 Winter energy budget

Value	Unit	Sept.	Oct.	Nov.	Dec.	Jan.	Feb.	Mar.
Daily global irradiance	Wh/m^2/day	3376	2133	990	646	806	1598	2882
Monthly electrical production	Ah/month	8.10	5.29	2.38	1.60	2.00	3.58	7.15
Monthly electrical consumption	Ah/month	2.43	2.51	2.43	2.51	2.51	2.27	2.51
Difference	Ah/month	+5.67	+2.78	−0.05	−0.91	−0.51	+1.31	+4.64
Battery charge state at end of month	(% of nominal capacity)	100	100	99	81	71	97	100
Winter deficit					$0.05 + 0.91 + 0.51 = \mathbf{1.47\ Ah}$			

The *daily production* (Ah) is

$$Q_{prod} = I_m \times E_{sol} \times C_1 \tag{5.29}$$

Monthly production is therefore equal to $N_{\text{days in the month}} \times I_m \times E_{sol} \times C_1$

Monthly production of panel = (number of days in the month) × (panel STC current) × (daily solar energy during October) × (current loss coefficient).

Example 1

During October, monthly production of the panel will be equal to $31 \times 100 \times 213 \times 0.081 = 5282\ \text{mAh} = 5.29\ \text{Ah}$

Monthly electrical production is calculated simply by multiplying the daily electrical consumption by the number of days in the month.

Example 2

In the month of September, total consumption reached $30 \times 81 = 2430$ mAh $= 2.43$ Ah

---

It will be noticed that for 3 months consumption is more than production, but that is acceptable. This should be normally limited to 3 months, because after 3 months, there is a risk of entering a chronic year-on-year deficit.

To ensure sufficient capacity to cover the winter deficit and the days of autonomy without solar input (to overcome successions of days with little sunshine), the battery's nominal capacity C_n must satisfy the following formula (C_u being the useful capacity):

$$C_n = \frac{C_u}{R_T} = \frac{D_W + N_{da} \times Q_{cons}}{R_T} \tag{5.30}$$

or in this case,

$$C_n = \frac{1.47 + 0.081}{0.75} = 3.04 \text{ Ah} \tag{5.31}$$

Therefore, in principle, out of the capacities available in the battery range chosen – 2.5 Ah, 5 Ah and 8 Ah – it is the 5 Ah model that will be used.

To be sure that the system is globally balanced, despite the 3-month episodes during which the battery 'helps' the panel with its reserve, we calculate the battery charge state at the end of the month during the critical period and beyond (last line of Table 5.21). To do this, we add the positive and subtract the negative balances at the end of the month (as in book keeping). We can then see that by the end of March, the battery is fully charged again.

Example

At the end of December, the battery will have supplied a deficit of 0.05 Ah in November and 0.91 Ah in December. It has thus lost 0.96 Ah. As a percentage of its nominal capacity, it is therefore at a charge level of 81%: $(5 - 0.96)/5 = 0.808$.

Charge regulation: To protect the battery being overcharged, an on–off shunt model with a Zener diode carefully aligned to the correct voltage is adequate for this low power (<2 W). There is no need for a boost charge or an equalisation charge with sealed batteries (see Section 5.1.1). The Zener diode and the anti-return diode will be installed on a small tropicalised circuit board that can be integrated into the output cable of the panel (Figure 3.29). As it is only acting as an overcharge protection, it is placed between the panel and the battery.

It is not necessary to install discharge protection, both because consumption is programmed and should not exceed forecasts, and because the data card includes a 'low battery' alarm.

So the telemetry equipment (data card, flowmeter and GSM transmitter) will be supplied directly by the battery via a fuse or a circuit breaker (electrical diagram in Figure 5.54).

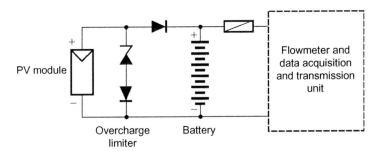

Figure 5.54   Electrical diagram of the solar supply

Cost of PV system:

12 V amorphous silicon solar panel 15 × 30 cm² (100 mA–15 V): €46
Overcharge limiter built-in to cable: €15
5 Ah–12 V sealed lead battery with gel electrolyte: €26
Panel fixing and waterproofing: €8
Total cost of PV solution: €95 excluding other expenses

5.6.1.3   Configurations adopted and installation

Balance of technical solutions

Three solutions are therefore technically valid and compatible with the technical specifications, and their advantages and costs are summarised in Table 5.22.

By reviewing these solutions with the constraints of sites number 1–12, it is fairly simple to decide which solutions are best.

Connection to the grid, even in the case of site number 10, which is only 50 m from the 230 V AC mains, is definitely the most expensive solution. This is obviously because of the very low energy requirements. The conclusion would have been different if the PV system had had to be bigger, because the grid connection solution does not have a bearing on the power consumed.

In the case we are describing, it is not economical and will not be adopted.

The PV solution is the cheapest, but it can only be applied if the site exposure is suitable. It cannot be generally used in woodland, even by installing a bigger surface of panels.

Table 5.22   Balance of technical solutions

	Connection to grid	**Battery**	**PV system**
Advantages	Simple and reliable	Applicable everywhere	Total autonomy
Disadvantages	Expensive in the present cases	– Space needed – Six years maximum	Need for good exposure
Cost	€2500/50 m of distance from grid	€137	€95
Site suitability	None	No. 11 and 12 in woodland	No. 1–9: isolated and well exposed No. 10: 50 m from electrified buildings and well exposed

Rather than making risky and expensive guesses of the losses through shading by trees, which also varies with the seasons, the clients prefer in this case to install batteries (zinc-air battery with small DC/DC converter).

The last line of Table 5.22 shows the solutions chosen for each site.

Installation of PV system
To summarise the equipment to be installed:

- one solar panel $30 \times 15 \times 3$ cm with integrated overcharge limiter;
- one sealed AGM lead battery of 5 Ah–12 V composed of two 6 V packs.

As the client wished, and according to the system calculation, the panel will be installed flat on the top of the cabinet. Its dimensions of 15×30 cm are compatible with the top of the box (Figure 5.55). As there is no frame, there will not be a risk of creating a collecting point for rainwater, and the wind will dry any water. Its output cable exits at the back on a contact block where the cable/panel soldering points had been sealed in resin. An opening will be made on the top of the cabinet for the cable and the module will be fixed on this surface with neutral silicon or butyl.

Inside the cabinet, on the bottom, are installed the appliances (electronic measuring and data cards, GSM transmitter), wired directly to the battery. The battery will be placed in the cabinet, taking care to avoid any likely causes of heating, which would reduce its life and encourage corrosion of the terminals. For increased safety, the terminals will be smeared with a protective grease of the silicon type. It should not be forgotten that the battery is the weak link of the system.

That being said, the region is not particularly hot, and the battery will last at least 3 years as specified, but probably, provided the above precautions are taken, it will not be necessary to change it for 5 or 6 years. The panel has a life expectancy of at least 8 years.

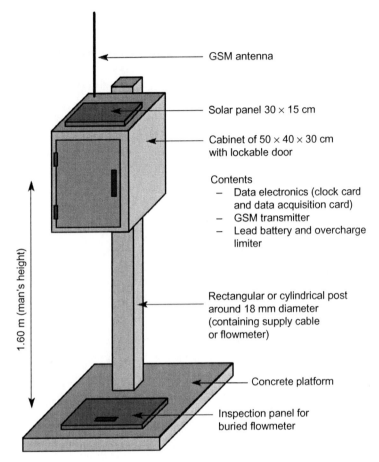

GSM antenna

Solar panel 30 × 15 cm

Cabinet of 50 × 40 × 30 cm
with lockable door

Contents
– Data electronics (clock card
  and data acquisition card)
– GSM transmitter
– Lead battery and overcharge
  limiter

Rectangular or cylindrical post
around 18 mm diameter
(containing supply cable
or flowmeter)

Concrete platform

Inspection panel for
buried flowmeter

1.60 m (man's height)

Figure 5.55   Installation of telemetry cabinet with its solar power supply

5.6.2   Chalet in Switzerland

The second case study takes us to the Swiss Alps where many chalet owners enjoy isolation amid beautiful natural surroundings. This is certainly the simplest case study, in which the plans can be made logically and without complex calculations, as we shall see below. This type of residence is often far from traditional electric lines, sunshine is favourable and requirements modest. PV energy is therefore suitable for supplying a minimum of comfort, and this type of installation has proved fairly successful.

5.6.2.1   Situation and requirements

General situation and occupation
The chalet in question is situated in the canton of Grisons in eastern Switzerland at an altitude of 1500 m. It has a good southerly outlook. It can be reached by car in

the summer, but preferably in a 4 × 4 as the road leading to it is not tarred. In winter, snowfall makes access more difficult: the owners get there with the help of cross-country skis and skimobiles. A family of four owns the house and occupies it regularly but only at weekends. Sometimes in summer they stay longer. The nearest mains electricity is 5 km away as the crow flies, but in any case they do not want it since they prefer the wildness of the site as it is.

The chalet is not far from the well-known resort of Davos, on which we will base our meteorological data. Its altitude is similar (1590 m), and this is important because the solar radiation conditions are much more favourable than in the valley (see Section 2.2.3).

Electrical consumption

The electrical appliances fall into four categories. The general idea is to have only a modest installation, firstly to keep the investment costs low, but also so as not to change the rustic character of the chalet. In principle, therefore, the installation could all be in 12 V DC.

Lighting

Eight 13 W lighting points are planned for the different rooms of the house. As there are four occupants, calculations are based on a maximum of one light switched on per person at a time.

In summer, 4 × 13 W × 3 h, or 13 Ah at 12 V; and in winter, 4 × 13 × 6 h, or 26 Ah at 12 V.

This is an average consumption that could be differently distributed to different lamps, which would make no difference as long as the energy supply is sufficient to meet the load.

Water supply

A natural spring is available, so all that is needed is a pump to take the water to a tank at the top of the house so that there is sufficient pressure in the taps. For a flow of 10 l/min, this pump consumes 6 A at 12 V. As there are four persons, and one reckons on 100 l/person/day (in total, for washing, cooking, etc.), the requirement is 400 l/day. The plant will therefore operate 40 min/day, giving a consumption of

$$6 \text{ A} \times 40/60 \text{ h} = 4 \text{ Ah at } 12 \text{ V}$$

Refrigeration

A refrigerator with a capacity of 110 l, with good thermal insulation, is used to keep food fresh, but only when the family is present. Also, it was decided to only use it in summer, since in winter any perishables can be left outside as the temperature rarely exceeds 5 °C. This equipment uses a 70 W compressor and has a consumption of 300 Ah/day or 25 Ah/day (in summer only).

Television

To avoid buying a television working on DC (there is very little choice in this area), a commercial 230 V AC model will be used with a power of 90 W. It is planned to connect it via a small inverter, with 90% efficiency, which will only be switched on at the same time as the television. For a projected use of 4 h/day, consumption will therefore be

$$90 \text{ W} \times 4 \text{ h}/0.9 = 400 \text{ Wh or } 33.3 \text{ Ah at } 12 \text{ V}$$

Summary of consumption

Table 5.23 summarises the consumption, by season. To obtain average consumption over time, which will enable us to size the panels, we need to take account of the fact that the chalet is only occupied 2 days/week:

- average daily consumption in summer: 75.3 Ah $\times$ 2/7 = 21.5 Ah;
- average daily consumption in winter: 63.3 Ah $\times$ 2/7 = 18.1 Ah.

Table 5.23   Electrical consumption of the chalet

	Summer (Ah/day at 12 V)	Winter (Ah/day at 12 V)
Lighting	13	26
Water	4	4
Refrigeration	25	0
Television	33.3	33.3
Total/day of occupation	75.3	63.3
Average/day	21.5	18.1

Note

This is a domestic application, subject to possible variations. It is quite different from our telemetry case study (see Section 5.6.1) where consumption was programmed and unlikely to vary. The users must therefore be careful in their energy consumption. But they will learn by experience and should be able to balance their consumption conveniently without exceeding the possibilities of their system. Accurate monitoring of the battery voltage is important as it enables one to keep track of the actual situation (with the help of an 11–14 V voltmeter on the regulator, for example).

To size the regulator, the peak current of the appliances must be calculated. If all the appliances are working at the same time, it would give 28.8 A. In practice, the load will be at its maximum when the refrigerator, television and two lamps are all on at the same time: it is easy to arrange not to operate the pump at the same time as the television. Thus, the peak power load will be: 70 + 100 + 26 = 196 W, which represents a current of 16.3 A.

5.6.2.2   Choice of equipment

Our Swiss family is using the local electrical installer, who is used to working in mountain locations and has already done several installations of this type. His experience will enable him to give effective advice and rapidly evaluate the technical solutions.

The PV components that he suggests are relatively standard, and available at a good ratio of quality to price:

- polycrystalline silicon 50 Wp–12 V PV modules measuring 800 × 450 mm providing 3 A–16.5 V STC, a voltage normally adequate in temperate climates for small installations;
- open 'solar' 220 Ah–12 V plate lead batteries, the largest 12 V batteries of this type. Their main disadvantage is a fairly low number of cycles, 250 at 80% discharge, but this is not critical in our case: even supposing one cycle per weekend, battery life would be around 5 years, which is quite satisfactory.

The installer will therefore build the system from these components. The calculations are fairly simple since we are only considering multiples of one, two or three panels. There is no need to be as accurate as when the system comprises more panels. The result is two proposals, one basic and the other a bit more ambitious. We will summarise them and then look to see how far they meet requirements.

Basic system
The system comprises

- three 50 Wp–12 V PV modules described in the earlier section (total surface area 1.1 m^2);
- a 220 Ah–12 V open solar lead battery;
- a 20 A–12 V charge/discharge series regulator with manual resetting (with boost charge option);
- a 400 VA TV type converter;
- eight 12 V DC terminal blocks.

The total price of these components is €2380.[34] The television set, the refrigerator, mounting accessories and installation are not included in this price.

The installer also suggested a more comfortable system, with low-energy lighting in AC, for aesthetic considerations, as there is a much wider choice. They are also less expensive but need to be supplied by the inverter that would slightly increase its consumption. For more safety and flexibility in use, an extra solar panel is added. The regulator is sized by the load of the appliances and is adequate for this new consumption as well as for the four panels (4 × 4 A = 16 A).

More comfortable system
This 'option 2' system comprises

- four 50 Wp–12 V PV modules (total surface area 1.5 m^2);
- a 220 Ah–12 V open solar lead battery;
- a 20 A–12 V charge/discharge series regulator with manual resetting (with boost charge option);

[34] Estimated retail price excluding tax early 2008.

- a 400 VA TV type converter;
- eight 13 W–230 V AC low-energy lamps.

The cost of this solution is €3290, again not including the television set, the refrigerator, mounting accessories or installation.

Balancing the equipment to requirements

Solar radiation for Davos is given in Appendix 2.

For 60° tilt south exposure, the radiation received is 3 kWh/m^2/day in winter and 4 kWh/m^2/day in summer.

Let us first consider the three panel solution. Their production in winter amounts to 3 × 3 A × 3 kWh/m^2/day = 27 Ah/day without loss coefficient, against an average consumption of 18.1 Ah/day. And in summer the production would be 3 × 3 A × 4 kWh/m^2/day = 36 Ah/day against an average consumption of 21.5 Ah/day.

Energy production is slightly in surplus, compatible with a current loss coefficient of 0.67 in winter and 0.60 in summer. Provided dirt on the panels does not cause a loss of more than 10% and the battery has an efficiency of 80% or more, the current loss coefficient will be 0.72. The effective production will therefore be 19.5 Ah (= 27 × 0.72) in winter and 26 Ah in summer.

With option 2, the lighting will be supplied via an inverter, which will slightly increase their consumption: instead of 13 Ah in summer, we should allow for 13/09 = 14.5 Ah. Similarly, in winter, consumption will increase from 26 to 29 Ah. This will only have a slight repercussion on average consumption, which will increase from 21.5 to 22 Ah in summer and from 18 to 19 Ah in winter. The four panel system will be considerably in surplus, but this is one of its objectives, to give a good margin to allow the occupants to enjoy more energy.

The margin can be estimated as follows: with four panels, production will be 4 × 3 A × 0.72 × 3 kWh/m^2 = 26 Ah/day in winter and 4 × 3 A × 0.72 × 4 kWh/m^2 = 34.5 Ah/day in summer.

The battery, with its 220 Ah nominal, will have a useful capacity of 176 Ah, with 80% of depth of discharge authorised. For maximum summer consumption (option 2: AC lighting), which is 22 Ah × 7/2 = 77 Ah/day effective consumption, this will provide an autonomy without solar contribution of 2.3 days, or slightly more than a full weekend. This should be sufficient since the risk of bad weather is low in summer. In winter, with a maximum consumption of 19 Ah × 7/2 = 66.5 Ah of effective daily consumption, there will be around 3 days' autonomy without solar input. We have not considered losses due to low temperatures because the battery is located in the house and when the occupants are there in winter they heat the house with a wood burning stove, so the temperature in the house is at least 15 °C. The battery should not be put in the cellar (where it would be colder) for this reason.

The regulators are intended to protect the battery against overcharging and overdischarging. This load-shedding function is particularly useful for intermittent occupation such as this: if an appliance (a lamp, for example) is left on inadvertently when leaving the house, it is better that the regulator should disconnect all the appliances rather than risking the deep discharge of the battery. With manual resetting, the occupants can turn the service back on when they return.

Also, since the battery is often unused, it is preferable to have a boost charge function to stir up the electrolyte from time to time (in particular after a major discharge), which will avoid stratification.

5.6.2.3   System installation

For the actual installation, some extra components will be needed, whichever option is chosen:

- a tilting wall frame for the modules,
- a container for the battery,
- a junction box to mount the modules in parallel, with anti-return diodes,
- wiring, switches, overload protectors, and some fuses and circuit breakers.

Physical installation

The tilting frame for the panels is an interesting solution that allows them to be tilted according to the season so as to maximise the collection of solar radiation. In winter, this will give the values previously calculated as the flux had already been optimised (60° tilt from the horizontal); however, in summer, extra energy can be gained provided that it is not situated just under the eaves, which might cast a shadow. It should also not be placed too low to avoid the risk of theft.

A further advantage of this frame is that when the chalet is unoccupied, the panel can be folded to the wall to protect the modules from rain and snowfall, and this will extend their life (Figure 5.56). On the other hand, if snow falls on the modules during their use, it will have to be removed (although it may slip off when the Sun is out).

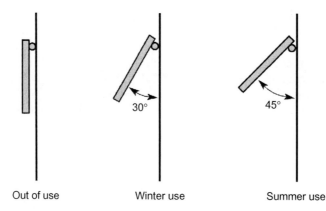

Out of use   Winter use   Summer use

Figure 5.56   Tilting frame to support the modules in different positions

The battery will be ideally located in a room on its own or somewhere in the middle of the chalet, in other words, in a place where temperature variations are at their lowest. The charge controller will be fixed on the wall at eye level so that it can be checked, and as close as possible to the battery so that the voltage measurement will be reliable.

As the inverter will only be used for the television set, it will be small and could be placed near the TV (remembering to switch it on just before the television set).

Electrical wiring
The overall electrical wiring diagram is shown in Figure 5.57.

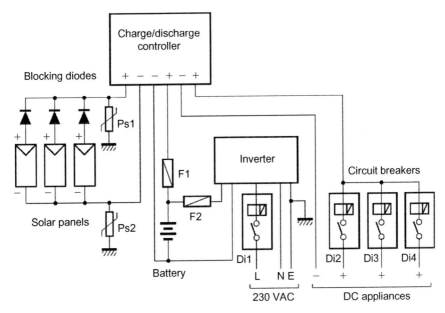

Figure 5.57   Electrical wiring diagram of the chalet

Solar panels
The three panels have their negative pole in common; one blocking diode is installed for each panel (if possible Schottky 30 V–5 A) in the junction box where they are running parallel (these boxes were described in Section 3.1.6, Figure 3.17).

The two overload protection elements Ps1 and Ps2 should be connected to an equipotential junction with a single earth point. The third overload protection device can be installed between the + and – terminals of the panels (see Section 5.1.4). Here the DC distribution section is not earthed but remains floating.

Charge controller
The regulator controls the charge from the panels and during discharge disconnects the DC appliances if the battery voltage falls too low. It is a series regulator, so blocking diodes are fitted with each panel. Check in the manufacturer's specifications that the blocking diode is not already integrated with the regulator, in which case it is unnecessary to add another one.

Battery

The battery should be encased in polystyrene insulation of the Bonisol type and placed in a PVC battery box with a lid. The Bonisol serves to wedge the battery in the box and as thermal insulation. The box cover protects the terminals and prevents them from being touched. The boxes have two large pre-pierced holes to the left and right to facilitate wiring; these holes also allow any gas to escape. The fuse F1 protects the connection between the panels and the DC appliances against accidental short circuits; they should be around 1.5 times the maximum current possible (calculated for the output of the regulator, 1.5 times $16 = 25$ A). Fuse F2 protects the connection to the inverter if this is not already integrated into the appliance.

Inverter

The inverter is equipped with its own regulator stopping it when the voltage falls too low. This is why it is wired directly to the battery and does not go via the output disconnect of the regulator. The circuit breaker Di1 enables the output to be disconnected; it can be replaced by a switch or omitted if the inverter has its own switch and its own protection. If the 230 V AC is used to supply the lamps as well, several circuit breakers should be mounted in parallel if certain floors or parts of the dwelling are to be disconnected. The remaining AC wiring must respect the local electricity distribution regulations.

AC appliances

The installation includes circuit breakers Di2, Di3 and Di4 to disconnect the various zones of the house: their power should be greater than the sum of the appliances installed on the line, for example, 6 A for four 13 W lamps. For the line feeding the pump, an individual circuit breaker is recommended of around three times the nominal current (16 A) of the pump.

For the wiring of the other DC appliances, ensure that the wiring section conforms with recommendations to minimise losses. In particular, for lamps, try to design the wiring to place the switches in series in the supply (star wiring) to avoid return wiring, which will increase the ohmic losses.

Balance of voltage losses

Now let us evaluate voltage losses and the adequacy of the panel voltage from this point of view. There is no loss due to extreme heat because the ambient temperature does not exceed 25 °C. On the other hand, the series regulator shows a loss of 0.24 V (typical value for 12 A) and the blocking diodes behind the panels 0.5 V.

As far as ohmic losses are concerned, as the components are fairly close to each other and the currents relatively low, these should not be critical and will only depend on a good choice of wiring sections. The panels have a total power of 9 A for the three panel option and 12 A for the four panel option. Assuming the distance between the output on the junction box and the battery is 8 m, one would choose a cable of 6 mm^2 section (for calculation, use the chart on Figure 5.42), which induces a loss of 6.6 mV/m for 1 A of current. Losses will thus be limited to 0.48 V for three panels and 0.63 V for four panels.

In total, the voltage losses between panels and battery will be 1.22 V (= 0.24 + 0.5 + 0.48 V) for the three panel system, and 1.37 V for the four panel system. As the STC voltage at maximum power point is 16.5 V, the battery input will be between 15.1 and 15.3 V, which is largely sufficient.

Between the battery and the regulator, if the distance is short (1–2 m), a 2 × 2.5 mm cable could be used, for example.

Life expectancy and maintenance

This system should be maintained according to the advice given in Section 5.5.7 and in Appendix 3. Maintenance is very limited, the main thing being to monitor the battery and probably change it after 4 or 5 years. The water pipes should be emptied in case of frost as in any other country house so as not to damage the pump or pipework.

5.6.3   Farm in Morocco

The third case study is completely different: a small farm in the south of Morocco. It is an example of rural electrification, because the system is entirely autonomous.

We will not attempt to cover all the challenges of this specific application of solar PV, which has been detailed in other works, but simply provide a modest example to highlight the following elements:

- PV enables vital needs to be met in remote areas: water supply, lighting and communication.
- Higher and more constant solar radiation than in Europe enables systems to be smaller to provide the same service.

As with the other case studies, we will approach the problem from the technical point of view and will discuss in order the client's requirements, the appliances to be supplied and the PV system to be used. Finally, we will describe the complete installation.

5.6.3.1   List of requirements

Situation and activity

The farm is part of an isolated hamlet, situated inland from Tarfaya, a coastal town opposite the Canary Islands. The Moroccan climate is both Mediterranean and Atlantic, with a hot dry season followed by a cool and humid season, the end of the hot period being ended by rainfall in October. The Atlantic Ocean is 150 km distant from the farm as the crow flies, and this maritime influence moderates the extremes of temperature that can be found in other parts of the country. The average temperature is around 22–24 °C in summer and 12–14 °C in winter, only occasionally falling below freezing. The maximum temperature, influenced by the proximity of the Sahara, can reach 35–40 °C.

A family of three lives there, engaged in market gardening and goat rearing (50 head). The goats provide milk, which is directly collected by a co-operative for local consumption or cheesemaking on another site. The vegetables are sold on the local market.

Goat farming is highly developed in Morocco, representing 25–30% of the agricultural GDP.

The farmers own a vehicle that they use to go to the village or the neighbouring town, where they can obtain diesel. Before developing the farming activities and considering a PV installation, they used a diesel generator for electricity. But it was noisy and unreliable. Their motivation to use solar energy was strengthened by the possibility of a regional aid package.[35]

Energy requirements
We will first list these requirements without prejudging the energy solution to be installed.

Water supply
Water is needed for the people, animals and crops. For domestic requirements – drinking, cooking and washing – some 100 l are needed per person per day. For livestock, 5 l are needed per head per day, or 250 l/day for 50 goats. The market gardening operation requires an average of 350 l/day during the warm season from March to September.

Table 5.24 summarises the water consumption.

Table 5.24   Summary of water requirements (l/day)

	March–September	October–February
People	300	300
Goats	250	250
Crops	350	0
TOTAL	**900**	**550**

Domestic electrical consumption
For the house and goat shed, 12 lighting points are needed, operating on average 5 h/day. Because of the heat, a ceiling fan will be fitted in the main room of the house. Additionally, there is a television set usually on for 4 h/day and a 140 l refrigerator for food storage. Occasionally, a portable power tool of 500 W–230 V AC is used for half an hour to one hour.

Agricultural electrical consumption
The only agricultural equipment on the farm requiring electrical energy (apart from the water pump) is a milking machine for the goats. Each goat produces 3–4 l/day, so this machine has to handle 150–200 l of milk per day (for 50 goats). The market garden activity only uses manual tools.

---

[35] We will not enter into financial detail, but simply point out that the investment required for the installation we will describe would certainly not be affordable for this family with its modest income. Programmes of rural electrification have existed in Morocco for some time and financial incentives are available.

Other electrical consumption

A computer consuming 150 W at 230 V AC is used 2 h every evening mainly for the farm accounts. In addition, to communicate with the outside world, the owners have a GSM telephone that is relayed by a station situated on high ground some kilometres distant. As the region is not too hilly, they do not have too many communication problems.

5.6.3.2 Choice of appliances and consumption

To consume as little energy as possible and reduce the size of the PV supply, low-energy appliances working on DC will be preferred.

Water pumping

The farm has a well situated 200 m from the house. The water table is 25 m deep. Because of the distance from the house, we will propose a solar pumping installation independent from the rest. We refer to this as 'pumping system' and to the other, which supplies the house and the goat shed, as 'main system'. The vegetable beds will be located close to the well, on a gentle slope so that trickle irrigation can be used. The water therefore has to be pumped to a height of 30 m (counting the tank and the depth of submersion of the pump), and stored in a tank at ground level. The size of this tank is calculated to allow storage of 4 days in summer, to allow for bad weather, so as not to interrupt supply. The resulting size is $4 \times 400 = 3600$ l minimum (or 4 m^3).

The occupants, who are used to drawing water from the well for their domestic requirements, do not want running water. They will come to fetch water from the tank in jerrycans for their domestic consumption. The goats will drink from a trough, which will be fed directly from the tank by gravity (Figure 5.59).

To fill this tank, a pump is needed, able to lift between 550 and 900 l/day to a height of 30 m.

Milking machine

A milking machine adapted to 24 V DC will be used on this farm. With a power of 1120 W, the machine can milk four or five goats at the same time, and uses less energy than a similar machine with a three-phase motor: it only requires 12 Wh/l of milk (see Section 5.4.3). As there are 50 goats, with each producing 3 or at most 4 l of milk per day, the electrical consumption of this machine will be a maximum of 2400 Wh/day.

Domestic appliances

For lighting, 10 W fluorescent lamps also at 24 V DC will be used, because they are available from a local retailer along with their replacement tubes – availability is an essential criterion of choice in remote areas with poor access to supplies. They are U-shaped fluorescent lamps (double PL) with an efficiency of 60 lm/W, which is correct for the requirement (see Section 5.2.2).

For ventilation, that essential standby in hot countries, a ceiling fan with large blades, driven by a DC motor with a speed regulator, is chosen. Its nominal power is 20 W and its likely use is 6 h/day.

The *television set* runs on 12 V DC and only consumes 60 W. If it is decided that the PV system should be 24 V DC, which is likely, a step-down converter 24 V–12 V DC will be needed, and the consumption of 60 W will be increased because of the 80% efficiency of the converter (see Section 5.1.3). This correction will therefore be introduced from the beginning (Table 5.25).

The *refrigerator* is chosen for its low consumption, with a compressor running on 24 V DC and with reinforced insulation for the storage compartment to reduce energy loss (see Section 5.2.3). Its electrical consumption is estimated at 500 Wh/day.

The *power tool* running on 230 V AC at a power of 500 W obviously needs an inverter to operate, and its efficiency will need to be taken into account. A sine wave inverter is selected so that the motor does not overheat, and the computer can also be supplied from it.

Office and telephone

The *computer* used is a classic desktop working on 230 V AC (it is not a laptop, which would be too expensive and complicated). It consumes a constant 150 W when switched on and its use should not exceed 2 h a day.

To power it, and the power tool mentioned above, an efficient sine wave inverter is chosen, with a minimum harmonic distortion (<3%) to protect the motor and a maximum efficiency (85–90% depending on the power output). As power of 500 W is needed for the tool, an inverter providing a permanent 800 VA is selected. This allows a small margin for any future more powerful equipment, and this model is compatible with the start-up power of the tool, which is higher than its running power. As with any appliance supplied through a converter, the total consumption of computer needs to take its efficiency into account.

Note

This computer will thus have two periods of use, and the time is chosen by the users. It will only be switched on when it is used and will never be left on standby. This is very important for total consumption, because standby power should never be neglected.

The *mobile telephone* needs to be charged from time to time. To avoid having another converter, it was decided to use a car charger that enables the phone to be charged from the car battery. The required energy is in any case small: 2–3 Wh/charge (600 mAh at 3.6 V). Otherwise, it would be possible to use a small step-down DC/DC converter with a 24 V DC input and an adjustable output of 3–12 V DC (see Section 5.1.3), and wire it to the main PV system. It could also be used for some other small appliances (rechargeable batteries, radio, etc.), provided their consumption remains small and does not affect the rest of the system.

Total consumption

Table 5.25 gives a summary of all the electrical appliances to be supplied by the main PV system. As mentioned earlier, the PV pump will be supplied by a separate system.

Table 5.25 Electrical consumption of the main PV system

	Voltage	Power in use (W)	Conversion efficiency	Power adjusted by efficiency (W)	Duration of use/day (h)	Electrical consumption (Wh)
Milking machine	24 V DC	1120	–	1120	–	2400
Lighting	24 V DC	120	–	120	5	600
Ventilation	24 V DC	20	–	20	6	120
Television	12 V DC	60	0.8	75	4	300
Refrigeration	24 V DC	70	–	70	–	500
Computer	230 V AC	150	0.85	188	2	376
Power tool	230 V AC	500	0.85	588	1/2	294
TOTAL				2181		**4590**

5.6.3.3 Sizing and installation of the PV systems

Solar radiation

The solar radiation data that we plan to use are those of Cape Juby, latitude 27.9° N, longitude 12.9° E. This is the nearest site to our installation for which meteorological data are available. The two curves in Figure 5.58 give the radiation for the horizontal position and at 40° tilt with an orientation due south. The influence of maritime cloudiness will be noticed for the horizontal radiation in June–July. It can also be seen that with an inclination of 40° due south a much better winter radiation will be afforded, but it will be slightly less in the summer.

Best exposure for the modules

There are two independent PV systems to be installed, the main system and the pump system. The modules for the two systems do not necessarily have to be oriented in the same direction, since they do not have the same requirements from this point of view.

The pump will need more energy in the summer because the need for water is greater at this period: it would therefore seem best to place the panels horizontally (or nearly) to maximise the energy received in summer. To promote the run-off of rainwater, it is preferable to mount the modules at a slight tilt (from around 5° to 10°), and this difference from the horizontal will only cause a very slight reduction in the radiation received.

The electrical consumption of the main system on the other hand (4.59 kWh/day) is constant all year around, so it is best to tilt the modules to maximise the minimum value received during the year. A tilt of 40° with orientation due south will achieve this result with a daily total radiation throughout the year of 4.4 kWh/m²/day or more. This is the optimal exposure for the latitude of our site (latitude of location +10°) for year-round use, and this is what needs to be used for our second system.

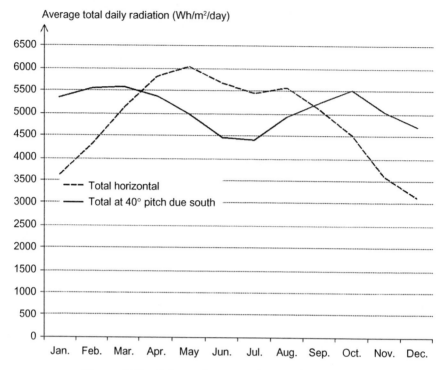

Figure 5.58   Solar radiation at Cape Juby, Morocco

Pumping system
Choice of equipment
As the well is 30 m deep, the pump must be of the submerged type (surface pumps are only generally suitable for wells of less than 6 m depth). Its PV system will be independent of the main system.

The pumping requirements are not very demanding: there are many suitable pumps available on the market. In fact there are PV pumps capable of pumping tens of cubic metres per day from a depth of 200 m (see Section 5.4.4).

The farm's requirements are more modest: the average daily volume required is 900 l/day from March to September and 550 l/day for the rest of the year, with the water being pumped from a depth of 30 m.

To meet these characteristics, and for the sake of simplicity, we would choose a DC pump to work directly off the Sun without a battery. The 4 m³ tank will act as a storage buffer.

The model chosen is a small volumetric pump (height 30 cm, diameter 9.5 cm), suitable for boreholes of 100 mm diameter. It is a three chamber diaphragm model, and the 24 V DC motor has a permanent magnet with thermal protection.

According to the manufacturer's specifications, with only 100 Wp installed, at 24 V DC, the volume pumped should reach 1.2 m³ for solar radiation of

4.5 kWh/m²/day and a depth of 30 m (from the pump to the tank feed). It is adequate for our daily requirement of 900 l. Between March and September, the horizontal total radiation is 5 kWh/m²/day (Figure 5.58). As the modules will be slightly tilted to avoid standing rainwater and as they are likely to get slightly dirty, it is good to have this margin. During the remainder of the year, the solar radiation will be between 3 and 5 kWh/m²/day. Bearing in mind the length of the days, it is probable that in winter the instantaneous radiation will rarely reach 1000 W/m². It will therefore be worthwhile to install uphill from the pump a booster that will enable it to start even in weaker sunshine, so as not to risk creating a water shortage. A booster also has the advantage of protecting the pump against accidental overloading or overvoltages.

The pumping system is thus made up of the following:

- a 100 Wp–24 V PV module (of the same type as those of the main system);
- a 24 V submerged volumetric pump 9.5 cm diameter;
- an electronic starter booster;
- a half-inch diameter water pipe;
- a waterproof feed cable;
- a 4 m³ tank mounted on the ground;
- an animal trough of around 500 l.

Figure 5.59 shows the positioning of these different elements.

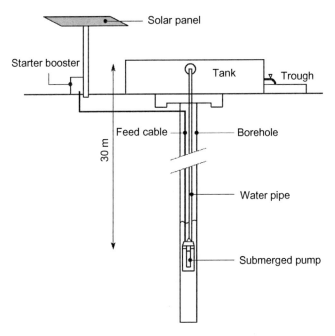

Figure 5.59   Pumping system

Main system

PV modules

First we should make an approximate calculation of the power in Wp to be applied (calculation described in Chapter 1).

The loss coefficient will be taken as 0.7 initially. The daily electrical consumption is 4590 Wh and the minimum total solar radiation is 4.4 kWh/day, therefore

$$P_p = \frac{4.59}{4.4 \times 0.7} = 1.49 \quad \text{kWp} \tag{5.32}$$

If we now refer to Table 5.18 on voltages recommended for PV systems (see Section 5.5.4), it is immediately clear that the system should be wired in 24 V. It is also clear that crystalline silicon technology should be chosen, because of the total power to be installed.

For this system, it is proposed to use 100 Wp–24 V modules each producing 2.85 A–35 V in STC conditions, with a surface of 0.7 m^2 (deficiency 14%).

The PV array would therefore have fifteen 100 Wp–24 V modules mounted in parallel, total peak power of 1500 Wp, charge current of 42.8 A (2.85 × 15), with a total surface of 11 m^2 (15 × 0.7 m).

Before confirming this choice, we will identify the other components of the system in order to calculate any losses.

Battery

The capacity of the battery will depend mainly on the need for autonomy without solar generation. Periods of bad weather in the region are mainly limited to October when there are often rainy spells that can last for 3 or 4 days. During the rest of the year, it is rare for the sky to be covered for more than two consecutive days. The duration of 3 days without Sun is therefore a fairly correct value that will considerably limit the risk of breakdown of the electrical equipment.

For the pumping system, it was decided to use a value of 4 days to ensure water supply without interruption throughout the year (in this case the storage is in the tank and not in a battery, which is much cheaper!).

The daily consumption of the main system is 191.2 Ah at 24 V (= 4590 Wh/24). With an initial coefficient of 0.7 and 3 days of autonomy without solar generation, the battery capacity needs to be

$$C_n = \frac{191.2 \times 3}{0.7} = 820 \quad \text{Ah} \tag{5.33}$$

With this capacity requirement, and bearing in mind that the premises are occupied, it is clear that the best solution is an open lead battery (with liquid electrolyte). The maintenance can easily be carried out regularly and there is no need to have recourse to a sealed battery. The technology chosen will be a model with tubular electrodes, which can deliver 500 discharge cycles at 80%.

There is a model of 800 Ah nominal capacity that would be compatible with the loss coefficient of 0.72.

Since the ambient temperature is between 15 and 25 °C most of the time, the temperature loss of capacity can be estimated to be 10% in winter. Still using the loss coefficient of 0.72, this gives an order authorised depth discharge of 80% ($0.72 = 0.8 \times 0.9$). This is correct because one can then have 500 cycles available.

We will therefore use an open lead battery composed of twelve 800 Ah–2 V elements with tubular electrodes. The size of each 2 V element is $200 \times 150 \times 650$ mm. If they are aligned in two rows of six elements, the battery bank will fill a space of 400×900 mm on the ground with a height of around 800 mm (for this type of arrangement, see Figure 5.50, and the accompanying explanations). Its charge current efficiency is at least 80%.

Charge controller
Taking into account the voltage and the power of the system required and the type of battery, it would seem best to have a series charge controller with equalisation charge. For this type of regulator, the maximum charge current should be taken as equal to 1.5 times the operating power of the modules in 24 V:

$$1.5 \times 2.85A \times 15 = 1.5 \times 42.8A = 64.1A \text{ for the 15 modules of 100 Wp}$$

The discharge current of the appliances depends on the power of all the appliances that will be installed on the 'users' output of the regulator. As in the case of the chalet, the inverter, with its own low battery protection, will be connected directly to the battery, as well as the milking machine equipped with the same device (it cannot operate when the battery is too low).

Taking the consumption of all the equipment except that of the milking machine and of those running on 230 V AC (the power tool and the inverter), the maximum load likely to be drawn from the charge controller is 285 W at 24 V (Table 5.25), which is only around 12 A. The series regulation must therefore be in 24 V and sized for an input of 70 A and an output of 12 A. With this fairly high current coming from the panels, in order to avoid massive cabling, it is simpler to put two regulators in parallel, each linked to a subgroup of panels. For 2×35 A input, the regulators could be installed with mercury power relays (there is a model that just fits 35 A). They are quite cheap and reliable in hot countries.

Wiring and installation of components
With these two regulators, the PV array is therefore to be divided into two subgroups of seven or eight panels. The power of a group of seven panels in parallel would be $7 \times 2.85 = 20$ A, and for another group of eight panels, $8 \times 2.85 = 23$ A. With the help of Table 5.19, the online losses between the components can be calculated. For 10 mm^2 section wiring, with 1 A current, 4 mV/m in double wiring are lost. Using this 2×10 mm^2 cable, for 23 A (eight panels), 92 mV/m would be lost between the modules and the regulator, or around 0.92 V if there are 10 m of wiring.

On the appliance side, power of the milking machine on its own is already 47 A ($= 1120$ W/24 V). If it is supplied from the battery with 10 m of 2×10 mm^2

cable, the loss will be 1.80 V: if the distance is reduced to 3 m, the loss will reduce to only 0.56 V.

The physical installation and the distance between the equipment will therefore be crucial in this project. In practice, the battery, the regulator and the inverter will be located in the goat shed with the milking machine, since it is the latter device that consumes the most. So these sensitive components will be under cover. Good ventilation is essential (for possible discharge of hydrogen in case of accidental overcharge). The PV array will be situated as near as possible to those so that there will be less than 10 m of cable length (general layout in Figure 5.60).

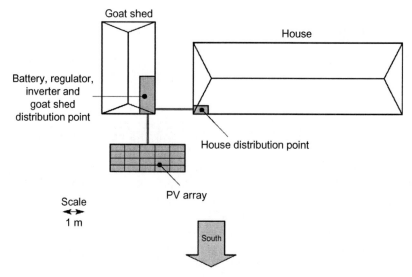

Figure 5.60   General situation of the farm and its main power distribution

The house will have its own 24 V DC domestic distribution wired from the regulator, plus a 230 V AC cable from the inverter to supply the computer.

Calculating losses
Voltage losses: Once the general plan has been established, we see that between the modules (connected to a junction box under the frame) and the battery, the actual distance will only be 6 m. This gives an ohmic loss in the wiring of 92 mV/m times 6 m = 0.55 V with 2×10 mm^2 wiring at 23 A (assuming a subgroup of eight panels).

The series regulator generates a loss of 0.44 V and the diodes at the output of each string of two modules have a voltage drop of 0.5 V.

To evaluate the loss incurred by rise in temperature, we will assume that the modules are permanently on NOCT (nominal operating cell temperature): 40 °C (see Section 3.1.4). The reduction being –0.4%/°C, there will be a drop in voltage of 6% for a temperature difference of 15 °C (40 °C instead of 25 °C). The voltage of the modules V_m being 35 V, the fall in voltage on account of temperature will be 2.1 V per parallel string.

The voltage losses per group are thus: $0.55 + 0.5 + 0.44 + 2.1 = 3.59$ V. The nominal voltage power group of panels is 35 V at 25 °C. In the conditions of operation, the available voltage at the battery would therefore be at least $35 - 3.59$ V $= 31.4$ V or 2.62 V/battery element (there are 12). It is sufficient to charge the battery: normally 2.5 V will be enough except in the case of a boost charge. The extra voltage is a good thing, which will enable an efficient charge even under instantaneous solar radiation lower than 1000 W/m². The voltage V_m will typically fall from 35 to 33 V under a radiation of 300 W/m²; hence, a charge voltage of $33 - 3.59 = 29.4$ V at the battery input, or 2.45 V/element.

Conclusion: The 100 Wp modules, with their STC charge point at 2.85 A–35 V are suitable, and the sizing of 1500 Wp installed can be confirmed.

Power losses: Dirt on the modules will not be a problem because the users will make sure they are kept clean, so only 5% of losses will be estimated on this account. The charge efficiency of the battery, which is 80%, must also be taken into account. We will therefore apply for the final calculation of the PV array charge current a current loss coefficient of

$$C_1 = 0.95 \times 0.8 = 0.76 \tag{5.34}$$

Any additional consumption must also be taken into account, in particular the consumption of the regulator(s). The technology chosen is economic in energy; its permanent consumption is 125 mA, which is equivalent to 3 Wh/day at 24 V. This is very low compared to the daily consumption of 4590 Wh, but even so it must be added on.

Final choices for the main system
The total load current for the system is therefore finally

$$I_m = \frac{4593}{4.4 \times 34 \times 0.76} = 40.4 \text{ A} \tag{5.35}$$

As the modules each deliver 2.85 A, the final requirement is

$$\frac{40.4 \text{ A}}{2.85 \text{ A}} = 13.8 \text{ modules}$$

So the number of modules to be installed will be 14 with a power of 1400 Wp (1.4 kWp). The total load current will be 14×2.85 A, or 40 A. The PV array will be divided into two subgroups of seven modules (20 A nominal per subgroup), each connected to a 35 A regulator. For reasons of convenience, one of the regulators will supply 24 V equipment in the goat shed (except the milking machine) and the other the domestic appliances (final diagram, Figure 5.61).

By careful evaluation of losses, we have saved on one module and defined the rules to respect in the physical and electrical installation of the system.

Final characteristics of the main PV system:

- fourteen crystalline silicon modules of 100 Wp–24 V, producing a total of 40 A–35 V in STC conditions (surface approximately 11 m^2);
- two 24 V series regulators, 35 A input with a mercury relay, output current 15 A for one and 3 A for the other;
- one open lead battery with tubular electrodes 800 Ah–24 V with 576 Ah useful charge (400 × 900 × 650 mm^3);
- one sine wave converter 24 V–800 VA with <3% harmonic distortion and minimum efficiency 85%;
- one 24 V DC/12 V DC converter with 80% efficiency.

Installation

Instructions for the installation are given in Section 5.5.7, with special attention being paid to the safety advice during the fitting of the modules and batteries. As there is no shortage of space, the frame for the modules will be built in such a way that there is no shading between the rows: the panels will be placed side by side and one above the other rather than some behind others. In each group, the seven modules will be wired in parallel to the junction box provided with 4 A anti-return diodes fitted in the middle. The wiring sections will be selected according to the length of connections with the help of Table 5.19 or the chart in Figure 5.42.

Figure 5.61 shows the layout and connections between the components. For the fuses, circuit breakers, overall voltage protection and installation precautions, see the description of wiring of the chalet. The domestic appliances (lighting, fan, refrigerator and television set) will be divided between the two regulators according to their location (one regulator for the house, another for the goat shed), and a distribution chart will be put up on the wall in each building. The DC/DC converter for the television will be put close to it so as to avoid doubling the amperage along the cable during the conversion to 12 V.

5.6.4   Wastewater treatment plant in the Vaucluse

A small agricultural commune in the Vaucluse Department of southern France built a wastewater treatment plant below the village far from any electricity lines. They planned to dig a trench to make a connection to the grid, but the owner of the field that had to be crossed refused permission. The plant was finished and only needed a three-phase supply of 3 × 400 V. The commune then decided to supply it using PV panels.

The technical specification was to drive a three-phase pump using 2.2 kW for 3–4 h everyday to transfer the effluent over a height of around 2 m.

Table 5.26 shows all the parameters of this stand-alone system. The first eight lines summarise the system characteristics (voltage, power, current, etc.). All these parameters are variable and will modify the balance of the system that must

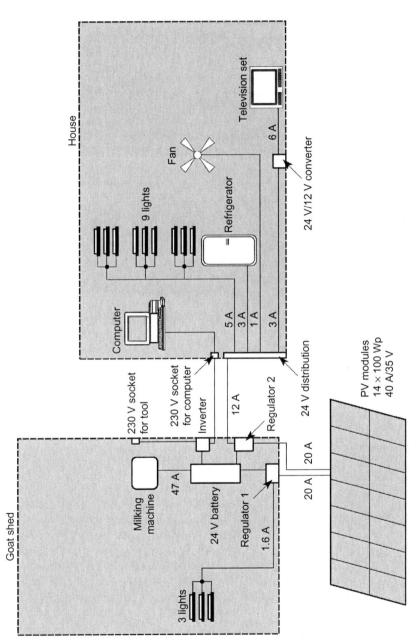

Figure 5.61 Detailed installation diagram of farm equipment (main system)

Table 5.26 Data for the Vaucluse treatment plant

Wastewater treatment plant – Vaucluse					
Pump	2200	W	Standby command:	12	W
I pumping	108	A		0.50	A
V system	24	V	System efficiency	0.85	Ah
I panel	7.23	A STC	Battery capacity	900	Ah/100 h
No. panels series	2		Consum./day	93	Ah
No. panels parallel	4		Installed power of panels	1000	W
I panel total	28.92	A STC			
Power 1 panel	125	W			

	Solar irradiance (kWh/m²/m) tilt 60° S	T day ave. (°C)	Duration/ day (h)	Duration/ month (h)	Consum. standby (Ah/m)	Consum. Total (Ah/m)	Power panel (at 29 V)	Production (Ah/month)	Capacity (Ah)
Jan.	102.6	6.4	0.75	23.3	372	2879	7.70	3160.4	900
Feb.	103.6	7.9	0.75	21.0	336	2601	7.70	3190.9	900
Mar.	152.8	11.5	0.75	23.3	372	2879	7.65	4676.6	900
Apr.	142.2	13.9	0.75	22.5	360	2786	7.62	4334.3	900
May	147.3	18.5	0.75	23.3	372	2879	7.55	4447.0	900
Jun.	155.4	22.8	0.75	22.5	360	2786	7.42	4612.3	900
Jul.	169.0	25.1	0.75	23.3	372	2879	7.00	4730.6	900
Aug.	173.0	24.6	0.75	23.3	372	2879	7.00	4843.4	900
Sept.	160.2	20.3	0.75	22.5	360	2786	7.42	4754.7	900
Oct.	127.1	16.7	0.75	23.3	372	2879	7.55	3838.4	900
Nov.	102.6	10.4	0.75	22.5	360	2786	7.65	3139.6	900
Dec.	87.1	7.0	0.75	23.3	372	2879	7.70	2683.0	704

produce enough energy to cover the pumping requirements. The calculations used an Excel spreadsheet in which each cell can be individually programmed. For example, by increasing the number of panels in parallel, the total charge current of the system is proportionately modified.

The following lines on the chart cover the sizing of the system: for each month, the average energy production and remaining battery charge are calculated. The solar data are taken from the Photovoltaic Geographical Information System (PVGIS) website[36] established by the solar laboratory of the European Commission, which gives solar irradiance data for locations in Europe and Africa based on statistics from 1981 to 1990. This website also gives average daytime temperatures, which enables the nominal charge account of the panels to be calculated for each month and temperature. We will take as a reference the current at 29 V that corresponds to an end of charge. We will not take the internal resistance of the battery into account, as this current is low (approximately 30 A) compared to the large capacity of the battery (900 Ah). The drops in voltage across the wiring and regulators are estimated at 1 V. Two 30 A regulators were installed in parallel to have some reserve in case of overload due to clouding and to be able also to add some extra panels if necessary.

All the energy calculations are carried out in Ah so as to get away from the charge and discharge voltages, which would be too complicated to model in this example. Alternatively, this calculation can be made in PVsyst, which will, in a more general manner, take account of the battery voltage. The charge/discharge efficiency of 85% in Ah is a conservative value that allows for some ageing of the system.

This simple calculation assumes that the daily solar production is uniform over the month: the monthly irradiance is divided by the number of days in the month. This procedure is practical, because with experience, we finish by knowing for a given region, what a panel can supply each day. In the example of the wastewater treatment plant, the solar charge increases from 2.8 kWh/m^2/day in December to 5.6 kWh/m^2/day in August when the panels are tilted at 60°.

The system could therefore consume twice as much energy at the peak of summer compared to the low point in winter. However, it should not be forgotten that a panel connected to a battery without an MPPT regulator remains at the battery voltage (plus the reduction in voltage due to the wiring and the working voltage of the regulator) and therefore charges on average at the corresponding current. In the case in point, it is often simpler to calculate the whole system in charge and discharge amperes and assimilate the energy to the Ah at the 'battery voltage'. To determine the nominal charge current of the panel, we take a maximum battery charge voltage (here 29 V) and find on the I/V curves of the panel the corresponding current for the daily temperatures during each month of the year. In this example, we have assumed that in July–August the panel is at its NOCT

[36] http://re.jrc.ec.europa.eu/pvgis/index.htm

temperature (daily temperature of around 25 °C), and for the less hot months, we have estimated the fall of NOCT to a minimum of 20 °C in winter.

The estimation of the efficiency of the system takes account of the charge/discharge efficiency in Ah of the battery and of the loss due to the inverters. As the system also functions by day, the efficiency is slightly higher than if all the energy had to pass through the battery (in the case of lighting, for example). We will take a value of 0.85, or 2% higher than the nocturnal value.

The calculation of power consumption is therefore expressed as

$$Cons = conp + stby \qquad (5.36)$$

where
  conp = consumption of the pump,
  stby = command standby.

For the solar production, we will calculate:

$$Prod = I_{rr} \times I_{pan} \times N_{pan} \times E_{ta} \qquad (5.37)$$

where
  I_{rr} = monthly irradiance,
  I_{pan} = current 'for the month', function of temperature;
  N_{pan} = number of panels in parallel;
  E_{ta} = efficiency of system = 0.85.

The charge state of the battery (capacity) is thus:

$$Cap = MIN[Ca; (Prod - Cons) + Cap(m - 1)] \qquad (5.38)$$

where
  Ca = battery nominal capacity,
  Cap $(m - 1)$ = final capacity of previous month.

As this calculation is circular throughout the year, we are assuming that the battery is fully charged during the best month of the year (here August); therefore, for this month we enter the nominal capacity Ca instead of the state of charge formula.

By varying the number of panels in parallel, we can quickly find the minimum power to be installed to meet the technical requirements.

Figures 5.62–5.64 show the electrical diagrams of this installation and review of the PV array.

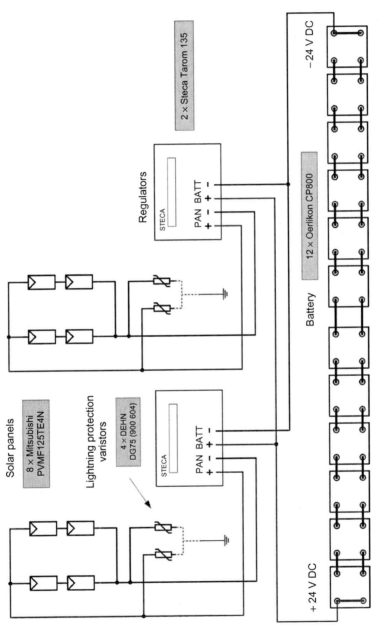

Figure 5.62 Electrical diagram of the wastewater treatment plant solar system

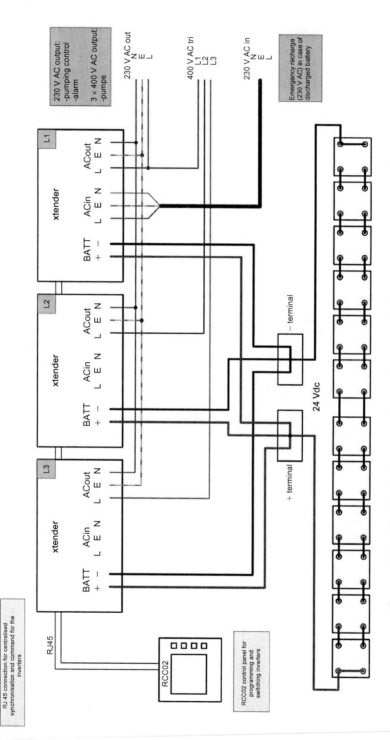

Figure 5.63 Electrical diagram of three-phase supply of wastewater treatment plant

The sizing of the battery needs to allow for 10 days of autonomy in case of bad weather. This period is not more because of the climatic conditions of the region where winter is often dry and sunny.

Figure 5.64   Wastewater treatment plant solar array

Appendix 1

Physical sizes and units

Electrical characteristics of an appliance

Instantaneous data

P: electrical power in watts (W)	1 kW = 1000 W = 100,000 mW
I: current consumed in amperes (A)	1 A = 1000 mA = 100,000 μA
V or U: operating voltage in volts (V)	1 V = 1000 mV

$$P = U \times I$$

$$(\text{W}) = (\text{V}) \times (\text{A})$$

Example: For appliance consuming 4 A at 12 V, electrical power $P = 48$ W.

Integrated data over a period of time

E: Energy consumed over a period of time

$$E = P \times N \text{ in watt-hours for a period of } N \text{ hours}$$
$$(\text{Wh}) = (\text{W}) \times (\text{h})$$

or for a constant system voltage (for example, 12 V)

$$E = I \times N \quad \text{in ampere-hours for a period of } N \text{ hours}$$
$$(\text{Ah}) = (\text{A}) \times (\text{h})$$

Note: This value in Ah is equivalent to the capacity (of a battery, for example).

Example: Energy consumed by a 4 A appliance at 12 V for 5 h:

$$E = 4 \text{ A} \times 5 \text{ h} = 20 \text{ Ah at } 12 \text{ V}$$
$$\text{or} \quad E = 48 \text{ W} \times 5 \text{ h} = 240 \text{ Wh}$$

Note: Divide the duration by 60 if it is expressed in minutes and by 3600 if it is expressed in seconds, to convert it into hours.

$$E = \frac{P \times n}{60} \quad \text{in watt-hours for a duration of } n \text{ minutes}$$

Example: Energy consumed by a 4 A appliance at 12 V for 10 min:

$$E = \frac{48\,\text{W} \times 10\,\text{min}}{60} = 8 \text{ Wh}$$

$$\text{or} \quad E = \frac{4\,\text{A} \times 10\,\text{min}}{60} = 0.667\,\text{Ah} = 667 \text{ mAh}$$

Note: The values W/h and A/h are meaningless.

Light radiation

Wavelength of light radiation

Micrometres (μm) = 10^{-6} m
Nanometres (nm) = 10^{-9} m

Example: Radiation of colour green: λ = 550 nm = 0.55 μm

Energy of a photon

$$E(\text{electron-volt, eV}) = h\nu = \frac{hc}{\lambda}$$

where h is Planck's constant, ν is the frequency, c is the speed of light and λ is the wavelength, or more simply $E = 1.24/\lambda$ with E in eV and λ in μm.

Example: Energy of photon with a wavelength of 550 nm:

$$E = \frac{1.24}{0.55} = 2.25 \text{ eV}$$

Intensity of solar radiation

Instantaneous solar radiation received on a surface:
watts per square metre (W/m^2)
1 W/m^2 = 0.1 mW/cm^2

Integrated (or cumulative) solar radiation over a period of 24 h (= energy):
watts × hour or kilowatts × hour per square metre per day
(Wh/m^2/day) or (kWh/m^2/day)

Example: Average cumulative solar radiation per day in Paris in December:

1.12 kWh/m^2/day

Other units of energy: langley (Ly), joule per square centimetre (J/cm^2) or kilo-calories per square metre (kcal/m^2)

$$1 \text{ J/cm}^2 = 2.9 \text{ kcal/m}^2 = 2.78 \text{ Wh/m}^2$$

$$1 \text{ Ly} = 1 \text{ cal/cm}^2 = 11.62 \text{ Wh/m}^2$$

Emission by a source of artificial light

Total emission:

lumens (lm)

Luminous efficacy of a lamp (quantity of lumens emitted compared to the electric power consumed):

lumens per watt (lm/W)

Illumination received on a surface within the sensitivity of the human eye (400–700 nm):

$$\text{lux(lx)} = \text{lm/m}^2 \text{(equivalent to W/m}^2\text{)}$$

Example: Illumination on a desk: 300 lx.

Appendix 2
Solar radiation data

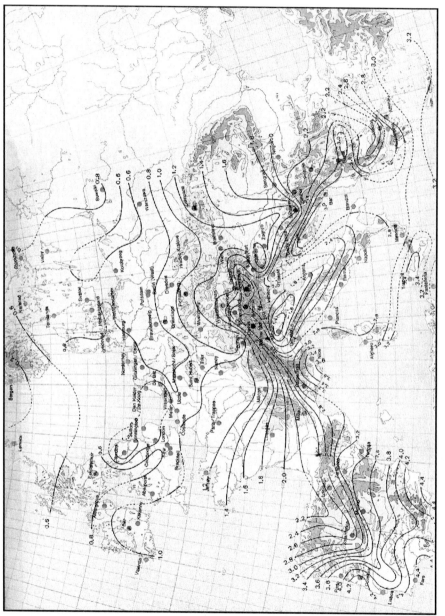

Figure A2.1 Solar radiation in Europe. Average values of global solar radiation (expressed over 24 h in December) for orientation due south and a tilt of 60° from the horizontal (in kWh/m²/day)

Table A2.1 Europe

Country	Place	Lat.	Long.		Jan.	Feb.	Mar.	Apr.	May	Jun.	Jul.	Aug.	Sept.	Oct.	Nov.	Dec.
Norway	Bergen	60.2° N	5.2° E	Gh	196	721	1708	3272	4134	4853	4145	3493	1857	938	302	118
				G 60° S	300	2380	2380	3660	3840	4190	3650	3580	2270	1470	510	170
Germany	Hamburg	53.5° N	10.0° E	Gh	521	2231	2231	3553	4688	5437	4820	4340	2786	1489	671	401
				G 60° S	900	2830	2830	3680	4200	4580	4170	4340	3290	2180	1160	800
	Dresden	51.7° N	13.3° E	Gh	721	2377	2377	3435	4416	5015	5047	4368	3075	1718	831	500
				G 60° S	1350	2910	2910	3430	3840	4080	4290	4220	3580	2420	1390	910
Belgium	Ostend	51.2° N	2.9° E	Gh	622	2624	2624	3968	4991	5795	5122	4370	3268	1887	849	534
				G 60° S	1000	3210	3210	3970	4310	4690	4270	4190	3830	2660	1390	960
Spain	Madrid	40.4° N	3.7° W	Gh	1728	4154	4154	5450	6169	6692	7224	6485	4801	3161	1985	1768
				G 60° S	3040	5260	5260	4970	4600	4690	5310	5870	5230	4370	3320	3800
France	Carpentras	44.1° N	5.1° E	Gh	1591	3750	3750	5295	6291	7047	7515	6229	4592	3210	1884	1461
				G 60° S	3130	4620	4620	5220	5180	5320	6000	5850	5310	4880	3450	3170
	Nice	43.7° N	7.2° E	Gh	1723	3931	3913	5356	6095	6789	7130	5916	4593	3270	1989	1645
				G 60° S	3540	4790	4790	5250	4980	5110	5680	5490	5260	4910	3760	3760
	Limoges	45.5° N	1.1° E	Gh	1190	3070	3070	4388	5192	5902	5959	4864	3858	2603	1390	1000
				G 60° S	2090	3590	3590	4190	4250	4520	4770	4430	4310	3690	2330	1940
	Nantes	47.1° N	1.4° W	Gh	983	3036	3036	4338	5157	5917	6132	4896	3611	2345	1260	868
				G 60° S	1956	3879	3879	4250	4181	4426	4752	4401	4174	3642	2433	1843
	Trappes (Paris)	48.4° N	2.0° E	Gh	823	2699	2699	4012	4824	5567	5575	4563	3475	2113	1049	663
				G 60° S	1350	3230	3230	3950	4090	4430	4600	4290	3980	2980	1730	1120

(Continues)

Table A2.1 *(Continues)*

Country	Place	Lat.	Long.		Jan.	Feb.	Mar.	Apr.	May	Jun.	Jul.	Aug.	Sept.	Oct.	Nov.	Dec.
Great Britain	Eskdalmuir	55.2° N	3.1° W	Gh	381	1117	2012	3240	3898	4669	4056	3421	2310	1292	651	338
				G 60° S	630	1890	2540	3330	3430	3880	3440	3260	2610	1880	1270	730
	Jersey	49.1° N	2.1° W	Gh	506	1598	2882	4296	5335	5991	5606	4549	3376	2133	990	646
				G 60° S	1350	2430	3590	4330	4600	4810	4660	4290	3880	3090	1640	1230
	London	51.3° N	0.0° W	Gh	555	1099	2074	3036	4122	4993	4383	3618	2709	1553	807	470
				G 60° S	880	1630	2550	3080	3660	4130	3730	3500	3170	2230	1380	880
Greece	Athens	37.6° N	23.4° E	Gh	1754	2620	3820	5149	6407	6838	6883	6179	4856	3383	2329	1694
				G 60° S	2760	3580	4260	4730	5000	4830	5120	5470	5320	4630	3870	3040
Italy	Milan	45.4° N	9.3° E	Gh	921	1705	3004	4490	5137	5828	6072	5258	3818	2452	1140	859
				G 60° S	1370	2300	3510	4320	4250	4500	4920	4820	4170	3340	1730	1340
	Messina	38.1° N	15.3° E	Gh	2068	2856	3872	5203	6089	6686	6955	6375	5051	3531	2535	1794
				G 60° S	3380	3780	4190	4650	4700	4740	5120	5480	5340	4710	4130	3070
	Rome	31.8° N	12.6° E	Gh	1695	2539	3780	4993	6025	6585	6861	6159	4693	3287	2020	1512
				G 60° S	2930	3580	4330	4670	4820	4860	5350	5580	5220	4610	3420	2830
Portugal	Faro	37.1° N	7.5° W	Gh	2224	3122	4355	5714	7219	7597	7688	6802	5508	3980	2710	2310
				G 60° S	3920	4360	4870	5310	5460	5220	5540	5860	5850	5380	4530	4740
Switzerland	Zurich	47.3° N	8.3° E	Gh	832	1614	2724	3933	5025	5454	5797	4558	3575	2009	1000	657
				G 60° S	1090	2210	3100	3750	4130	4180	4660	4120	3950	2640	1410	880
	Davos	46.5° N	9.5° E	Gh	1551	2454	3987	5233	5656	5479	5558	4736	4089	2876	1671	1351
				G 60° S	3030	3850	4880	4650	4540	4100	4330	4220	4630	4330	3220	3080

Gh, global horizontal plane (in Wh/m²/day); G 60° S, global 60° tilt due south (in Wh/m²/day).

Table A2.2 Rest of the world

Capital	Place	Lat.	Long.		Jan.	Feb.	Mar.	Apr.	May	Jun.	Jul.	Aug.	Sep.	Oct.	Nov.	Dec.
AFRICA																
Morocco	Casablanca	33.4° N	7.7° W	Gh	2903	4064	5225	5806	6386	5806	6386	5806	5225	4644	3483	2903
				G 40° S	4647	5707	6099	5650	5484	4750	5330	5337	5647	6154	5443	4897
	Cap Juby	27.9° N	12.9° W	Gh	3599	4296	5109	5806	6038	5689	5457	5573	5109	4528	3599	3135
				G 40° S	5348	5561	5589	5392	4983	4474	4419	4913	5224	5514	5055	4708
Egypt	Guizeh	30.0° N	31.2° E	Gh	3367	4412	5840	6769	7211	7594	7443	6967	6142	4865	3588	3089
				G 40° S	5155	5934	6637	6425	5996	5920	5991	6230	6501	6167	5243	4848
Senegal	Dakar	14.7° N	17.4° W	Gh	4586	5353	6258	6584	6456	6258	5295	4435	4819	4842	4656	4226
				G 20° S	5416	5974	6480	6283	5798	5456	4738	4167	4814	5220	5383	5027
Ivory Coast	Abidjan	5.0° N	4.0° W	Gh	4651	4883	5232	5813	5582	4767	4069	4185	4651	5348	4767	4651
				G 10° S	4963	5075	5263	5633	5253	4438	3847	4035	4615	5504	5045	4994
Saudi Arabia	Riyadh	24.0° N	50.0° E	Gh	4064	4644	5806	6386	7547	6967	5806	6967	5806	5806	4644	3483
				G 30° S	5517	5674	6298	6106	6534	5790	4993	6385	5983	6924	6181	4742
ASIA																
China	Hong Kong	22.3° N	114.2° E	Gh	3716	3646	3588	3808	4528	4819	5527	4656	4343	5388	4621	3808
				G 20° S	4853	4251	3721	3603	3988	4079	4722	4262	4373	6261	6005	5133
India	Bombay	18.9° N	72.8° E	Gh	5341	6038	6734	7083	7315	5341	4528	4296	5341	5922	5806	5225
				G 30° S	7059	7243	7094	6545	6126	4389	3857	3877	5305	6738	7512	7097
Japan	Tokyo	35.7° N	139.8° E	Gh	2160	2624	3112	3553	3901	3449	3820	3843	2891	2299	2102	1927
				G 40° S	3397	3510	3479	3442	3447	2959	3315	3568	3004	2750	3047	3095
Philippines	Quezon City	14.4° N	121° E	Gh	3576	3507	5062	6015	5341	5028	4087	4226	4122	4226	4099	4029
				G 10° S	3887	3688	5188	5942	5137	4776	3933	4139	4156	4418	4436	4443
AMERICAS																
Canada	Montreal	45° N	72° W	Gh	1471	2438	3472	4401	5295	5620	5817	4784	3727	2229	1277	1091
				G 60° S	3130	4163	4365	4193	4198	4122	4414	4186	4168	3183	2210	2266
Chile	Santiago	33.4° S	70.7° W	Gh	5573	5922	5922	6270	6270	6038	5573	5225	5806	6038	5806	5341
				G 45° N	6201	6865	6326	5653	3770	3123	3337	3855	4266	5688	5516	6191
Brazil	Manaus	3.13° S	60° W	Gh	4296	4180	4180	4064	4296	4644	4993	5457	5457	5341	4877	4528

Gh, global horizontal plane (in Wh/m²/day); G 60° S, global 60° tilt due south (in Wh/m²/day).

Appendix 3

System monitoring: checklist

Solar panels

Clean the surface with plain water: remove dirt, sand, any spider's webs or insects (also in the junction box).

Cut back any vegetation around the panels.

Check the appearance of the modules: no brown cells, no water leakage, no other damage.

Check the stands: check for any corrosion, tighten the mechanical fixings.

Note: For a grid-connected system, always disconnect the PV array from the inverter and work with insulated tools and gloves, observing all the safety rules when dealing with high voltages.

Charge controller

Check the fixing of the charge controller and fuses.

Check the state of charge: the indicators should accurately reflect the voltage state of the battery at 'battery full', 'battery charging' or 'battery low = load shedding'.

Tighten the terminal clamps.

Batteries

The recommended checks should be done at least once a month for small systems. In the case of larger systems, monitoring should be planned and organised by the supplier of the components with on-the-spot training of the users.

Note: The checks and maintenance operations on the batteries should always be carried out with insulated tools and observing safety regulations to avoid any short circuit.

Open batteries

Voltage measurement of each element and noting of the values in a logbook.

Check appearance: case normal, not distorted, clean terminals (if not clean them), no visible deposit at the bottom of the plates (which indicates a loss of actor material, visible when the case is transparent).

Check the connections: terminal clamps tight, cables in good condition, fuses in place and clean for all batteries.

Electrolyte level: plates well covered (top with distilled water if necessary), no deposit or suspect masses between the grids.

Measure the electrolyte density of each 2 V element: note the values in the same logbook as the voltages and compare the readings; if one or several readings are very different, carry out the measurements again after an equalisation charge and once more after several days in an intermediate charge state. If the densities remain very different, battery replacement should be considered if the capacity is no longer sufficient.

Sealed batteries

Check the voltage.

Check appearance: case normal, not distorted, clean terminals (if not clean them), the safety valves not distorted, no traces of electrolyte on the surface (which would indicate a high overcharge with loss of acid).

Lamps

The checking of appliances is usually done regularly as they are used. However, *fluorescent tubes age* with time and the number of times they are switched on or off. When a DC fluorescent tube does not start easily, check that the wear of the electrodes (black halo) is the same at each end of the tube, otherwise turn the lamp in its socket (U-shaped tube) or reverse the ends (straight tube) and try again to start it. The black halo appearing at the ends comes from the degradation of the electrodes, which lose some matter at every cold start or if the wave is not symmetric (DC component). It is recommended to have spare fluorescent tubes that are often difficult to obtain in traditional sales outlets. If the ambient temperature is very low (in a chalet in winter, for example), wait until it has risen if the lamps cannot be started – it is also possible to take the tube out and heat it somewhere else (this can be done without problem up to 50 °C) if light is absolutely necessary in a cold room. If starting still proves impossible, do not keep trying, to avoid overheating the ballast, which, in general, has to dissipate more energy at start-up.

For the maintenance of low-energy bulbs operating on 230 V AC, there is no access to the tube, so when they do not start, they should be changed. However, the remarks above about low temperatures remain valid and it is also possible to heat the lamp before turning it on in the cold if it does not start immediately.

Bibliography

General works

Séverine Martrenchard-Barra, *Lumière Matière*, CNRS Éditions, Centre de vulgarisation de la connaissance, coll. 'Nature des Sciences'.

C. Vauge et M. Bellanger, *L'aube des énergies solaires*, Hachette, 1984.

A. Ricaud, *Photopiles solaires – De la physique de la conversion photovoltaïque aux filières, matériaux et procédés*, Cahiers de Chimie, Presses polytechniques et universitaires romandes, 1997 – http://www.ppur.com

Anne Labouret, Michel Villoz, *Cellules solaires: les bases de l'énergie photovoltaïque*, Dunod, coll. 'ETSF', 4th édition, 2005.

Jean-Paul Louineau, *Guide pratique du solaire photovoltaïque, dimensionnement, installation et maintenance*, éditions Systèmes solaires, 2nd édition, 2005.

Magazines and sources of economic and technical information

Revue Systèmes solaires – Le journal des énergies renouvelables – 146 rue de l'Université 75 007 Paris et le baromètre photovoltaïque d'OBSERVER – http://www.energies-renouvelables.org

Photon International – The Photovoltaic Magazine, Solar Verlag GmbH (Allemagne) – http://www.photon-magazine.com

Cythélia, *La lettre du solaire*, Savoie-Technolac, Bât. Aero, 73 370 Le Bourget du Lac, April 2009, vol. 9, no 4.

World PV Industry Report Summary, *Solarbuzz reports world solar photovoltaic market grew to 5.95 gigawatts in 2008* – http://www.solarbuzz.com/Marketbuzz2009-intro.htm

Renewable Energy World – http://www.renewableenergyworld.com/rea/magazine

National documentation (France)

Dossier de presse du ministère de l'écologie, de l'énergie, du développement durable et de l'aménagement du territoire: *Grenelle Environnement: Réussir la transition énergétique: 50 mesures pour un développement des énergies renouvelables à haute qualité environnementale* – 17 November 2008 – http://www.developpement-durable.gouv.fr

Critères d'éligibilité des équipements de production d'électricité photovoltaïque pour le bénéfice de la prime d'intégration au bâti – http://www.industrie.gouv.fr/energie/electric/pdf/guide-integration.pdf

European standards and directives

Commission électrotechnique internationale – International Electrotechnical Commission – http://www.iec.ch

TUV (Certification organisation for PV modules) – http://www.tuv.com/de/en/pv_module_certification.html

Restriction of Hazardous Substances (RoHS): Directive 2002/95/CE – http://europa.eu.int

Swiss low-energy homes – http://www.minergie.ch

Laboratories

Institute for Applied Sustainability to the Built Environment (ISAAC), Switzerland – http://www.isaac.supsi.ch

Laboratoire photovoltaïque et couches minces électroniques, Institut de Microtechnique, Université de Neuchâtel (Suisse).

Laboratoire de physique des interfaces et des couches minces, École Polytechnique, Palaiseau (France).

Laboratoire de physique et applications des semi-conducteurs PHASE/CNRS, Strasbourg (France).

Solar radiation data and sizing software

W. Palz, *European Solar Radiation Atlas*, 3rd revised edition, Berlin: Springer–Verlag; Heidelberg: GmbH & Co. K; 1996.

http://www.meteotest.ch/

http://www.pvsyst.com

http://eosweb.larc.nasa.gov/sse

http://www.retscreen.net

ISCCP (International Satellite Cloud Climatology Project) – http://eospso.gsfc.nasa.gov

PVGIS – http://re.jrc.ec.europa.eu/pvgis/index.htm

Solar panel technologies

For a directory of all manufacturers – http://www.solarbuzz.com

Photovoltaic International, 2nd edn., 2008 – http://www.pv-tech.org

P.-J. Verlinden, R.-M. Swanson and R.-A. Crane, 'High efficiency silicon point-contact solar cells for concentrator and high value one-sun applications', *Proceedings 12th EC Photovoltaic Solar Energy Conference*, Amsterdam, Apr 1994, pp. 1477–80.

Comparison of panel technologies

K.-W. Jansen, S.-B. Kadam and J.-F. Groelinger, 'The advantages of amorphous silicon photovoltaic modules in grid-tied systems', *Photovoltaic Energy Conversion, Conference Record of the 2006 IEEE 4th World Conference*, May 2006, vol. 2, pp. 2363–6.

S. Adhikari, S. Kumar and P. Siripuekpong, 'Comparison of amorphous and single crystal silicon based residential grid connected PV systems: case of Thailand', Technical Digest of the International PVSEC-14, Bangkok, Thailand, 2004.

Connection to the grid (France)

Rhonalpénergie-environnement, *Guide Perseus* – édition 2007, téléchargeable sur – http://www.raee.org/administration/publis/upload_doc/20071121090955.pdf
Documentation détaillée sur les démarches, les aides et conditions de rachat – http://www.hespul.org

Standalone systems and rural electrification

Directives générales pour l'utilisation des EnR dans l'Électrification rurale décentralisée (Directives ERD), June 1997.
X. Vallvé, G. Gafas, *Problems related to appliances in stand-alone PV power systems*, http://www.iea-pvps.org
IEA PVPS Task 3, *Managing the Quality of Stand-Alone Photovoltaic Systems – Recommended Practices* – http://www.iea-pvps.org
Christophe de Gouvello et Yves Maigne (dir.), *L'Électrification rurale décentralisée, une chance pour les hommes, des techniques pour la planète*, éditions Systèmes solaires, 2000.
Fondation Énergies pour le Monde (dir.), Hubert Bonneviot (Consultant indépendant), *Adduction d'eau potable avec pompe photovoltaïque – Pratiques et recommandations de conception et d'installation*.
The Cold Chain Product Information Sheets, SUPDIR 55 AMT 5, Expanded Programme on Immunization, World Health Organization, 1 211 Geneva 27, Switzerland.
X. Vallvé, *et al.*, 'Key parameters for quality analysis of multi-user solar hybrid grids (MSGs)', *17th European Solar Energy Conference*, Munich, Oct 2001.
J. Agredano, J. Huacuz and J. Flores, 'The Mexican Experience', *Proceedings of PV Hybrid Power Systems Conference, Hybrid Systems*, Aix-en-Provence, 7–9 Sep 2000.

Batteries, regulators, inverters, protections

IEA PVPS Task 3, *Management of Batteries Used in Stand Alone PV Power Supply Systems* – http://www.iea-pvps.org
D. Berndt, 'Valve-regulated lead-acid batteries', *Journal of Power Sources*, 2004;**100**:29.
IEA PVPS Task 3, *Guidelines for Selecting Lead-acid Batteries in Stand-Alone PV Power Systems* – http://www.iea-pvps.org
http://www.itpower.co.uk/investire/home.html
L. Torcheux, P. Laillier, 'A new electrolyte formulation for low cost cycling lead acid batteries', *Journal of Power Sources*, 2001;**95**:248–254.

R.-H. Newnham, W.-G.-A. Baldsing, 'Benefits of partial-state-of-charge operation in remote-area power-supply systems', *Journal of Power Sources*, 2002;**107**:273–9.

M. Perrin, H. Döring, K. Ihmels, A. Weiss, E. Vogel and R. Wagner, 'Extending cycles life of lead-acid batteries: a new separation system allows the application of pressure on the plate group', *Journal of Power Sources*, 2002;**105**:114–119.

IEA PVPS Task 3, *Recommended Practices for Charge Controllers,* PVPS Task 3 – http://www.iea-pvps.org

G. Moine, *Protection contre les effets de la foudre dans les installations faisant appel aux énergies renouvelables,* Ademe, June 2001.

Organisations and associations

France

ADEME (Agence de l'environnement et de la maîtrise de l'énergie)
http://www.ademe.fr

CLER (Comité de liaison des énergies renouvelables)
93 – Montreuil
http://www.cler.org

Association HESPUL
Efficacité énergétique et énergies renouvelables, photovoltaïque raccorde au réseau.
Espace Info Énergie du Rhône
69 – Villeurbanne
http://www.hespul.org

Enerplan
Association professionnelle de l'énergie solaire
13 – La Ciotat
http://www.enerplan.asso.fr

SER (Syndicat des énergies renouvelables)
Le Syndicat des énergies renouvelables et sa commission photovoltaïque SOLER, le groupement français des professionnels du solaire photovoltaïque
75 – Paris
http://www.enr.fr

CIDFER (Centre d'information, de documentation et de formation sur les énergies renouvelables)
146, rue de l'Université – 75 007 Paris
http://www.energies-renouvelables.org/centre_ressources.asp

FONDEM (Fondation Énergie pour le monde)
146, rue de l'Université – 75 007 Paris
http://www.energies-renouvelables.org/accueil_fondation.asp

Technosolar
Association des installateurs photovoltaïciens et thermiciens solaires 66 – Ria
http://www.technosolar.fr/

Switzerland

OFEN (Office fédéral de l'énergie)
CH-3003 Berne
http://www.bfe.admin.ch/index.html?lang=en [English]

Swissolar
Association suisse des professionnels de l'énergie solaire
CH-8005 Zurich
http://www.swissolar.ch/fr/ [French]

BiPV Competence Centre (Building integrated PhotoVoltaics)
CH-6952 Canobbio
http://www.bipv.ch/base_e.asp

Swissgrid
Operator managing purchases of renewable electricity.
CH-5070 Frick
http://www.swissgrid.ch/

Europe

EPIA (European Photovoltaic Industry Association)
The world's largest photovoltaic industry association, representing about 95% of the European photovoltaic industry and 80% of the worldwide photovoltaic industry.
http://www.epia.org

Index

Page numbers followed by "*f*" indicate figure; those followed by "*t*" and "*n*" indicate table and note respectively.